Applying Drone Technologies and Robotics for Agricultural Sustainability

Pethuru Raj
Reliance Jio Platforms Ltd., Bangalore, India

Kavita Saini
Galgotias University, India

Vinicius Pacheco
Office of the Prosecutor General, Brazil

IGI Global
PUBLISHER of TIMELY KNOWLEDGE

A volume in the Advances in Environmental Engineering and Green Technologies (AEEGT) Book Series

Published in the United States of America by
IGI Global
Engineering Science Reference (an imprint of IGI Global)
701 E. Chocolate Avenue
Hershey PA, USA 17033
Tel: 717-533-8845
Fax: 717-533-8661
E-mail: cust@igi-global.com
Web site: http://www.igi-global.com

Copyright © 2023 by IGI Global. All rights reserved. No part of this publication may be reproduced, stored or distributed in any form or by any means, electronic or mechanical, including photocopying, without written permission from the publisher. Product or company names used in this set are for identification purposes only. Inclusion of the names of the products or companies does not indicate a claim of ownership by IGI Global of the trademark or registered trademark.

Library of Congress Cataloging-in-Publication Data

Names: Raj, Pethuru, editor. | Saini, Kavita, 1976- editor. | Pacheco, Vinicius Feitosa, editor.
Title: Applying drone technologies and robotics for agricultural sustainability / Pethuru Raj, Kavita Saini, Vinicius Pacheco, editors.
Description: Hershey, PA : Engineering Science Reference, [2023] | Includes bibliographical references and index. | Summary: "As digital intelligence and transformation technologies continue to emerge and evolve, this book is attempts to demystify how the pioneering digital technologies impact the agricultural field deeply and decisively"-- Provided by publisher.
Identifiers: LCCN 2022031080 (print) | LCCN 2022031081 (ebook) | ISBN 9781668464137 (h/c) | ISBN 9781668464144 (s/c) | ISBN 9781668464151 (eISBN)
Subjects: LCSH: Aeronautics in agriculture. | Sustainable agriculture. | Micro air vehicles--Industrial applications.
Classification: LCC S494.5.A3 .A67 2023 (print) | LCC S494.5.A3 (ebook) | DDC 632/.94--dc23/eng/20220822
LC record available at https://lccn.loc.gov/2022031080
LC ebook record available at https://lccn.loc.gov/2022031081

This book is published in the IGI Global book series Advances in Environmental Engineering and Green Technologies (AEEGT) (ISSN: 2326-9162; eISSN: 2326-9170)

British Cataloguing in Publication Data
A Cataloguing in Publication record for this book is available from the British Library.

All work contributed to this book is new, previously-unpublished material. The views expressed in this book are those of the authors, but not necessarily of the publisher.

For electronic access to this publication, please contact: eresources@igi-global.com.

Advances in Environmental Engineering and Green Technologies (AEEGT) Book Series

Sang-Bing Tsai
Zhongshan Institute, University of Electronic Science and Technology of China, China & Wuyi University, China
Ming-Lang Tseng
Lunghwa University of Science and Technology, Taiwan
Yuchi Wang
University of Electronic Science and Technology of China Zhongshan Institute, China

ISSN:2326-9162
EISSN:2326-9170

Mission

Growing awareness and an increased focus on environmental issues such as climate change, energy use, and loss of non-renewable resources have brought about a greater need for research that provides potential solutions to these problems. Research in environmental science and engineering continues to play a vital role in uncovering new opportunities for a "green" future.

The **Advances in Environmental Engineering and Green Technologies (AEEGT)** book series is a mouthpiece for research in all aspects of environmental science, earth science, and green initiatives. This series supports the ongoing research in this field through publishing books that discuss topics within environmental engineering or that deal with the interdisciplinary field of green technologies.

Coverage

- Industrial Waste Management and Minimization
- Contaminated Site Remediation
- Alternative Power Sources
- Green Transportation
- Air Quality
- Cleantech
- Sustainable Communities
- Green Technology
- Pollution Management
- Waste Management

IGI Global is currently accepting manuscripts for publication within this series. To submit a proposal for a volume in this series, please contact our Acquisition Editors at Acquisitions@igi-global.com or visit: http://www.igi-global.com/publish/.

The Advances in Environmental Engineering and Green Technologies (AEEGT) Book Series (ISSN 2326-9162) is published by IGI Global, 701 E. Chocolate Avenue, Hershey, PA 17033-1240, USA, www.igi-global.com. This series is composed of titles available for purchase individually; each title is edited to be contextually exclusive from any other title within the series. For pricing and ordering information please visit http://www.igi-global.com/book-series/advances-environmental-engineering-green-technologies/73679. Postmaster: Send all address changes to above address. Copyright © 2023 IGI Global. All rights, including translation in other languages reserved by the publisher. No part of this series may be reproduced or used in any form or by any means – graphics, electronic, or mechanical, including photocopying, recording, taping, or information and retrieval systems – without written permission from the publisher, except for non commercial, educational use, including classroom teaching purposes. The views expressed in this series are those of the authors, but not necessarily of IGI Global.

Titles in this Series

For a list of additional titles in this series, please visit: www.igi-global.com/book-series/advances-environmental-engineering-green-technologies/73679.

Implications of Nanoecotoxicology on Environmental Sustainability
Rafiq Lone (Central University of Kashmir, Ganderbal, Jammu and Kashmir, India) and Javid Ahmad Malik (Guru Ghasidas Vishwavidyalaya, Bilaspur Chhattisgarh, India)
Engineering Science Reference • © 2023 • 300pp • H/C (ISBN: 9781668455333) • US $250.00

Food Sustainability, Environmental Awareness, and Adaptation and Mitigation Strategies for Developing Countries
Ahmad Ni'matullah Al-Baarri (Diponegoro University, Indonesia) and Diana Nur Afifah (Diponegoro University, Indonesia)
Engineering Science Reference • © 2023 • 315pp • H/C (ISBN: 9781668456293) • US $110.00

Biomass and Bioenergy Solutions for Climate Change Mitigation and Sustainability
Ashok Kumar Rathoure (M/s Akone Services, India) and Shankar Mukundrao Khade (Ajeenkya D.Y. Patil University, India)
Engineering Science Reference • © 2023 • 412pp • H/C (ISBN: 9781668452691) • US $250.00

Transcending Humanitarian Engineering Strategies for Sustainable Futures
Yiannis Koumpouros (University of West Attica, Greece) Angelos Georgoulas (University of West Attica, Greece) and Georgia Kremmyda (University of Warwick, UK)
Engineering Science Reference • © 2023 • 315pp • H/C (ISBN: 9781668456194) • US $270.00

Handbook of Research on Building Greener Economics and Adopting Digital Tools in the Era of Climate Change
Patricia Ordóñez de Pablos (The University of Oviedo, Spain)
Engineering Science Reference • © 2022 • 391pp • H/C (ISBN: 9781668446102) • US $350.00

Improving Integrated Pest Management for Crop Protection and Food Security
Muhammad Haseeb (Florida A&M University, USA) Lambert H.B. Kanga (Florida A&M University, USA) and Jawwad A. Qureshi (University of Florida, USA)
Engineering Science Reference • © 2022 • 315pp • H/C (ISBN: 9781799841111) • US $215.00

Geoscientific Investigations From the Indian Antarctic Program
Neloy Khare (Ministry of Earth Sciences, India)
Engineering Science Reference • © 2022 • 305pp • H/C (ISBN: 9781668440780) • US $250.00

IGI Global
PUBLISHER of TIMELY KNOWLEDGE

701 East Chocolate Avenue, Hershey, PA 17033, USA
Tel: 717-533-8845 x100 • Fax: 717-533-8661
E-Mail: cust@igi-global.com • www.igi-global.com

Table of Contents

Detailed Table of Contents

Agriculture v4.0 revolves around digital technologies such as artificial intelligence, data analytics, cellular agriculture, and sensors. These latent technologies can help us meet sustainable development goals. By end of 2020, the global digital agriculture-based market was valued at USD10 billion, and is projected to grow at 15% between 2021-2026 to be valued at USD 22.88 billion by 2026. Rising global temperatures, shrinking natural resource base, and mushrooming human population have increased the pressure on limited resources for enhanced productivity. Ubiquitous digital technologies in the near future will help transform agriculture into a precise, productive, and sustainable sector. Smart farming technologies involve precise and efficient data acquisition, data management, decision making, and field management. Big data management has been a key hurdle in application of digitalized agriculture. The issue of data management has been resolved to a great extent using IoT. Mass field application of futuristic technologies awaits developing and under-developed countries.

This chapter explores the use of emerging ICT technologies for agriculture in Zimbabwe, a developing country in sub-Saharan Africa. It investigates the concept of ICT technologies and their potential impact on agriculture. The chapter then brings the background and context of the study. The chapter is premised on a qualitative approach which fuses a literature review to situate the chapter and eight (8) in-depth interviews and a ten (10) member focus group discussion to gather empirical data. The findings reveal that there are emerging ICT technologies in Zimbabwe which include artificial intelligence, drone technology, IoT, robots, and big data. However, the application and use of these technologies are still

A healthy plant has a significant impact on the environment and human health. To put it another way, early discovery of plant diseases results in a momentous increase in the quality of the finished product. Using IoT-based agriculture, deep learning, and image processing techniques are essential to developing an effective way for detecting and controlling the spread of disease in agricultural products. Leaf drop is a common symptom of many diseases, as is the disappearance of the plant or plants nearby. This means that early detection and prevention are critical to ensuring healthy grape growth and yield. A method for detecting grape disease utilizing a CNN methodology and real-time data from environmental characteristics collected to predict illness likelihood. The image is subjected to multiple layers of CNN processing. The shuffled frog-leaping technique, a meta-heuristic algorithm, is used to optimize the CNN's learning rate (SFLA). The results suggest that the proposed SFLA-CNN outperforms existing approaches in terms of performance.

The barren land detection management using manual methods has proven to be of great difficulty in terms of time and workforce. Also, several other methods such as vibration-based models, etc. have been costly to be implemented, as well as their questionable accuracy is along with their limitations. This system aims at designing and developing a drone that mainly focuses on addressing the issue of deforestation with the integration of digital intelligence and drone technology.

Drone technology brings a revolutionary change in every sector of production, supply, and service. It came as a new revolution in the field of agriculture also. Being a light weight, easy-to-handle, and smart device, drone devices along, with IOT and sensors, help make the smart agriculture to rise; it can also be considered as the agriculture 4.0 revolution. This chapter demonstrates the potential application of drone technology that are currently in use at the countries like US for smart cultivation. Usage of drone

by day in real-time execution which can be addressed using current trends of technology like machine learning and deep learning. Machine learning is a subset of artificial intelligence. ML is functioning for image & speech recognition, mail filtering, Facebook tagging mechanism, and many others. Deep learning is an advanced technology that is a subset of machine learning with the capacity to learn more intelligently on a large set of data. Deep learning works with multiple hidden layers to produce the predicted outcomes. Deep learning algorithms include convolutional neural networks, recurrent neural networks, long short-term memory networks, stacked auto-encoders, deep boltzmann machines &, etc.

The previous work is two versatile 3D methodologies for independent horticultural UAVs to find impact-free ideal ways in time with the interaction costs in various conditions, containing different obstructions. These strategies were created, roused by two metaheuristic calculations. This concentrate likewise involved an instrument for snagging the board. It observes the best way between the beginning and target stations of each UAV. In way arranging techniques, as a rule, either the robots haphazardly move, or expensive cycles are attempted in tracking down an advanced way. This interaction is performed steadily between the two stations. The UAV attempts to track down a method between two places. It enhances haphazardly made ways to observe the ideal direction. On account of this technique, robots can likewise effectively keep away from impediments. The amount of every one of the tuples' expenses is determined, and the expense of the way is acquired. The suggestion implements some security measures to make the system better.

Machine learning is the branch of artificial intelligence which makes the computers capable enough to learn from the input data using statistical models and algorithms to imitate the human learning process, gradually becoming more accurate at outcome prediction without being explicitly programmed to do so. Once the algorithm learns what task it has to do with the data, it works on it automatically. Deep learning, also referred to as deep neural network or deep neural learning, on the other hand, is the branch of machine learning, which is just a neural network with many layers to optimize and upgrade for accuracy.

Cellular technology has undergone remarkable development over the last four decades. With the commercial introduction of 5G, humans are again going to take a considerable leap in the fields of IoT, augmented and virtual reality, and tactile internet. Like LTE, the 5G new radio technology is also based on orthogonal frequency-division multiplexing, but unlike the previous generations, it supports a large number of different use cases because of its ability to operate over both high and low-frequency bands.

Jigna Bhupendra Prajapati, Acharya Motibhai Patel Institute of Computer Studies, Ganpat University, India
Akash Kumar, Center for Health and Applied Sciences, Ganpat University, India
Jhilam Pramanik, Center for Health and Applied Sciences, Ganpat University, India
Bhupendra G. Prajapati, Shree S. K. Patel College of Pharmaceutical Education and Research, Ganpat University, India
Kavita Saini, Galgotias University, India

Advancements of the last decade in edge computing, edge IoT, and edge artificial intelligence now allow for autonomous, efficient, and intelligent systems to be proposed for various industrial applications. Intelligence agricultural solutions allow farmers to achieve more with less while improving quality and providing a rapid go-to-market approach for produce. Using AI is an effective technique to detect any crop health concerns or nutrient inadequacies in the field. Plant diseases affect the food system, economy, and environment. This chapter covers intelligent agriculture & challenges in front of technology. It focuses AI application using machine learning, artificial neural network (ANN), and deep learning. The various AI applications in agriculture for land monitoring, crop and varietal selection, smart irrigation or automation of irrigation, monitoring of crop health, crop disease detection, predictive analytics, weed control, precision agriculture, harvesting, yield estimation and phenotyping, supply chain management, and food quality.

Esha Saxena, Dr. Akhilesh Das Gupta Institute of Technology and Management, India
Preety Shoran, Dr. Akhilesh Das Gupta Institute of Technology and Management, India
Meenakshi Yadav, Dr. Akhilesh Das Gupta Institute of Technology and Management, India

Artificial intelligence is facilitating the future of smart systems and their associated services. The world is witnessing the rapid growth in smart systems with artificial intelligence (AI) algorithms. Services must include a set of features to handle future trends, intelligent decisions, security, adaptivity, scalability, user-friendliness, time-saving ability, and adaptability to new architectural models. To satisfy these objectives, smart systems and services need advancements in working models of IoTs and applications of cloud computing. The AI and IoT create a great combination in various fields; health systems, agriculture, weather forecasting, manufacturing units, prediction, etc. Some of its applications include smart homes, smart cities, and smart businesses. This chapter also discuss about the role of IoT and cloud computing in smart systems, followed by smart city development, education system, environment protection, healthcare, water reservation, security, etc.

Developed by 3GPP, 5G is the technology standard for cellular broadband networks. Introduced in July of 2016 and commercialized almost 3 years later, it is seen as the successor to 4G networks. 5G uses device-to-device communication, dividing the service area into small geographical locations called cells. These cells are connected intensely through localized antennas. The connectivity of these antennas, despite being small, amounts to high-speed downloads ranging up to 10 Gbps. This chapter includes a brief introduction of 5G network; concise notes on the history of its predecessors such as 2G, 3G, and 4G; as well as a thorough look through the architecture of the 5G network model.

Preface

Having understood the strategic significance, every industry vertical is meticulously embarking on digital transformation. There are several renowned digital transformation technologies such as the Internet of Things (IoT), artificial intelligence (AI), 5G communication, digital twins, cloud-native computing, serverless and edge computing, microservices and event-driven architecture, streaming analytics, cybersecurity, etc. With the faster maturity and stability of these technologies and tools, businesses, government organizations and institutions across the globe are keenly embracing these techniques to be right and relevant to their growing constituents, clients and consumers. The aspect of digital transformation has the power to bring forth a bevy of multifaceted opportunities and possibilities for industrial houses in enhancing their revenue by exploring new avenues, in visualizing and realizing sophisticated services to retain their loyal customers, in attaining fresh business partners and prospects, in envisaging new trendsetting ways and means of achieving heightened productivity, and in offering premium and hitherto unknown capabilities. Precisely speaking, with the growing power of breakthrough technologies, the digital transformation goal is being neatly and nicely implemented so that the much-anticipated customer delight can be fulfilled sooner than later.

Agriculture is an indispensable domain of study and research for the whole world. As the world population is growing steadily, the field of agriculture has to equally grow in order to meet up the internal and external challenges with clarity and alacrity. Agriculture processes are being rationalized and optimized. Besides process excellence, infrastructure optimization and upgrade are being meticulously looked into. Connectivity and cognition get utmost fillip. All kinds of physical, mechanical and electrical systems being widely leveraged in agriculture environments are being systematically digitized through a host of digitization and edge technologies. The agriculture data is being gleaned and subjected to a variety of deeper and decisive investigations in order to emit out actionable insights, which comes handy in continuously working in empowering agriculture offerings, outputs and operations.

This book is being planned and prepared to tell all about the various information-processing, communication and digital transformation technologies to achieve better and bigger things in agriculture, sericulture and horticulture. Thus, the arrival of evolutionary and revolutionary technologies is being seen as the grandiose sign and signal to energize the struggling agriculture sector. We have incorporated 16 chapters in this book and each chapter is closely associated with digital transformation technologies and how they individually and collectively contribute for the betterment of the vital phenomenon of garniture.

There is a chapter on smart farming ingredients: the IoT sensors and actuators, software-defined cloud environments, high-bandwidth connectivity, real-time capture and crunching of agriculture field data, the growing contributions of AI-enabled robots and drones, etc. The chapters focus on how AI models are being immaculately designed, developed and deployed for complex agriculture problems.

The drone technology has gained a greater say in monitoring, measuring, managing and maintaining agriculture fields. Thus, the overwhelming idea of how digital transformation methods in arriving at highly optimized agriculture processes is gaining a lot of importance. We have incorporated all the 16 chapter abstracts in order to clearly articulate and accentuate the core ideals of each chapter.

Chapter 1 focuses on Agriculture v4.0 and how it revolves around digital technologies such as artificial intelligence, data analytics, cellular agriculture and sensors. These latent technologies can help us meet Sustainable Development Goals. By end of 2020, the Global digital agriculture-based market was valued at USD10 billion and is projected to grow at 15% between 2021-2026 and will be valued at USD 22.88 billion by 2026. Rising global temperatures, shrinking natural resource base, and mushrooming human population have increased the pressure on limited resources for enhanced productivity. Ubiquitous digital technologies in near future will help transform agriculture into a precise, productive, and sustainable sector. Smart farming technologies involves precise and efficient data acquisition, data management, decision making and field management. Big data management has been a key hurdle in application of digitalized agriculture. The issue of data management has been resolved to a great extent using IoT. Mass field application of futuristic technologies awaits developing and under-developed countries.

Chapter 2 explores the use of emerging ICT technologies for agriculture in Zimbabwe a developing country in sub-Saharan Africa. The concept of ICT technologies and their potential impact on agriculture. The chapter then brings the background and context of the study. The chapter is premised on a qualitative approach which fuses a literature review to situate the chapter and eight (8) in-depth interviews and a ten (10) member focus group discussion to gather empirical data. The findings reveal that there are emerging ICT technologies in Zimbabwe which include artificial intelligence, drone technology, IoT, Robots, and Big Data. However, the application and use of these technologies are still very localized and limited to big and highly capacitated commercial farmers with very few smallholder farmers using the technology because of challenges that impede the adoption and use of these emerging ICT technologies for agriculture by the bulk of farmers. The chapter submitted measures that could be implemented to try and increase the adoption and increased use of these technologies.

Chapter 3 focuses on Smart Farming. Smart Farming uses Information and Communication Technologies in various fields of agriculture. Smart farming refers to the use of information and data management technologies in agriculture. Smart Farming leads towards high productive and sustainable agricultural production. Smart Farming provides the farmer many advantages for decision making for better management. Smart farming technologies collect precise measurements of factors that determine farming outcomes. It enables agriculture more reliable, predictable and sustainable. It also improve crop health, reduce the ecological footprint of farming, help feed the increasing global population, provide food security in climate change scenarios, Achieve higher yields while reducing operating costs. It's also needed to meet the needs of the growing population. There are many technological devises as IoT, Software support, Connection, Data Analytics, Robots, Drones, GPS is useful to enhance the quantity and quality of Agriculture production with minimizing labor.

Chapter 4 examines the use of drone technology in agriculture to predict the occurrence of pest and disease outbreaks. The purpose of the study is to comprehend competition analyses and determine the best entry tactics for the aerial imaging and advice service market. They can determine the current state of drone technology and create strategies based on the different companies providing drone services to farms by studying how farmers perceive the drone employed in predictive pest and disease analyses. These observations are highly valuable for comprehending how drone technology can affect Indian agriculture. A virtual survey gives a useful and accurate image, whether or not they utilise it for predictive forecasting and identification and if they depend on the advice given by various sources or not and most importantly, how best the drone service available in the market, whether farmers can afford for it.

A healthy plant has a significant impact on the environment and human health. To put it another way, early discovery of plant diseases results in a momentous increase in the quality of the finished product. Using IoT-based agriculture, deep learning and image processing techniques are essential to developing an effective way for detecting and controlling the spread of disease in agricultural products. Leaf drop is a common symptom of many diseases, as is the disappearance of the plant or plants nearby. This means that early detection and prevention are critical to ensuring healthy grape growth and yield. A method for detecting grape disease utilizing a CNN methodology and real-time data from environmental characteristics collected to predict illness likelihood. The image is subjected to multiple layers of CNN processing. The Shuffled Frog-Leaping Technique, a meta-heuristic algorithm, is used to optimize the CNN's learning rate (SFLA). The results studied in Chapter 5 suggest that the proposed SFLA-CNN outperforms existing approaches in terms of performance.

The barren land detection management using manual methods has proven to be of great difficulty in terms of time and workforce. Also, several other methods such as vibration-based models, etc. have been costly to be implemented as well as their accuracy is questionable along with their limitations. This system of Chapter 6 aims at designing and developing a Drone that mainly focuses on addressing the issue of deforestation with the integration of Digital Intelligence and Drone Technology.

Chapter 7 focuses on how drone technology brings a revolutionary change in every sector of production, supply and services. It came as a new revolution in the field of agriculture also. Being a light weight, easy to handle and smart device, drone devices along with IOT and sensor help make the smart agriculture to come up, it can also be considered as the Agriculture 4.0 revolution. This chapter demonstrates about the potential and applications of drone technology that are currently in use at the countries like US for smart cultivation. Usage of drone technology not only helps to improve the production; it can also help to avoid the wastage of the crops and fruits grown at the places that are harder to be achieved by the human being due the geographical conditions. Considering the same problem author proposes a smart drone-based harvester that can be used to pluck the fruits and vegetable at high altitude or difficult to reach places. This harvester is also able to measure the ripening condition of the fruits in order to avoid any type of wastage.

Since the advent of technology, it is growing by leaps and bounds. The thirst to be connected to the world has led to the emergence of 5G in 2019. Mobile technologies are evolving rapidly to address various concerns related to speed, efficiency, connectivity, power consumption, bandwidth, and latency. As compared to its predecessors, 5G provides a greater speed and higher bandwidth to exchange data, in lesser time. Thus, it can cater to many advanced communication applications and fully supports the current modern technology and solves its intended purpose. 5G seems promising in connecting any device, object, or human to any other node, at any location and at all times. It is expected that by 2030, the network traffic and dependency on wireless communication will experience a massive increase and 5G is ready to support such a huge volume of traffic. Chapter 8 will present a systematic and detailed study of 5G, a comparative analysis of 5G with its predecessor technologies, its potential applications, and current use cases of 5G, with a special focus on the agricultural sector.

Chapter 9 is devoted to the Study of the Legal Regulation of the use Of Artificial Intelligence In Deep Learning, Machine Learning And Services. A New Service System Classification that requires minimum Human Interaction but Works especially based on things Interaction. Efficient results can be acquired in the Fields of Information Technology, IoT, Security, Healthcare, Business, Agriculture, etc. by using AI in Smart Systems And Services. We can use AI algorithms and techniques to solve the issues surrounding regular smart based services, as the future is a clear sky of AI. The goal is to investigate the issues and drawbacks of existing methods and to improve the service's efficiency in order to achieve the best results.

In Chapter 10, each primary, secondary, tertiary, quaternary, and the quinary sector has huge or very huge incremental data from large-scale, small-scale industries, medium industries, or cottage industries. The data associated with each of them are very crucial from every point of view. The complex problems are increasing day by day in real-time execution which can be addressed using current trends of technology like Machine learning and Deep Learning. Machine learning is a subset of Artificial Intelligence. ML is functioning for image & speech recognition, mail filtering, Facebook tag mechanism, and many others. Deep learning is an advanced technology that is a subset of machine learning with the capacity to learn more intelligently on a large set of data. Deep learning works with multiple hidden layers to produce the predicted outcomes. Deep Learning Algorithms include Convolutional Neural networks, Recurrent Neural Networks, Long Short-Term Memory Networks, Stacked Auto-Encoders, Deep Boltzmann machines, etc.

In Chapter 11, the previous work is two versatile 3D methodologies for independent horticultural UAVs to find impact-free ideal ways in time with the interaction costs in various conditions, containing different obstructions. These strategies were created, roused by two metaheuristic calculations. This concentrate likewise involved an instrument for snagging the board. It observes the best way between the beginning and target stations of each UAV. In way arranging techniques, as a rule, either the robots haphazardly move, or expensive cycles are attempted in tracking down an advanced way. This interaction is performed steadily between the two stations. The UAV attempts to track down a method between two places. It enhances haphazardly made ways to observe the ideal direction. On account of this technique, robots can likewise effectively keep away from impediments. The amount of every one of the tuples' expenses is determined, and the expense of the way is acquired. The suggestion implements some security measures to make the system better.

In Chapter 12, we learn machine learning is the branch of artificial intelligence, which makes the computers capable enough to learn from the input data using statistical models and algorithms to imitate the human learning process, gradually becoming more accurate at outcome prediction without being explicitly programmed to do so. Once the algorithm learns what task it has to do with the data, it works on it automatically. Deep learning, also referred to as deep neural network or deep neural learning, on the other hand, is the branch of machine learning, which is just a neural network with many layers to optimize and upgrade for accuracy.

Chapter 13 looks at how cellular technology has undergone a remarkable development over the last four decades. With the commercial introduction of 5G, we are again going to take a considerable leap in the fields of IoT, Augmented and Virtual Reality, and Tactile Internet. Like LTE, the 5G New Radio Technology is also based on Orthogonal Frequency-Division Multiplexing but unlike the previous generations, it supports a large number of different use cases because of its ability to operate over both, high and low-frequency bands.

Advancements of the last decade in Edge Computing, Edge IoT, and Edge Artificial Intelligence now allow for autonomous, efficient, and intelligent systems to be proposed for various industrial applications. Intelligence agricultural solutions allow farmers to achieve more with less while improving quality and providing a rapid go-to-market approach for produce. Using AI is an effective technique to detect any crop health concerns or nutrient inadequacies in the field. Plant diseases affect the food system, economy, and environment. Chapter 14 covers intelligent agriculture and challenges in front of technology. It focuses AI application using machine learning, artificial neural network (ANN) and deep learning. The various AI Applications in agriculture for land monitoring, crop and varietal selection, smart irrigation or automation of irrigation, monitoring of crop health, crop disease detection, predictive analytics, weed control, precision agriculture, harvesting, yield estimation and phenotyping, supply chain management and food quality.

Artificial Intelligence is facilitating the future of smart systems and their associated services. The world is witnessing the rapid growth in smart systems with Artificial intelligence (AI) algorithms. Services must include a set of features to handle future trends, intelligent decisions, security, adaptivity, scalability, user-friendliness, time-saving, and adaptability to new architectural models. To satisfy these objectives, smart systems and services need advancements in working models of IoTs and applications of cloud computing. The AI and IoT create a great combination in various fields; health systems, agriculture, weather forecasting, manufacturing units, prediction, etc. Some of its applications include smart homes, smart cities, and smart businesses. Chapter 15 also discuss about the role of IoT and Cloud Computing in smart systems, followed by smart city development, education system, environment protection, healthcare, water reservation, security etc.

Developed by 3GPP, 5G is the technology standard for cellular broadband networks. Introduced in July of 2016, and commercialized almost 3 years later, it is seen as the successor to 4G networks. 5G uses device-to-device communication, dividing the service area into small geographical locations called cells. These cells are connected intensely through localized antennas. The connectivity of these antennas, despite being small, amounts to high-speed downloads ranging up to 10 Gbps. Chapter 16 includes a brief introduction of 5G network, concise notes on the history of its predecessors such as 2G, 3G, and 4G, as well as a thorough look through the architecture of the 5G network model.

Pethuru Raj
Reliance Jio Platforms Ltd., Bangalore, India

Kavita Saini
Galgotias University, India

Vinicius Pacheco
Office of the Prosecutor General, Brazil

Chapter 1
Digital Technologies and Tools for the Ensuing Digital Era

Tarun Sharma
CSK Himachal Pradesh Krishi Vishvavidyalaya, India

Akashdeep Singh
https://orcid.org/0000-0003-1473-8746
CSK Himachal Pradesh Krishi Vishvavidyalaya, India

Naveen Kumar
CSK Himachal Pradesh Krishi Vishvavidyalaya, India

Garima Chauhan
CSK Himachal Pradesh Krishi Vishvavidyalaya, India

ABSTRACT

Agriculture v4.0 revolves around digital technologies such as artificial intelligence, data analytics, cellular agriculture, and sensors. These latent technologies can help us meet sustainable development goals. By end of 2020, the global digital agriculture-based market was valued at USD10 billion, and is projected to grow at 15% between 2021-2026 to be valued at USD 22.88 billion by 2026. Rising global temperatures, shrinking natural resource base, and mushrooming human population have increased the pressure on limited resources for enhanced productivity. Ubiquitous digital technologies in the near future will help transform agriculture into a precise, productive, and sustainable sector. Smart farming technologies involve precise and efficient data acquisition, data management, decision making, and field management. Big data management has been a key hurdle in application of digitalized agriculture. The issue of data management has been resolved to a great extent using IoT. Mass field application of futuristic technologies awaits developing and under-developed countries.

INTRODUCTION

From time immemorial, agriculture has been the backbone of civilizations which makes innovations in

DOI: 10.4018/978-1-6684-6413-7.ch001

Copyright © 2023, IGI Global. Copying or distributing in print or electronic forms without written permission of IGI Global is prohibited.

the sector a necessity. As the global population surges, demands for food production also intensify which stresses the current agricultural production systems. Besides the burgeoning population, other stressors such as the environmental effects of the production systems having undesirable consequences and consumers demanding more nutritious and safe food have led the scientific community to think of alternatives to improve agricultural production as well practices that are also within the scope of sustainability.

From a perspective, one can imagine the task at hand to be intimidating and insuperable but this has long been the commandment. For society to sustain and prosper, innovation in agricultural practices is obligated. It is safe to say, presently, novel tools revolve around digital technologies such as sensors, artificial intelligence, big data, and smartphone applications which when implemented effectively in agricultural production systems offer unparalleled contributions to tackling the challenges mentioned earlier. Digital agriculture is information and communication technology and data ecosystems to support the development and delivery of timely, targeted information and services to make farming profitable and sustainable while delivering safe nutritious, and affordable food for all (ICRISAT, 2022). In the current setting, farmers around the globe are encountering altering climates resulting in increased temperatures, variations in rainfall patterns, extreme weather events, and reductions in water availability.

The farming community must continue to evolve to sustain and improve productivity to meet the demand. Digital technologies have the potential which can help farmers overcome the vagaries and seize the opportunity for growth. They offer the farming communities the possibility of curtailing the inputs while maximizing returns. The world is treading the path toward smart agriculture which encompasses all the information communication technologies (ICTs) which help increase productivity and sustainability.

A STEP TOWARDS DIGITALIZATION

A significant transformation could be observed in all the sectors with the introduction of ICTs but within the agriculture sector, its adoption has been rather slow. The factor primarily affecting the promotion and adoption of such technologies is the technical know-how followed by social influence (Fox et al., 2021). It is the trust of the farmers in the technology that impacts the overall performance of any digital technology. Digital agriculture offers technologies that will help transform agricultural practices and provide sustainable solutions to the problem of delivering food to the swelling global population. For eons, humans have played a significant role in improving agriculture as a system. It is fascinating to watch the years of experience be linked to mathematical equations resulting in complex algorithms which form the foundation for artificial intelligence or computer modeling systems (Mitton et al., 2022).

ICTs continue to pervade the agriculture sector with smart agriculture, digital agriculture, and e-agriculture terms used frequently in the scientific community and used interchangeably, all leading to enhanced productivity and sustainability. They comprise a wide spectrum of technologies that need to be harnessed according to the needs of the present and the future to sustain agriculture production.

Technologies for Digital Agriculture

Digital agriculture is still in its infancy with current technologies mainly constituting positioning systems, remote sensing through spectral imaging systems, decision support systems (computing and its related software), and robotics in agriculture. Agricultural practices were centered around witnessing the prevailing conditions in a field which ultimately impacted the overall crop production. With the advancement in technologies, data became the source of regulating field operations. The extent of profit from the farm is directly linked to the data provided for the technology to function aptly. The digitization of the agriculture system would increase the capability of the farmers to improve crop production, resource and farm management along with a better understanding of the agricultural ecosystem (Bellon-Maurel et al.,2022).

Agricultural Technologies for Digitization

The fourth industrial revolution revolves around reshaping every possible sector with the use of intuitive technologies such as sensors, artificial intelligence (AI), big data, unmanned aerial vehicles (UAVs), and unmanned terrestrial vehicles (UTVs), etc. The integration of such technologies in the agriculture sector is now regarded as Agriculture 4.0 or smart agriculture or digital agriculture or digital farming. Through this chapter, the authors have tried to cover the emerging technologies that find their way into the agriculture sector.

Unmanned Terrestrial Vehicles in Agriculture

At times during peak season of agricultural operations, there is a shortage of labor combined with the high cost of labor makes the farmers look for suitable alternatives. Robots operating on the ground are typical examples of UTVs. The prime gears of UTVs are a platform for locomotive apparatus, sensors for navigation, supervisory control systems, an interface for the control systems, communication links for information exchange between devices, and system architecture for integration between hardware and software agents. The UTVs can be semi-autonomous (human intervention using the interface) or fully autonomous (with the use of sensors and artificial intelligence). They are also referred to as unmanned ground vehicles (UGVs). They are capable of performing various agricultural operations like weed control, spraying of pesticides, seed sowing, harvesting operations, and even pollination (Abbasi et al., 2022).

Unmanned Aerial Vehicles in Agriculture

The global drone market is estimated to reach US$ 28.47 billion in the financial year 2021-2022, with India likely to account for a 4.25% share (IBEF, 2022). UAVs depending on the technology employed are either fixed-wing type or rotor copter type (Fig. 1). Among the roto copter type, drones with four or six rotors are used more often in the agriculture sector. This is mainly because they are easy to use. Depending on the autonomy level of the UAVs, they can either be teleoperated by the pilot by giving certain reference points to the machine or telecommand in which the aircraft relies on an automatic controller.

The UAVs are mounted with appropriate sensors such as vision, infrared, multispectral and hyperspectral, etc. which allows farmers to receive data pertaining to vegetation, leaf area, disease incidences, etc. These vehicles have made the monitoring of the field easy and comprehensive but the issue lies with the battery pack of the drones and copters. Increasing the size of the battery also increases the weight and reduces the carrying capacity of the UAVs affecting their efficiency.

Figure 1. Left side showing rotocopter which can move in all directions for spraying of either pesticides or growth hormones. On the right is a fixed-wing type of UAV that can move in a single direction.

REMOTE SENSING

Remote sensing (RS) refers to the collection of data about an object from a distance. Such dimensions are subjected to gravitational and electromagnetic forces. It classically comprises data acquisition, data processing, and data analysis system. They have been systematically employed to monitor vegetation, soil, water, snow, and cloud parameters with the aim to optimize crop production. Among various digital technologies, RS is the most widely used component but its use is generally restricted to weather forecasting.

RS employs a global positioning system (GPS) and geographical information system (GIS) to identify the area of interest (Gebeyehu, 2019). It is through RS technology UAVs and UTVs are able to detect a specific location in the field. It is the myriad of systems that work together to give the desired effects. The sensors mounted on UAVs and UTVs using RS gather a large amount of dataset which is further processed in the hardware-software system thereby enabling various operations such as disease monitoring, field water status, yield estimation, etc. (Adão et al., 2017). RS applications include crop acreage using Norval Difference Vegetation Index (NDVI), crop yield forecasting, flood monitoring, pest infestation, and water resource management (Shanmugapriya et al., 2019; Gebeyehu, 2019; Khanal et al., 2020).

Sensors in Agriculture

Sensors are the devices that are spatially arranged or distributed in the open field or protected systems that monitor, the physical conditions in which the crop plants grow, in specified iterations.

They are also capable of temporarily storing the data and transmitting the collected data to a specific location. These sensors may be connected through a wireless network system in order to facilitate the easy transmission of data from anywhere. Once the sensors have acquired the data, it is their job

Figure 2. Various types of sensors that find their way into the agriculture production systems

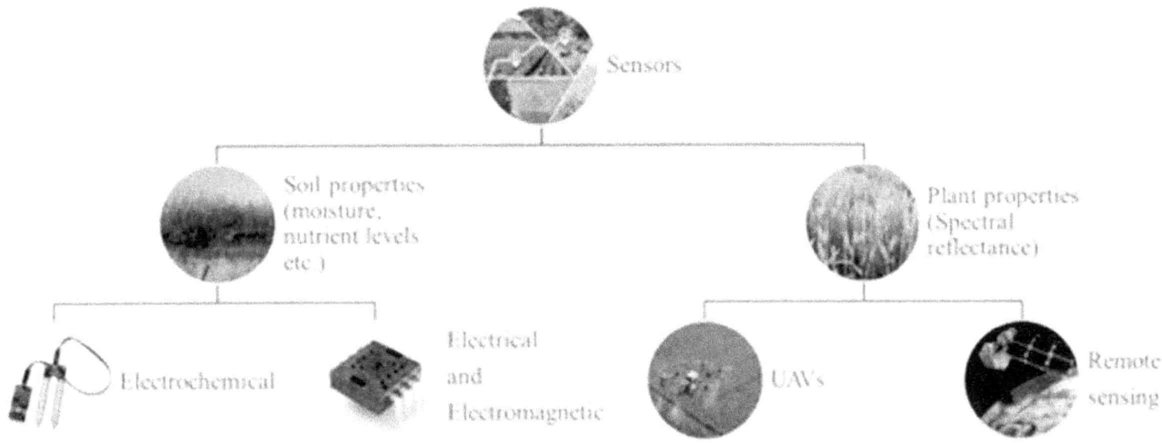

to transmit the same without any loss. The sensors are generally fitted with wireless communication devices so that they are able to transmit data from far-off regions. Such sensors/devices feed on certain energy sources which are often limited (*e.g.* Lithium-ion batteries etc.). Under such circumstances, it becomes essential to filter the type of data that must be transmitted because the data transmission requires the most energy (Mitton et al., 2022). There are various types of sensors such as terrestrial wireless sensor networks (TWSNs), wireless underground sensor networks (WUSNs), underwater wireless sensor networks (UWSNs), wireless multimedia sensor networks (WMSNs), and mobile wireless sensor networks (MWSNs) (Abbasi et al., 2022). Apart from these sensors, the broad categories of sensors used to measure soil properties are optical sensors which use light to measure the soil properties like organic matter content, moisture content, and clay content in the soil, electrochemical sensors measure the pH and soil nutrient status, and dielectric sensors primarily find their way in irrigation management (Queiroz et al., 2020). Mapping of the field based on the images received from the sensors enables the monitoring of the field fast and effectively. Sensors such as Light Detection and Ranging (LIDAR) can distinguish between plants based on their height (Peteinatos et al., 2013). Such sensors can be used in technologies to identify weeds or in spraying systems on UTVs.

A project was designed to monitor moisture and temperature using sensors in citrus plants. The sensors in question were air humidity sensor, air temperature sensor, soil temperature sensor, and soil humidity sensor. The sensors collected the data every minute in real-time and transmitted it to the decision support system which would then compare the data with the standard sample in the database controlling the irrigation system (Zhang et al., 2017). The system demonstrated its aptitude to make scientific decisions with the only shortcoming, high cost. The site-specific management practices not only cut the costs of cultivation but also conserves resources and make the system more sustainable. The farmers' dependence on pesticides for enhancing crop production has been gaining ground. Exposure to elevated levels of pesticides can result in cancer, asthma, bronchitis, autism, and infertility (Shah, 2020). This is controlled by the use of sensor systems that can be mounted on UAVs or UTVs. Consequently, herbicides are applied only where are detected, and with variable rate technology onboard, it is now possible to adjust the quantity of herbicide to be applied to the identified weed (Lopez-Granado, 2011; Frasconia et al., 2017; Lopez-Granada et al., 2016). The problem identified few groups with the UAVs

Figure 3. Working of AI in the field of agriculture using the data from various sensors as an input

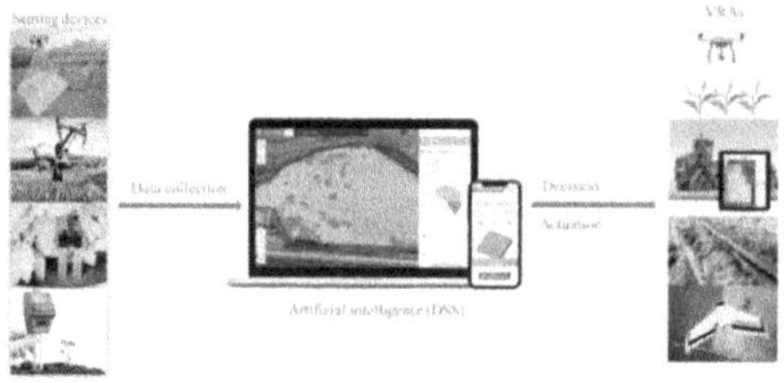

sensor is that they fail to identify weed infestation, which is generally due to low-cost imaging sensors *viz.,* visible (RGB) and near-infrared (NIR) (Castaldi et al., 2017; Adão et al., 2017). Subsequently, the use of higher spectral resolution sensors such as multispectral and hyperspectral sensors overcome the inadequacy (Adão et al., 2017).

ARTIFICIAL INTELLIGENCE AND MACHINE LEARNING

Modern agriculture systems have to tackle a plethora of problems such as increasing food demands due to population explosion, natural resource management, climate changes and sustainability. This calls for immediate steps that must be taken in order to lessen the burden on the agriculture production system, simultaneously, reducing the environmental burden. The chaos in the system has driven the farming community to precision agriculture which has the potential to increase productivity and maintain sustainability. AI is the study of the computer system that models the situation at hand and applies human intelligence to get results (Saxena et al., 2020). It simulates the intelligent behavior of humans to do cognitive computing and respond to different situations. In agriculture, AI has immense potential to provide information to the farmers regarding the time of sowing, the technique of sowing, scheduling of pesticide sprays, irrigation scheduling, optimizing soil nutrient levels, and yield forecasting (Kim et al., 2008; Talaviya et al., 2020). Machine learning (ML) is a tool to strengthen the AI which in turn monitors and manages various on-farm operations and forecasting. The algorithms are usually the talk when discussing AI.

ML and deep learning are concerned with algorithms making up the entire Artificial Neural Networks (ANNs) which are similar to neuron networks of the brain. Deep learning (DL) can hence solve more complicated problems by feeding on data in the form of images and videos resulting in the identification of deficiencies, diseases, etc. ML in the agriculture sector is associated with weather and yield forecasting, weed identification, disease, and pest detection, fertilizer gradient, and its determination. Combining the ML datasets results in the progress of deep learning. DL involves the extraction of information from a large mass data set. This enables the machine to truly think like a human. When a machine is fed with hundreds and thousands of images of say, *Galinsoga quadrifolia* it will be able to identify that weed on its own in any condition due to DL algorithms. The future of farming is a mix of technological and biological skills which would improve the quality while minimizing the losses and workforce. The central reason for the push toward AI is the efficiency and precision of robotics which is unmatched when compared to conventional farming practices (Dursun and Ozden, 2011). They are capable of performing multiple farm operations such as weeding, irrigation, spraying, fertilizer application, etc., and

Figure 4. Digital technologies for modern agriculture system (Trendov et al., 2019)

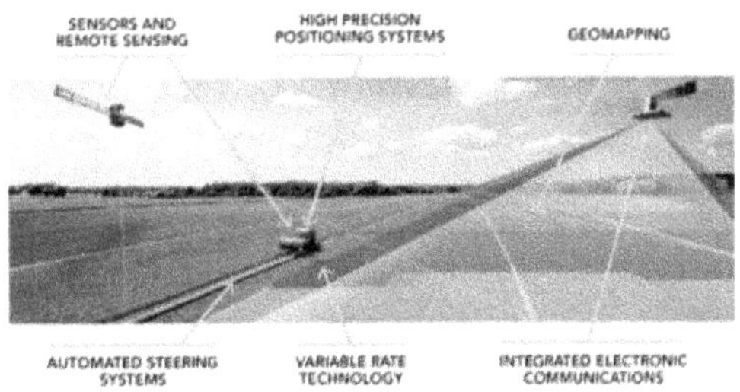

also ensuring the sustainability of the environment and ecosystem, which we as humans have failed, without compromising the crop production. A farm called "NatureFresh Farms" has got itself a robotic lens mounted on UAVs that zooms on the flower of a tomato crop and images of that are fed to the AI algorithms. The AI in turn predicts the time required for the flower to turn into a ripe tomato. Knowing the number of tomatoes available from the farm, the processing of orders becomes easier. The farm is located between Ontario and Ohio in Canada (Talaviya et al., 2020).

In the changing climate, it is pertinent to conserve resources and obtain maximum returns to sustain life on the planet. The role of AI will become an integral part of the agricultural systems in the near future to realize sustainability. The ANN is analogous to the human brain where synaptic endings are replaced by artificial synapsis. This enables AI to make decisions more efficiently and accurately. In Turkey, the use of an Artificial Neural Network was able to conserve as much as 20.46% water and 23.90% energy. The ANN system was fed with data pertaining to soil type and moisture content in order to schedule the time and amount of irrigation (Khare and Seth 2020).

Big Data and Analytics in Agriculture

The advancement in technology, particularly in the Internet of Things (IoT) has brought a fourth revolution in the farming sector, generating an enormous amount of data pertaining to meteorological information, soil-related information, and land uses with the use of multiple sensors thus making the agriculture system increasingly data-driven and data-enabled (Zhai et al., 2020). The machines when equipped with various sensors would measure different parameters in their respective environment that is aimed at particular behavior. This seemingly immense data is now referred to as big data, which includes contents either in the form of text or multimedia (*i.e.,* photographs, video, audio). The process of probing the data to find likings of customers, market inclinations, and other useful information is big data analytics.

The concept of big data includes three elements *i.e.,* network management, stakeholder network and business processes (Lambert and Cooper, 2000). These elements are interlinked to each other and influence big data analytics. Within the concept, business processes are primarily focused on the generation and use of big data for the manipulation of farming processes. It comprises data chain, farm management and farm processes (Davenport, 1993; Beer, 1981; Miller and Mork, 2013). The use of sensors helps the decision support system within the AI in managing the farm operation through the input data. The stakeholder network comprises of people who are either the users of big data or the managers of data. Finally, there is network management, typically comprising organizational and technological structures. They control the management of the data chain and the whole working of the processes. They are in-

Figure 5. Conceptual framework of Big Data from a socio-economic perspective (Lambert and Cooper, 2000)

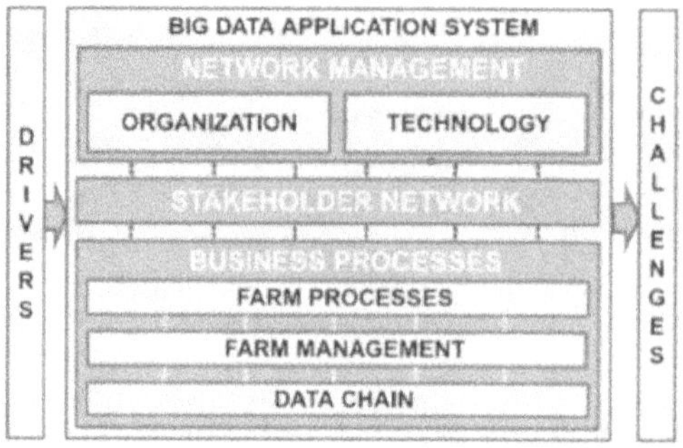

volved in the governance and transfer of technology for the use of big data in agriculture systems (van der Vorst et al., 2005).

Big data is characterized by the five V's *viz.,* volume (amount of data collected from sources), velocity (the speed at which the data is generated, collected, transmitted, and processed), variety (textual or multimedia data collected), veracity (quality, reliability, accuracy of data) and value (identifying the useful information from the dataset) (Abbasi et al., 2022; Sonka, 2021). This is too much data for an individual to handle and transform the data into practical knowledge. Therefore, a need for decision support systems (DSS) arises in order to make use of data precisely. In the words of a few researchers, DSS is defined as "a specific class of computerized information system, enabling to manage decision-making activities" (Yazdani et al., 2017).

CONCLUSION

In the current scenario, where the climate has been changing dramatically, it is the farming practices that have been embarking upon substantial difficulties like changes in temperature, erratic rainfall, groundwater shortage, poor irrigation, etc. This has raised concerns about global food security and intensive cultivation practices affecting sustainability. At this time, digital or smart farming practices could be the vanguard that would increase crop yields, curtail costs, and cut waste. Large-scale research is underway, with few of the technologies available to serve the farming community. In regards to agriculture, AI or robotics or autonomous decision-making systems are still in their infancy. Another important aspect when it comes to digital technologies is their exorbitant costs. There is a need for industries to discover a path to make these technologies more affordable so that they can be readily accepted by the farming community. The current chapter has covered the basics of digital technologies. These technologies work synchronously among themselves to establish what we can call a digital or smart farm.

REFERENCES

Abbasi, R., Martinez, P., & Ahmad, R. (2022). The digitization of agricultural industry–a systematic literature review on agriculture 4.0. *Smart Agricultural Technology*, 100042.

Adão, T., Hruška, J., Pádua, L., Bessa, J., Peres, E., Morais, R., & Sousa, J. J. (2017). Hyperspectral imaging: A review on UAV-based sensors, data processing and applications for agriculture and forestry. *Remote Sensing*, *9*(11), 1110. doi:10.3390/rs9111110

Beer, S. (1981). *Brain of the Firm*. Wiley.

Bellon-Maurel, V., Brossard, L., Garcia, F., Mitton, N., Termier, A., & Barbier, S. (2022). Agriculture and Digital Technology.

Castaldi, F., Pelosi, F., Pascucci, S., & Casa, R. (2017). Assessing the potential of images from unmanned aerial vehicles (UAV) to support herbicide patch spraying in maize. *Precision Agriculture*, *18*(1), 76–94. doi:10.100711119-016-9468-3

Davenport, T. H. (1993). *Process innovation: reengineering work through information technology*. Harvard Business Press.

Dursun, M., & Ozden, S. (2011). A wireless application of drip irrigation automation supported by soil moisture sensors. *Scientific Research and Essays*, *6*(7), 1573–1582.

Fox, G., Mooney, J., Rosati, P., & Lynn, T. (2021). AgriTech Innovators: A Study of Initial Adoption and Continued Use of a Mobile Digital Platform by Family-Operated Farming Enterprises. *Agriculture*, *11*(12), 1283. doi:10.3390/agriculture11121283

Frasconi, C., Raffaelli, M., Emmi, L., Fontanelli, M., Martelloni, L., & Peruzzi, A. (2017). An automatic machine able to perform variable rate application of flame weeding: design and assembly. *Chemical Engineering Transactions, 58.*

Gebeyehu, M. N. (2019). Remote sensing and GIS application in agriculture and natural resource management. *International Journal of Environmental Sciences & Natural Resources*, *19*(2), 45–49. doi:10.19080/IJESNR.2019.19.556009

IBEF. (2022). *Indian drone industry reaching the skies. India Brand Equity Foundation. Ministry of Commerce and Industry*. Government of India.

ICRISAT. (2021). Digital Agriculture: Pathway to Prosperity. In: International Crops Research Institute For The Semi-Arid Tropics. *ISCRISAT.* https://www.icrisat.org/digital-agriculture/

Khanal, S., Kc, K., Fulton, J. P., Shearer, S., & Ozkan, E. (2020). Remote sensing in agriculture—Accomplishments, limitations, and opportunities. *Remote Sensing*, *12*(22), 3783. doi:10.3390/rs12223783

Khare, V., & Seth, A. (2020). The case for AI in smallholder farmer water management. In AI holds water: A case for AI-enabled water management in agriculture. [White Paper, Wadhwani Institute for Artificial Intelligence, India].

Kim, Y., Evans, R. G., & Iversen, W. M. (2008). Remote sensing and control of an irrigation system using a distributed wireless sensor network. *IEEE Transactions on Instrumentation and Measurement*, *57*(7), 1379–1387. doi:10.1109/TIM.2008.917198

López-Granados, F. (2011). Weed detection for site-specific weed management: Mapping and real-time approaches. *Weed Research*, *51*(1), 1–11. doi:10.1111/j.1365-3180.2010.00829.x

López-Granados, F., Torres-Sánchez, J., Serrano-Pérez, A., de Castro, A. I., Mesas-Carrascosa, F., & Peña, J. M. (2016). Early season weed mapping in sunflower using UAV technology: Variability of herbicide treatment maps against weed thresholds. *Precision Agriculture*, *17*(2), 183–199. doi:10.100711119-015-9415-8

Miller, H. G., & Mork, P. (2013). From data to decisions: A value chain for big data. *IT Professional*, *15*(1), 57–59. doi:10.1109/MITP.2013.11

Mitton, N., Brossard, L., Bouadi, T., Garcia, F., Gautron, R., Hilgert, N., & Bellon Maurel, V. (2022). Foundations and state of play. In *Agriculture and Digital Technology: getting the most out of digital technology to contribute to the transition to sustainable agriculture and food systems*. INRAE, p. 30-75.

Peteinatos, G. G., Weis, M., Andújar, D., Rueda Ayala, V., & Gerhards, R. (2014). Potential use of ground-based sensor technologies for weed detection. *Pest Management Science*, *70*(2), 190–199. doi:10.1002/ps.3677 PMID:24203911

Queiroz, D. M. D., Coelho, A. L. D. F., Valente, D. S. M., & Schueller, J. K. (2021). Sensors applied to Digital Agriculture: A review. *Revista Ciência Agronômica*, 51.

Saxena, A., Suna, T., & Saha, D. (2020). Application of Artificial Intelligence in Indian Agriculture. In Souvenir: 19 national convention–artificial intelligence in agriculture: Indian perspective. RCA Alumni Association, xvi.

Shah, R. (2020). Pesticides and Human Health. In *Emerging Contaminants*. IntechOpen.

Shanmugapriya, P., Rathika, S., Ramesh, T., & Janaki, P. (2019). Applications of remote sensing in agriculture-A Review. *International Journal of Current Microbiology and Applied Sciences*, *8*(1), 2270–2283. doi:10.20546/ijcmas.2019.801.238

Sonka, S. T. (2020). Digital Technologies, Big Data, and Agricultural Innovation. *The innovation revolution in Agriculture*, 207.

Talaviya, T., Shah, D., Patel, N., Yagnik, H., & Shah, M. (2020). Implementation of artificial intelligence in agriculture for optimisation of irrigation and application of pesticides and herbicides. *Artificial Intelligence in Agriculture*, *4*, 58–73. doi:10.1016/j.aiia.2020.04.002

Trendov, M., Varas, S., & Zeng, M. (2019). Digital technologies in agriculture and rural areas: status report. *Digital technologies in agriculture and rural areas: status report.*

Van der Vorst, J., Beulens, A., & van Beek, P. (2005). 10 Innovations in logistics and ICT in food supply chain networks. *Innov. agri-food Syst,* 245.

Yazdani, M., Zarate, P., Coulibaly, A., & Zavadskas, E. K. (2017). A group decision making support system in logistics and supply chain management. *Expert Systems with Applications*, *88*, 376–392. doi:10.1016/j.eswa.2017.07.014

Zhai, Z., Martínez, J. F., Beltran, V., & Martínez, N. L. (2020). Decision support systems for agriculture 4.0: Survey and challenges. *Computers and Electronics in Agriculture*, *170*, 105256. doi:10.1016/j.compag.2020.105256

Zhang, X., Zhang, J., Li, L., Zhang, Y., & Yang, G. (2017). Monitoring citrus soil moisture and nutrients using an IoT based system. *Sensors (Basel)*, *17*(3), 447. doi:10.339017030447 PMID:28241488

Chapter 2
Emerging ICT Technologies for Agriculture, Training, and Capacity Building for Farmers in Developing Countries:
A Case Study in Zimbabwe

Samuel Musungwini
https://orcid.org/0000-0002-9315-6252
Midlands State University, Zimbabwe

Petros Venganai Gavai
Midlands State University, Zimbabwe

Briget Munyoro
Midlands State University, Zimbabwe

Anyway Chare
https://orcid.org/0000-0002-1317-6633
Marondera University of Agricultural Sciences and Technology, Zimbabwe

ABSTRACT

This chapter explores the use of emerging ICT technologies for agriculture in Zimbabwe, a developing country in sub-Saharan Africa. It investigates the concept of ICT technologies and their potential impact on agriculture. The chapter then brings the background and context of the study. The chapter is premised on a qualitative approach which fuses a literature review to situate the chapter and eight (8) in-depth interviews and a ten (10) member focus group discussion to gather empirical data. The findings reveal that there are emerging ICT technologies in Zimbabwe which include artificial intelligence, drone technology, IoT, robots, and big data. However, the application and use of these technologies are still very localized and limited to big and highly capacitated commercial farmers, with very few smallholder farmers using the technology because of challenges that impede the adoption and use of these emerging ICT technologies for agriculture by the bulk of farmers. The chapter submitted measures that could be implemented to try and increase the adoption and increased use of these technologies.

DOI: 10.4018/978-1-6684-6413-7.ch002

Copyright © 2023, IGI Global. Copying or distributing in print or electronic forms without written permission of IGI Global is prohibited.

INTRODUCTION

The advent of Information and communication technology (ICT) has transformed the general operational landscape and it is now business unusual. This transformation has been felt across different facets from government operations (Twinomurinzi et al., 2017), health (Simbarashe & Coleman, 2018), education (Gocha, 2014), and production areas like agriculture (Batani et al., 2019) changing at a rapid pace. This has resulted in the coining of such terms as ICT for development, Mobile for development (M4D), and sustainable development goals (SDGs) (Msendema & Nyirenda, 2019). These digital technologies enable people from all lifestyles to connect from their remote locations to any part of the world across the globe at high speeds at any time (Musungwini & Zyl, 2017). These ICTs are continuously evolving since the advent of the computer and the SDGs framework is premised on harnessing ICTs to attain the world we want. In the last decade, literature is awash with research articles, dissertations, and policy documents on the adoption, application, and use of ICTs in various areas including agriculture (Ayim et al., 2020; Musungwini et al., 2022). Historically, the agriculture sector has been the mainstay of the Zimbabwean economy, which resulted in the country being acclaimed as the breadbasket of Africa.

However, in the last three decades from 1990 to 2020, the sector has undergone a rapid decline because of the crumbling economy. Technology is continuously evolving and its tilting towards an Artificial Intelligence (AI)-enabled smart precision agriculture as posited by Loures et al., (2020). The use of digital smart connected farm culture in which technology is the heart and soul, which dictates and empowers a farmer in a truly rewarding way, is now the order of the day (Linaza et al., 2021). In recent years, agriculture practitioners and farmers alike have been increasingly fighting all sorts of external and internal factors such as unpredictable hostile weather, pest problems, water shortage, and even drought. Advanced applications of ICTs like AI, Sensors, Drones, IoT, and Big data analytics, have emerged as a window of opportunity for the farming community (Psirofonia et al., 2017). This smart technology can ensure that every plant is scanned for health and growth tracking such that any slight change in the condition of the plant is identified and notified to the farmer in real-time (Ben Ayed & Hanana, 2021). The government has acquired the High performance-computing infrastructure, which can be used to process large amounts of data in a short time frame and to make sense of large quantities of structured and unstructured data.

BACKGROUND AND CONTEXT OF THE STUDY

In Sub-Saharan Africa, the agriculture sector plays a fundamental role in the development of the economy (Jayne, 2012; Musungwini, 2016). The general population is swelling at an astronomical pace; this increase comes with the ever-increasing demand for food. The global population is projected to reach 9.7 billion and the bulk of these will be urbanised (Rahmann et al., 2021). The general disposable incomes of the general populace are rising and this is coupled with the changing consumption habits of the growing modern population (OECD/FAO 2016, 2016). As a result, the farmers are under immense pressure to meet the ever-increasing demand, and as a result, they need a way to increase their agricultural productivity yields (Sachitra & Padmini, 2021). Generally, more people are required to be fed but the amount of farming land available is limited and even shrinking (Mugandani et al., 2012). The bulk of farmers in Sub-Saharan Africa including Zimbabwe is Smallholder farmers. These farmers are usually under-resourced and ill-equipped hence they perennially under-produce (Musungwini, 2018). If the agriculture activities of these Smallholder farmers in Zimbabwe are to change there will be a need to move beyond their traditional farming ways and embrace emerging modern technologies for their agricultural activities (Musungwini et al., 2022).

Therefore, harnessing the emerging modern technologies for agriculture automation has since become the fundamental concern for the contemporary global world as asserted by Talaviya et al. (2020). The traditional methods, which have been used by the Smallholder farmers over the years, have since been superseded by the evolving events. As a result, new automated methods of farming are required to meet the new obtaining food requirements of the growing billions of people (Rahmann et al., 2021). Thus, emerging technologies like Artificial Intelligence, the Internet of Things, Drones, Robots, and Big Data in agriculture have brought an agriculture revolution to the developed world (Talaviya et al., 2020) and have the potential to transform Smallholder agriculture in Zimbabwe. Smallholder farmers to safeguard their crop yields from various factors like climate changes, population growth, and food security problems can use these emerging technologies.

In this book chapter, we explore the current level of application of Artificial Intelligence, IoT, Drones, and Big Data in Agriculture in Zimbabwe and their impact. It also explores current opportunities and challenges of Agriculture digitalization in Zimbabwe. Agricultural technologies in mechanization generally are promptly changing in the direction of a new pattern which includes but is not limited to digitalization, automation, and artificial intelligence. Mechanization plays a major role in crop production, including weeding, pest control, and other agricultural activities (FAO, 2020). Food insecurity in developing countries specifically African countries has become a major challenge as the number of people in these countries requires food. Consequently, more than a few African countries still need and rely on food assistance and backing from other continents. These untiring food insecurities are greatly due to unsustainable farming approaches, limited mechanization efforts, limited skills capacity, and ineffective farming innovative technologies hence there is a need to adopt precision agriculture to expand Africa's job construction and guarantee adequate food security.

Therefore, harnessing and adopting emerging technologies for sustainable smart farming through the use of such technologies as AI, drones, IoT technologies as well as robots will help farmers to reduce generated waste and enhance their agricultural productivity (Saiz-Rubio & Rovira-Más, 2020). The use of emerging technologies simply results in a high-tech agriculture system of growing food that is clean and sustainable for the masses.

METHODOLOGY

The study used a qualitative research approach to provide new data in line with Teddlie & Tashakkori (2009). The qualitative research approach is famed for its methods that give primacy to the research subject's opinion. As a result, this approach emphasizes the need for extensive observation of research subjects in their everyday lives to provide a 'rich' and 'depth' account as recommended by Mays and Pope (1995). That is why in qualitative research, there is a need to give careful thought to the larger contextual settings in which the research is situated and the problem is to be integrated. This is because the researchers can only comprehend the obtained data if they understand the data from the viewpoint of a greater contextual setting (Rendle et al., 2017).

Given (2008) asserts that the qualitative research approach enables the researchers to observe the experiences, ideas, behaviors, and attitudes of the research respondents. Because the qualitative research techniques were developed in the social sciences arena, hence the approach allows researchers to investigate societal and cultural issues, which is the reason we used it in this work. Although there are many different types of qualitative research approaches which encompass action research, grounded theory, and case study research, this study used a grounded theory approach. Observation and participation observation (fieldwork), focus group discussions (FGD), interviews, records, publications, and the researchers' perceptions and responses are some of the sources of qualitative data. Because this study also included a review of literature, an FGD, and in-depth interviews, therefore, the literature study served as a foundation for this chapter's positioning and footing of emerging technologies in agriculture in a developing context. The in-depth interviews with emerging technologies specialists and farmers offered a mechanism to assess the degree of understanding and perspectives on problems relating to challenges in the adoption and use of emerging technologies in agriculture in Zimbabwe. A key technique frequently utilised by qualitative researchers is interviewing chosen persons (Gubrium & Holstein, 2001).

The participants' thoughts, attitudes, and sentiments towards emerging ICT technologies for agriculture, the challenges faced in the adoption and use of these technologies as well as training and capacity-building needs of farmers were ascertained by the researchers through in-depth interviews with the participants. Eight (8) of the ten (10) participants in the FGD farmers while two were experts in emerging ICT technologies for agriculture. The researchers used literature review business documents and newspaper articles as well as a focus group discussion and in-depth interviews for data collection and data was analysed in line with the key tenets suggested by Saldana (2009).

LITERATURE REVIEW

What are the Emerging ICT Technologies for Agriculture?

Technology has been the main backing factor in how people across the globe live, communicate, move around, and network in present society and all industries, especially agriculture. Agriculture is described as the instrument that is at the center of livelihood improvement and economic progression not only in developing countries but across the entire globe (Adam, 2007; Friha et al., 2021). Literature has also reflected that agriculture is the principal source of food in the universe and vital to the development of civilizations throughout history. The increase in the global population points to the high demand for food and in that same vein if agricultural production mechanisms are to become more sustainable, then technology, research, and development must be used to the maximum levels possible to realize Sustainable Development Goal (SDG) 4.

As agriculture has been evolving over years, there have also been emergencies of new technologies across evolutions to support various agricultural activities. According to Friha et al. (2021), there are four distinct revolutions ranging from agriculture 1.0 which featured human and animal power to agriculture 4.0 smart agriculture featuring emerging technologies.

The technologies that have emerged today under agriculture 4.0 serve different purposes. These technologies according to Sadiku et al. (2020) include but are not limited to the following:

- **Smart farming:** It has been regarded as the future of modern-day farming. It involves the application of IoT solutions making use of predictive analytics to make smarter decisions in farming by gathering real-time data on issues that concerns farmers. It makes use of information and com-

Figure 1. Agricultural Revolutions
Source: (Friha et al., 2021)

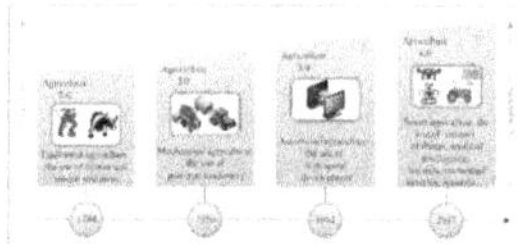

munications technologies (ICT) to augment and automate agricultural processes and operations. Smart farming technologies shelter all these facets of precision agriculture. As an emerging technology, it is postured to upsurge the competence of some if not all farming activities(Sadiku et al., 2020; Subramanian et al., 2019).

- **Machine learning:** It is an emerging branch of artificial intelligence that has started to be applied to the agriculture sector across the globe. It refers to the automated monitoring of the use of agricultural resources such as water, nutrients, and energy in a way that would benefit the farmers by detecting meaningful patterns in available data. Machine learning in agriculture goes beyond monitoring of resources but also more accurate disease diagnosis and crop disease prediction to maximise production by making use of AI and ML algorithms to assist farmers to go through data to arrive at powerful insights to aid them (Yahya, 2020).
- **Big data:** Nowadays huge volumes of information are generated, collected, and managed to add value to people's daily lives and among them farmers. Currently, farmers are deploying vast amounts of data generated by networked devices to have an overall understanding of what is going on on their farms. The big data revolution is revamped the farming industry at a stride we have never perceived before and will continue to revolutionise how farmers can manage their farm routines (Coble et al., 2018; Ribarics, 2016; Sadiku et al., 2020; Yahya, 2020).
- **Internet of Things (IoT):** It is an idea that is positively disrupting the agriculture industry by attempting to connect devices, at any time, and in any place through the internet. It is a network of devices that are connected to facilitate the exchange of information among themselves. It is being deployed across the whole world to deal with the most pressing issues not only in agriculture but across all sectors of the economy. It assists farmers to control the activities of the devices through connectivity (Alam, 2020; Kanchana, 2018).
- **Agbots:** These are robots that are used in agricultural setups. This involves the use of robots or automated technologies as a substitute for humans to perform physical tasks in various agricultural activities. Farmers use agbots to automate activities that include but are not restricted to harvesting, picking of fruits, weeding, planting, irrigation, and spraying of chemicals just to mention but a few. Robots have the potential to improve productivity and reduce labour-associated costs (Honrubia, 2021).

Additional emerging technologies that farmers can embrace include artificial intelligence, biotechnology, GPS, GIS, genomics, regenerative agriculture, satellite greenhouses, aquaponics, conservation technology, carbon farming, microbial farming, hyper precision, remote monitoring systems, nuclear magnetic resonance, and nanotechnology. These emerging technologies can be utilized wisely to meet the growing demands and trials of contemporary agriculture

Figure 2. Application of Agbots
Source: (Honrubia, 2021)

Challenges Faced in the Adoption and Use of Emerging ICTs for Agriculture in Developing Zimbabwe

The agricultural sector in Zimbabwe is faced with a plethora of challenges across the agriculture value chain. Natural factors that support agricultural activities such as fertile land and water are gradually turning into a constraint (Trendov et al., 2019). That trend is a major threat to food security. To avert such challenges, agricultural transformation is, therefore, a top priority for Zimbabwe where the economy is largely agro-based. Technology, as it has been introduced and used in various sectors of the economy, is a vital cog in the transformation and development of the agricultural sector in Zimbabwe. The adoption of modern technologies could go a long way in driving productivity and alleviating natural factors affecting agriculture (Ayim et al., 2020). A digitally-driven agricultural transformation can see the country achieving sustainable growth. Digitising agriculture could promote greater cooperation in agriculture and bring about employment opportunities along the value chain (Tsan et al., 2019). Research has shown that many of the challenges in the adoption of emerging ICTs for agriculture may be due to such factors as limited awareness, cost, and infrastructure (Kamutuezu et al., 2021).

Access to agricultural information can be enhanced through ICTs such as the Internet and mobile applications. However, the adoption of emerging digital technologies particularly in rural farming communities is still a challenge due to a lack of awareness among the farming population about digital technologies and the benefits they bring if harnessed for agricultural activities (Kamutuezu et al., 2021). Studies have revealed that the level of education correlates to the level of awareness of emerging ICTs and their influence on agricultural production (Ugochukwu et al., 2020). Basic literacy, as well as specialized technical knowledge and abilities, are also required to use digital technologies. People who lack these skills may find themselves on the peripheries of increasingly digitized societies. Knowledge of how technology can help farming activities is essential to attract farmers towards adopting the technology for their farming activities. Further, the quality of education in rural areas is frequently hampered by a lack of adequate facilities and resources. As a result, learning is less effective and attendance is lower, and there is a high chance of students dropping out. Furthermore, in many rural areas, young people walk long distances to school and are sometimes expected to work, leaving little time for school. That results in low educational attainment in agricultural communities (Trendov et al., 2019). According to UNESCO (2017) literacy remains poor in many rural parts of least developed countries, especially among women. A lack of fundamental literacy and numeracy is therefore a substantial impediment to the uptake and usage of digital technologies in agriculture.

Infrastructure such as general information communication technology infrastructure is often hardly available in remote rural communities and areas with large indigenous populations (Musungwini, 2021). The lack of basic telecommunication infrastructure that enables access to data communication services is a huge impediment to making information available to rural farmers (Mutale, 2018). A large number of small-scale farmers in Zimbabwe reside in rural remote areas where telecommunication service is too basic and erratic and implementing emerging ICTs in such areas is close to impossible.

Modern technology is perceived as a very exorbitant investment by farmers in developing countries (Gavai et al., 2018). Cost is perceived as a huge impediment to farmers' adoption of emerging technologies owing to their small revenues (Victor et al., 2021). Investment in emerging technologies such as drone technology and other precision agriculture technologies seems like a luxury for farmers. Lower costs of emerging ICTs has a positive effect on the adoption of these ICTs and higher costs are a barrier to adoption (Kante et al., 2019).

Policies and frameworks are a big part of a plethora of initiatives that drive digitalisation. They are significant in creating a conducive environment for implementing a digital value chain, particularly in agriculture. Fakhoury (2018) points out that there is a need for top-level administrative capacity to harness digital technologies in government programs and many developing countries have made little progress in that regard. The Zimbabwean government's telecommunications licensing framework and spectrum allocation efficiency can have a big impact on promoting private sector investment in mobile network service in rural agricultural areas. Greater deregulation of the telecommunications sector enables other players to enter the industry and improve national network coverage. Lower deployment costs can also be realised by mobile network operators where there is efficient spectrum management, resulting in increased access to ICT services for end users. Government e-services development has also been notably slow in the agriculture sector, with very few e-Agriculture services (Trendov et al., 2019). Research has revealed that there is a rampant lack of prioritization, implementation, and monitoring of ICT policies and that seriously affects the adoption of ICTs in agriculture, particularly those that are meant for rural farming communities (Ayim et al., 2020).

What are the Training and Capacity-building Needs for Farmers in Developing Countries?

Recent efforts link cutting-edge technologies (typically tailored to agricultural situations) to centralized digital platforms that may store and analyse information gathered through farmer surveys, aerial monitoring, and/or ground sensors. Farmers can use the findings of such an analysis to guide their decision-making, sometimes through direct feedback systems and sometimes through organizations that offer agronomic guidance. To implement digital agriculture, a large volume of data must be collected, necessitating the use of agricultural data science, which can open up new analytical possibilities, create new types of advisor decision-support tools, and connect new services to digital technologies like social media, mobile apps, drones, and robots. Therefore, a process of digitalization of agricultural knowledge provider organizations is induced (Rijswijk et al., 2019)

The adoption of new technology is influenced by several factors, according to Padilla et al., (2020): the technology must meet farmers' needs; building relevant Co-operators enhances adoption; understanding the socioeconomic context, farmers' skills, and farming systems is crucial; and participatory involvement of the rural communities enhances adoption. For instance, giving farmers advice on which technologies to employ on their farms and how to use them is one operational responsibility. Setting and developing the needs for future technologies, and making sure they provide value to agriculture, would be another responsibility. To give farmers greater strategic guidance and help when selecting and implementing technologies, it was determined that it requires increasing one's knowledge of digital technology. On the other hand, the industry bodies appear to be more concerned with farmers' capacity to deal with digitalization than with their skills and competencies regarding digital technologies. Farmers need more advanced skills; thus, industry bodies' procedures and services need to adapt. They stated they would need to learn more about the technical requirements and expertise of farmers. Additionally, this would result in a new method of connecting with farmers, such as through custom websites and apps. Furthermore, they believed that to uphold their objectivity or impartiality, it is their duty to evaluate the added value and relevancy of a wide range of technologies and brands (Rijswijk et al., 2019).

The extension has shifted over the past forty years from its traditional focus on technology transfer and farm management information provided by the public sector to a broader public and private advisory service mode, addressing subjects like marketing, environmental sustainability, pest diagnostics, and risk management (Norton & Alwang, 2020). For any agritex extension program, strong in-service training is essential, but it is now even more crucial as ICT technologies evolve just as quickly as agricultural advances. Maintaining close ties to the most recent discoveries produced by both public and commercial research is essential for successful agritex extension.

Innovations are characterized by uncertainty, particularly in the context of agriculture, where factors such as the farmer's human capital and regional (agronomic and meteorological) variables determine the applicability and significance of new technology for a certain farm. Most of the time, the farmer is unable to predict how the new technology would perform in the circumstances in which he or she is working. The optimum way to apply the technology could also be unclear, especially if it is combined with other operational aspects (e.g., fertilizers). In this situation, farming expertise and human talents may also be important. These elements make the farmer's potential future income unclear. Future prices are also likely to be unpredictable, particularly how decisions concerning widespread adoption may affect prices.

Increased adoption of smart farming may result in new client engagement strategies for advisors, which in turn will require new methods of gathering and disseminating information, the development of new skills for advisors in the curation and management of remote sensing data, and strengthening networks with technology providers and R&D in smart farming. Advisors can help farmers learn about the nature of digital data and how to interpret it to manage the uncertainty associated with implementing a smart farming innovation, which can affect its final adoption. Additionally, technology providers should play a bigger advising support role for farmers. The 'translators' of knowledge at the interface between technology and farm management have also been identified as technology suppliers. Participatory training is necessary for farmers and their advisors since it is crucial for learning about data and technology, such as in promoting the use of seasonal climate information (Nettle & Rue, 2019).

Table 1. Interviewee profiles

INTERVIEWEE	AGE	INTERVIEW METHOD	DATE	DURATION (Minutes)
FARMER1	48	Face to face	21 April 2022	37
FARMER2	52	Face to face	23 April 2022	35
FARMER3	39	Face to face	27 April 2022	29
FARMER4	46	Face to face	28 April 2022	38
FARMER5	47	Face to face	30 April 2022	40
EXPERT1	50	Face to face	2 May 2022	31
EXPERT2	45	Telephone	5 May 2022	29
EXPERT3	41	Telephone	8 May 2022	36

To enable advisers to experiment (e.g., trialing various technologies or data interpretation procedures), engage in collaborative learning with actors like farmers and technology vendors, and pursue formal training, innovative learning models may be needed (e.g., in data analytics). Cooperation between farmers and advisers might occasionally be useful for both parties to comprehend how the relationship is evolving (Nettle & Rue, 2019).

RESEARCH PARTICIPANTS

The researcher conducted eight (8) interviews, five (5) of which were conducted with farmers, and three (3) were conducted with experts with various backgrounds who met the criteria relevant to the research subject domain. The interviews with farmers were conducted face-to-face while those with experts were conducted telephonically. Table 1.1 presents the interviewees' profiles and the interview method that was used to interview each participant and the duration of the interview.

The researchers conducted a focus group discussion with twelve (10) FGD participants. The FGD was conducted in sync with the dictates of conducting focus group discussions set by Dzino-Silajdzic, (2018). The FGD participants were seven (7) farmers and three (3) ICT agriculture technology experts in Zimbabwe. The researchers facilitated the FGD meeting and it lasted 59 minutes.

RESEARCH FINDINGS

The Emerging ICT Technologies for Agriculture in Zimbabwe

Research institutions, non-profit organizations, and farmer cooperatives all play a critical role in extension, knowledge sharing, farmer training, and enabling change. The research established that there are many emerging ICT agricultural technologies that are arising in the Zimbabwean agricultural and ICT landscapes, but their use is just budding and still confined to those highly resourced commercial and highly skilled farmers. There are many challenges dogging the adoption of the emerging technologies as a result this requires the training and capacity building of the farmers to capacitate them and increase

Table 2. Focus group discussion participants' profiles.

Participants	Age	Position Profile	Agricultural Knowledge and Experience	Emerging ICTs for Agricultural
FGDP1	52	Farmer	Very high	Low
FGDP2	39	Farmer	High	Moderate
FGDP3	47	Farmer	Very high	High
FGDP4	41	Farmer	Very high	Moderate
FGDP5	43	Farmer	Very high	High
FGDP6	36	Farmer	High	High
FGDP7	42	Agriculture Informatics Officer	High	High
FGDP8	39	ICT Agriculture expert	Moderate	Very High
FGDP9	41	ICT Agriculture engineer	High	Very High
FGDP10	49	Agriculture Extension services officer	High	High

their chances of adopting the emerging ICTs to boost agricultural production. Therefore, there is a need for a multi-stakeholder approach for the realisation of the high adoption levels of emerging ICTs for agriculture in Zimbabwe. Therefore, the research institutions, non-governmental organisations (NGOs), non-profit organizations, agricultural extension services, and farmer cooperatives all play a critical role in extension, knowledge sharing, farmer training, and enabling change.

The research participants suggested that there is a need for strengthening institutional as well as human capacity in agricultural settings.

The country needs to focus on training of farmers and capacity building and this should be done at the individual farmer level, community, as well as institutional levels. [...] the capacity building with farmers, with trainers, and with research institutions in the country (FGDP7).

The research participants out that there are emerging technologies like IoT which is an open and comprehensive network of intelligent objects that can self-organize exchange and share valuable information, data, and other valuable resources, reacting accordingly in line with the intelligence gathered about the situations and changes in the environment.

In the recent years the applications like IoT which have become a fused reality which is composed of an assortment of a network of objects that are connected in a dynamic environment have started to straddle the Zimbabwean landscape and some farmers with capacity are taking advantage of it (FGDP9).

This is usually asynchronous in nature, and this enables the possibility to provide a massive amount of information to feed machine-learning algorithms.

Participants pointed out that it has been practically proven beyond reasonable doubt that emerging ICTs have had a significant contribution to the growth and socio-economic development in different countries and sectors where they have been well deployed. The effective integration of these emerging technologies in the developed world's agricultural sectors has benefitted immensely from these emerging technologies. This has led to remarkable enhancement in agriculture value chain efficiency and productivity in those developed countries.

Precision agriculture has been a different set up in Zimbabwe where the levels of application of precision agriculture are still very low [...] but for those who have adopted these agricultural technologies are reaping the dividends as they are realizing larger crop yields (EXPERT2).

Even among smallholder farming households but also expressed concern with the low rate at which these technologies are being taken up, especially in developing countries like Zimbabwe.

According to research participants, the Government of Zimbabwe has shown some commitment by paving way for the adoption of technologies such as drones, especially in agriculture.

The government has begun the process of drawing up a drone framework which was expected to be gazetted by then in February 2019 spearheaded by the Civil Aviation Authority of Zimbabwe (CAAZ) [...] CAAZ is the responsible authority in regulating the country's aerospace did not pay dividends as it had minimal effect in spurring majority of farmers to apply such technologies as Artificial Intelligence, IoT, Drones and Big Data in their agricultural activities.

However, the researchers further argued that the adoption of general ICTs in agriculture remains a challenge in Zimbabwe.

All the research participants agreed that the use of emerging ICTs driven approaches to sustainable agriculture practices and precision farming has resulted in the increasing adoption and use of drone technology for agricultural purposes.

The Unmanned Aerial Vehicles (UAVs) are being used in conjunction with GPRS, infrared cameras and enhanced sensors as well as programmable controllers for data collection (EXPERT2).

However, the research participants indicated that this is increasingly being practiced by commercial farmers in Zimbabwe especially tobacco farmers, cotton farmers, and sugarcane farmers when spraying herbicides and on a larger scale, for agricultural studies, and in research, but they are isolated cases where the smallholder farmers are using these technologies.

The practice of using larger drones for spraying chemicals on the large tracts of land also enables the reduction of labour cost (FGDP8).

Participants suggested that the Zimbabwean government through its CAAZ arm has resulted in the promulgation of drone operation regulations and licensing systems, which are not so complex.

The crafted legal framework requires that individuals and organisations who want to operate drones should acquire a license for them to legally operate a drone in Zimbabwe as guided by Statutory Instrument (SI) 271 of 2018 (EXPERT3).

Although the research participants indicated that the license is valid for 2 years, however, the bulk of smallholder farmers are still acquiring and using this technology in Zimbabwe. Focus group participants pointed out that farmers with large heads of cattle and owning large farms are using drones in livestock health monitoring using sensors and imaging.

These farmers are using the drones that are fitted with picture and video cameras and with enabled thermal imaging capabilities to track and monitor their livestock movement and their health conditions (FGDP3).

The research participants suggested that the issue of drones is also an evolving aspect of emerging ICTs for agriculture in Zimbabwe and this involves the use of powerfully equipped drones in livestock monitoring and management. The continual evolution and development of AI applications and infrared cameras and their application in herds monitoring allow tasks that are even more sophisticated.

These drones can detect the sick or injured animals as well as those straying [...] The drones are also being used to detect cattle rustlers and predators before they prey on the farmer's livestock (EXPERT1).

In addition to that drone technology was also said to be used together with Global Positioning System and Remote Sensing (GPRS), systems for field and plant performance analyzes the drones' collected data on the state of plant health and field nutrients, presence of weeds, and diseases. The fusion of advanced crop aerial information, which is acquired using drones fused with data from other sources such as weather forecasts and soil maps, can help to refine the final information and enable the farmer to take full advantage of the farm and maximize the yields to their natural limits. The collection and analysis, as well as monitoring of plant health, enables farmers to make informed choices which can lead to high-quality products like high-quality tobacco which can then go on to fetch a higher price and enable the farmers to survive competition in the international market.

According to agronomy experts, drones can be used for agriculture crop productivity enhancement purposes by all farmers. Such activities include:

- **Crop health:** emerging ICTs like drones can help detect damage made by pests and colour change due to pest infection.
- **Water needs:** emerging ICTs like drones can help detect stressed parts of the field. Soil analysis emerging ICTs like drones can help detect nutrient availability for plant nutrient management.
- **Plant height:** emerging ICTs like drones can help detect plant height and density.

The use of GPS receivers is ground-based precise positioning technologies that assist in the detection of field heterogeneity which contains automated samplers, various sensors, and measuring systems, among others is still very limited in Zimbabwe. The use of agricultural robots to handle essential agricultural tasks such as harvesting crops at a higher volume and faster pace than human laborers or harvesting machines with automatic accounting of the harvest. Precision farming technology necessitates the use of remote sensing systems for crops and animals. For images and positioning, aerial drones and satellites are used. Internet connections via satellite, mobile wireless, fixed wireless, and Digital fixed line (DSL). In the agriculture sector, the Internet of Things (IoT) and data analytics (DA) are used to improve operational efficiency and productivity.

The Challenges Faced in the Adoption and use of Emerging ICTs for Agriculture in Zimbabwe

This research established that the adoption and use of emerging ICTs in agriculture in Zimbabwe are nascent and progressing feebly with those farmers who have adopted the technology enjoying good yields. However, the research established that there are challenges that hinder the adoption and use of these emerging ICTs technologies in Zimbabwe. The research participants cited such issues as erratic power supplies characterising the country as technology use requires energy to power their uses. There is a high capital requirement for setting up adequate ICT infrastructure to be able to function with the emerging ICT technologies for agriculture in Zimbabwe. The other issues cited are the very low incomes obtained by the bulk of farmers as the produce market prices are said to be usually depressed. The research participants also cited the lack of favourable policies to enhance ICT development in rural areas hence these are marginalised from the rest of the country's urban areas. The agricultural and ICT experts highlighted a lack of essential skills among the bulk of the intended beneficiaries who are the farmers to use these highly efficient and productive emerging technologies.

The research participants suggested that while the integration of emerging digital technologies for agriculture and smart platform solutions can help to bring about digital transformation in the agricultural sector in Zimbabwe, there are challenges faced in the full-scale adoption by the farmers. The experts suggested that there is a need to address some issues therein.

There are some basic requirements of agriculture digital transformation like the consumption sensors mounted on moving machines that measure while driving, or the use of multispectral sensors on a tractor's sprayer boom to estimate the nitrogen demand of the crop to be sprayed and adjust the dosage, are examples of modern agricultural machines that are controlled by an on-board computer and can perform differentially agro-technical operations (FGDP10).

Training Needs and Capacity Building of Farmers in Zimbabwe

The research participants pointed out that there is a need to do the following:

Understanding the Farmers' Technological Challenges and Contextualizing Training for Farmers

The research participants indicated that there is a need to view the farmers as individuals as well as collective communities as this would enable learning from the farmer's contextual collective needs and design tailor-made training solutions and address capacity-building needs for farmers rather than just prescribe blanket training. Obtaining a macro picture of the training and capacity-building requirements with micro-level details of the situation allows the formulation of training that empowers the farmers to engage directly and easily with the emerging ICT technologies for agriculture in Zimbabwe. The training programs should be wholesome to the effect that they can be structured in such a way that they may not be limited to the emerging ICT technologies for agriculture but also encompass the training of other needs that are identified therein in a given area. By employing a "bottom-up technique," which involves gathering data about the farmers and assessing the awareness levels of the farmers to situate and develop the program, it is possible to additionally solve other skills lacky areas of farmers' entrepreneurial illiteracy, consumer literacy, and sustainability literacy among others which are critical to the development of farmers.

Create Accessible Education for Trainers

In that respect, the training can be formulated and conducted with farmers, but this training can also be recorded on videos and saved on internet platforms accessible to agricultural extension services, farmers, and other experts. These videos can also be delivered to farmers' mobile phones using technology applications like WhatsApp. The WhatsApp videos should be brief like Tik Tok videos and they should seek to engage audiences in comprehending reasons why it is crucial to adopt and use the emerging ICT technologies for agriculture in their agricultural activities in the first place.

There is a need for the introduction of targeted mentor-guided emerging ICT technologies for agriculture training programs for farmers with a certificate of completion like those being done with agricultural extension services Master farmer and Advanced farmer programs. By carefully selecting their research to apply specifically to a specified location, which they have found beneficial in fostering interaction with their audience, the training content should be locally adaptable and contextually relevant. The training can also make a significant effect on the farmers by utilising a "train the trainer" strategy that focuses on the agricultural extension services officers as well as other farming community trainers and gatekeepers who provide agricultural knowledge to farmers in Zimbabwe.

DISCUSSION

The results of our research suggest that the issue of emerging ICTs in agriculture in Zimbabwe is still a budding phenomenon and is currently being used by very limited numbers of farmers. There are many of these emerging ICT technologies for agriculture which include precision farming, Artificial Intelligence, the Internet of Things, Drones, Robots, and Big Data which have brought an agriculture revolution. But the research participants highlighted the major challenges affecting the adoption and use of these emerging ICT technologies for agriculture by farmers in Zimbabwe. The identified challenges include the precondition to educating, training, capacity building, and provision of refresher courses for farmers, agricultural extension officers, and key stakeholders responsible for educating and training farmers in Zimbabwe. The research participants also cited the need to develop and establish ICT and other technology infrastructures in rural areas where most farmers are located in Zimbabwe. In addition to that, there may be a need to develop other supporting infrastructure like electricity generation and supply and road network infrastructure.

There is a need for a purposeful policy that promotes the growth and adoption of novel infrastructures and services for rising ICT in agriculture. It is necessary to foster partnerships and synergies between various players from the public and corporate sectors and Non-Governmental Organisations (NGOs) to ensure that this possibility can be realised. According to the experts whose opinions were sought, accessibility is essential if the adoption of emerging ICT technologies for agriculture are to be adopted and have a footprint on the agricultural landscape in Zimbabwe and have a lasting impact on agricultural development for the country. However, at the point in time when this research was being conducted, just very few farmers were using emerging ICT technologies for agriculture in Zimbabwe. The future of these technologies and other farmers, particularly the smallholder farmers remains bleak.

Therefore, to ensure that all farmers adopt and use emerging ICT technologies for agriculture in Zimbabwe, it is imperative to depart from the current perspective on agriculture modernisation, which is predicated on a top-down approach to the digital transformation of agriculture. The implementation of ICT technologies in agriculture must take a holistic approach that guarantees that the poor and smallholder farmers are not left behind. A technologically transformed agricultural sector in Zimbabwe may lead to the revitalising of agricultural productivity to the levels of yesteryear when the country was eulogised as the breadbasket of Africa. This may result in the revitalising of the Zimbabwean industry and the economy at large. Therefore, developing countries like Zimbabwe cannot experience real transformation, which encompasses all farmers unless all the key stakeholders come together the private sector mainly the ICT technologies for agriculture and other associated industries, the NGOs and government, and the farmers themselves.

While the emerging ICTs for agriculture have the potential to transition Zimbabwean agriculture towards educing its past glory but it cannot necessarily conjure miracles on its own unless there are deliberate steps undertaken to ensure that it thrives. The emerging ICT agriculture technologies have the potential to boost agricultural productivity in Zimbabwe but that is presently heavily localised to very few farmers hence, this calls for the need for a determined endeavour to ensure that all farmers can embrace these emerging technologies. Because of this, we contend in this chapter that there is a need for the ministry of agriculture, ministry of ICT, and courier services in Zimbabwe, non - governmental organisations, and businesses should work together. This will enable the creation of policies that promote the development of ICT infrastructure support services in the rural areas where the bulk of farmers are located. The chapter advocates for sapient, complex, and nuanced training and capacity development interventions customised in sync with the distinctive characteristics of each farmer and geographical context.

CONCLUSION

To conclude our chapter, we have established that there are emerging ICT technologies in Zimbabwe which include artificial intelligence, drone technology the Internet of Things, Robots, and Big Data. However, the application and use of these technologies are still very localized and limited to big and highly capacitated commercial farmers with very few smallholder farmers using the technology. Several challenges that impede the adoption and use of these emerging ICT technologies for agriculture by the bulk of farmers were identified. The research proffered measures that could be implemented to try and increase the adoption and increased use of these technologies.

The measures include the crafting of purposeful policy that promotes the growth and adoption of novel infrastructures and services for intensifying ICT in agriculture. It is necessary to foster partnerships and synergies between various players from the public and corporate sectors and Non Governmental Organisations (NGOs) to ensure that this possibility can be realised. The need for customised training for farmers that is regionally and contextually specific. A holistic approach to training and instruction that understands the constraints, accessibility, and values of a community will more effectively build improved capacity for the farmer. There is a need for tailor-made capacity-building programs that situates the farmers and those responsible for the training of farmers at the epicentre. This is because, to create positive change and deepen the impact of our programme, we must ensure that farmers and farm workers have access to the necessary tools, training, and support that they need to continuously improve their practices.

REFERENCES

Adam, L. (2007). *Information and Communication Technologies.* Knowledge Management and Indigenous Knowledge.

Alam, E. N. (2020). IoT in Agriculture Industry. *Jurnal Sistem Cerdas*, *3*(1), 36–42. doi:10.37396/jsc.v3i1.57

Ayim, C., Kassahun, A., Tekinerdogan, B., & Addison, C. (2020). Adoption of ICT innovations in the agriculture sector in Africa: A Systematic Literature Review. *Agriculture and Food Security, 11*(22).

Batani, J., Musungwini, S., & Rebanowako, T. G. (2019). An Assessment of the Use of mobile phones as sources of Agricultural information by tobacco Smallholder farmers in Zimbabwe. *International Journal of Systems Integration*, *3*(1), 1–21. doi:10.20470/jsi.v10i3.375

Ben Ayed, R., & Hanana, M. (2021). Artificial Intelligence to Improve the Food and Agriculture Sector. *Journal of Food Quality, 2021*(Ml). doi:10.1155/2021/5584754

Coble, K. H., Mishra, A. K., Ferrell, S., & Griffin, T. (2018). Big data in agriculture: A challenge for the future. *Applied Economic Perspectives and Policy*, *40*(1), 79–96. doi:10.1093/aepp/ppx056

Dzino-Silajdzic, V. (2018). *Practical Guide; Focus group discussions*. 40. https://www.crs.org/sites/default/files/tools-research/fgds_april_24_final_lo_res_.pdf

FAO. (2020). Agriculture 4.0: Start Agricultural robotics and automated equipment for sustainable crop production. *Integrated Crop Management*, 24.

Friha, O., Ferrag, M. A., Shu, L., Maglaras, L., & Wang, X. (2021). Internet of Things for the Future of Smart Agriculture: A Comprehensive Survey of Emerging Technologies. *IEEE/CAA Journal of Automatica Sinica, 8*(4), 718–752. doi:10.1109/JAS.2021.1003925

Gavai, P., Musungwini, S., & Mugoniwa, B. (2018). A Model for the Adoption and Effective Utilization of ICTs in Commercial Agriculture in Zimbabwe. *Journal of Systems Integration*, *9*(4), 40–58. doi:10.20470/jsi.v9i4.356

Given, L. M. (2008). Naturalistic Inquiry. In *Encyclopedia of Research Design.* SAGE Publications, Inc., doi:10.4135/9781412961288.n262

Honrubia, M. (2021). *4 Agriculture technology projects to transform the industry.*

Jayne, T. S. (2012). Managing food price instability in East and Southern Africa. *Global Food Security*, *1*(2), 143–149. doi:10.1016/j.gfs.2012.10.002

Kamutuezu, E. U., Winschiers-Theophilus, H., & Peters, A. (2021). An Exploration of Factors Influencing the Adoption of ICT Enabled Entrepreneurship Applications in Namibian Rural Communities. *Adoption of Rural Entrepreneurship Digital Services in Namibia.*

Kanchana, S. (2018). IoT in Agriculture : Smart Farming. *International Journal of Scientific Research in Computer Science. Engineering and Information Technology*, (May), 181–184. doi:10.32628/CSEIT183856

Kante, M., Oboko, R., & Chepken, C. (2019). An ICT model for increased adoption of farm input information in developing countries: A case in Sikasso, Mali. *Information Processing in Agriculture*, *6*(1), 26–46. doi:10.1016/j.inpa.2018.09.002

Linaza, M. T., Posada, J., Bund, J., Eisert, P., Quartulli, M., Döllner, J., Pagani, A., Olaizola, I. G., Barriguinha, A., Moysiadis, T., & Lucat, L. (2021). Data-driven artificial intelligence applications for sustainable precision agriculture. *Agronomy (Basel)*, *11*(6). doi:10.3390/agronomy11061227

Loures, L., Chamizo, A., Ferreira, P., Loures, A., Castanho, R., & Panagopoulos, T. (2020). Assessing the effectiveness of precision agriculture management systems in mediterranean small farms. *Sustainability (Switzerland)*, *12*(9), 1–15. doi:10.3390u12093765

Mays, N., & Pope, C. (1995). Qualitative Research: Rigour and qualitative research. *BMJ (Clinical Research Ed.)*, *311*(6997), 109–112. doi:10.1136/bmj.311.6997.109 PMID:7613363

Msendema, M. B., & Nyirenda, G. B. (2019). Unravelling Development Capabilities Enhanced by ICT in Informal Micro-Finance Settings. Case of Women Village Banks in Blantyre, Malawi. *2019 IST-Africa Week Conference. IST-Africa*, *2019*, 1–8. doi:10.23919/ISTAFRICA.2019.8764871

Mugandani, R., Wuta, M., Makarau, A., & Chipindu, B. (2012). Re-Classification of the Agro-Ecological Regions of Zimbabwe in Conformity With Climate Variability and Change. *African Crop Science Journal*, *20*(S2), 361–369. doi:10.4314/acsj.v20i2

Munyaradzi Gocha, S. M. and L. C. (2014). The Use Of Information Communication Technology By A-Level Teachers In Selected Secondary Schools In Gweru, Zimbabwe. *Journal of Technology Management.*, *1*(1), 1–21.

Musungwini, S. (2016). A model for harnessing the power of the Mobile Phone Technology to improve Smallholder Agriculture in Zimbabwe. *Proceedings of the 5th International Conference on M4D Mobile Communication Technology for Development: M4D 2016, General Tracks, December*, 237–252.

Musungwini, S. (2018). Mobile Phone Use by Zimbabwean Smallholder Farmers : A Baseline Study. *The African Journal of Information and Communication*, *22*(22), 29–52. doi:10.23962/10539/26171

Musungwini, S. (2021). *Inclusive Digital Transformation for the Marginalized Communities in a Developing Context.* IGI Global, 95–122. doi:10.4018/978-1-6684-3901-2.ch005

Musungwini, S., & Van Zyl, I. (2017). ` Mobile Technology for Development ' Experiences from Zimbabwe Vending Markets a Naturalistic Enquiry. *International Journal of Business and Management Studies.*, *06*(01), 101–111.

Musungwini, S., Zyl, I. van, & Kroeze, J. H. (2022). *The Perceptions of Smallholder Farmers on the Use of Mobile Technology: A Naturalistic Inquiry in Zimbabwe*. 530–544. doi:10.1007/978-3-030-98015-3_37

Mutale, M. (2018). Challenges and opportunities of information and communication technologies (ICTs) in Binga District, Zimbabwe. *African Journal of Social Work*, *8*(2), 55–62.

Nettle, C. E. M. A. R., & Dela Rue, B. (2019). Making sense in the cloud: Farm advisory services in a smart farming future NJAS-Wageningen J. *Life Sciences.*

Norton, G. W., & Alwang, J. (2020). Changes in Agricultural Extension and Implications for Farmer Adoption of New Practices. *Applied Economic Perspectives and Policy, 42*(1), 8–20. doi:10.1002/aepp.13008

OECD/FAO. (2016). Agriculture in Sub-Saharan Africa. Prospects and challenges for the next decade. *Agricultural Outlook 2016-2025, 181*(November 1947), 59–93. doi:10.1787/agr_outlook-2016-en

Padilla, N. E., Payne, J. A. G., Simbulan, V. S., Lapastura, R. J. S., & Cadeliña, E. J. A. (2020). Capacity building for dairy farmers towards commercialization of green corn silage, haylage and urea molasses mineral block. *Universal Journal of Agricultural Research, 8*(6), 233–240. doi:10.13189/ujar.2020.080604

Psirofonia, P., Samaritakis, V., Eliopoulos, P., & Potamitis, I. (2017). Use of Unmanned Aerial Vehicles for Agricultural Applications with Emphasis on Crop Protection: Three Novel Case - studies. *International Journal of Agricultural Science and Technology, 5*(1), 30–39. doi:10.12783/ijast.2017.0501.03

Rahmann, G., Azim, K., Brányiková, I., Chander, M., David, W., Erisman, J. W., Grimm, D., Hammermeister, A., Ji, L., Kuenz, A., Løes, A. K., Wan-Mohtar, W. A. A. Q. I., Neuhoff, D., Niassy, S., Olowe, V., Schoeber, M., Shade, J., Ullmann, J., & van Huis, A. (2021). Innovative, sustainable, and circular agricultural systems for the future. *Organic Agriculture, 11*(2), 179–185. doi:10.100713165-021-00356-0

Rendle, K. A., Garrett, S. B., & Halley, M. C. (2017). *Beyond Exploratory : A Tailored Framework for Assessing Rigor in Qualitative Health Services Research.* PMC.

Ribarics, P. (2016). Big Data and its impact on agriculture. *Ecocycles, 2*(1), 33–34. doi:10.19040/ecocycles.v2i1.54

Rijswijk, K., Klerkx, L., & Turner, J. A. (2019). Digitalisation in the New Zealand Agricultural Knowledge and Innovation System: Initial understandings and emerging organisational responses to digital agriculture. *NJAS Wageningen Journal of Life Sciences, 90–91*(1), 100313. doi:10.1016/j.njas.2019.100313

Sachitra, V., & Padmini, C. (2021). COVID-19 Challenging Period and Agriculture Sector in Sri Lanka: Way to Lead. *Asian Journal of Advances in Agricultural Research*, (July), 21–34. doi:10.9734/ajaar/2021/v15i430160

Sadiku, M. N. O., Ashaolu, T. J., & Musa, S. M. (2020). Emerging Technologies in Agriculture. *International Journal Of Scientific Advances, 1*(1), 1–5. doi:10.51542/ijscia.v1i1.6

Saiz-Rubio, V., & Rovira-Más, F. (2020). From smart farming towards agriculture 5.0: A review on crop data management. *Agronomy (Basel), 10*(2), 207. doi:10.3390/agronomy10020207

Saldana, J. (2009). An Introduction to Codes and Coding. *The Coding Manual for Qualitative Researchers., 2006*, 1–31. doi:10.1519/JSC.0b013e3181ddfd0a

Subramanian, B., Chen, S. S., & Reddy, K. R. (2019). *Emerging Technologies for Agriculture and Environment.* Springer.

Talaviya, T., Shah, D., Patel, N., Yagnik, H., & Shah, M. (2020). Implementation of artificial intelligence in agriculture for optimisation of irrigation and application of pesticides and herbicides. *Artificial Intelligence in Agriculture*, *4*, 58–73. doi:10.1016/j.aiia.2020.04.002

Teddlie, C., & Tashakkori, A. (2009). *Foundations of mixed methods research*. SAGE Publications.

Trendov, N. M., Varas, S., & Zeng, M. (2019). *Digital Technologies In Agriculture And Rural Areas*. FAO.

Tsan, M., Totapally, D. S., Hailu, D. M., Addom, B. K., & CTA. (2019). *Digitalisation for Agriculture Report* [Data set].

Twinomurinzi, H., Schofield, A., Hagen, L., Ditsoane-Molefe, S., & Tshidzumba, N. A. (2017). Towards a shared worldview on e-skills: A discourse between government, industry and academia on the ICT skills paradox. *South African Computer Journal*, *29*(3), 215–237. doi:10.18489acj.v29i3.408

Ugochukwu, N. C., Carlu, W., & Abiodun, O. (2020). *Rural Smallholder Farmers' Awareness and Use of ICT-based market information Sources in South Africa. May*. doi:10.20944/preprints202005.0159.v1

Victor, O., Nic, J. L., & Xiaomeng, L. (2021). Factors affecting the adoption of mobile applications by farmers: An empirical investigation. *African Journal of Agricultural Research*, *17*(1), 19–29. doi:10.5897/AJAR2020.14909

World Bank. (2011). ICT in agriculture: Connecting amallholders to knowledge, networks, and institutions. *WORLD (Oakland, Calif.)*, *64605*, 424. 64605

Yahya, A. (2020). *Emerging Technologies in Agriculture, Livestock, and Climate*. Springer. doi:10.1007/978-3-030-33487-1

Chapter 3

Smart Farming Ingredients:
IoT Sensors, Software, Connectivity, Data Analytics, Robots, Drones, GIS-GPS

Jigna Bhupendra Prajapati
https://orcid.org/0000-0003-4058-8622
Ganpat University, India

Roshani Barad
Kantaben Kashiram Institute of Agriculture Sciences and Research, Ganpat University, India

Meghna B. Patel
AMPICS, Ganpat University, India

Kavita Saini
Glgotias University, India

Dhvanil Prajapati
LDRP Institute of Technology and Research, India

Pinalkumar Engineer
Sardar Vallabhbhai National Institute of Technology, India

ABSTRACT

Smart farming uses information and communication technologies in various fields of agriculture. It refers to the use of information and data management technologies in agriculture. Smart farming leads towards high productive and sustainable agricultural production. Smart farming provides the farmer many advantages for decision making for better management. Smart farming technologies collect precise measurements of factors that determine farming outcomes. It enables agriculture more reliable, predictable, and sustainable. It also improves crop health, reduces the ecological footprint of farming, helps feed the increasing global population, provides food security in climate change scenarios, and achieves higher yields while reducing operating costs. It's also needed to meet the needs of the growing population. There are many technological devices, such as IoT, software support, connection, data analytics, robots, drones, and GPS, which is useful to enhance the quantity and quality of agriculture production with minimizing labor.

DOI: 10.4018/978-1-6684-6413-7.ch003

Copyright © 2023, IGI Global. Copying or distributing in print or electronic forms without written permission of IGI Global is prohibited.

INTRODUCTION

Smart Farming (Smart Farming Cycle)

Agriculture was ranked second in the world in 2014, providing more than 13 percent to India's Gross Domestic Product (GDP). It also gives a job to more than 50% of the Indian people. One of India's most varied industries, agriculture contributes to the nation's the social structure, the economics, and culture. The technique and practises applied in the existing agricultural system are inherently flawed. We should pose some study questions about food safety and agricultural output, because by 2050, it's predicted that there will be close to 10 billion people on the planet (Patil et al., 2012). A 60 percent increase in agricultural production is required to accomplish these targets. The old agricultural method is insufficient and inconsistent in achieving this. Furthermore, some specialists are already beginning to incorporate innovations into agricultural practise in an effort to reduce expenses and boost performance (Mohamed et al., 2021). As a result, IoT and data analytics are important in agriculture, resulting in innovations like wireless sensor networks (WSNs) (Navarro, Costa, and Pereira, 2020; Ouafiq et al., 2021; Vikranth, 2021). IT agriculture, often known as smart farming, is the practise of using technological breakthroughs and worldwide networking to help agricultural solve issues methodically. In this scenario, sensors are particularly crucial since they collect a variety of crop-related data, including moisture, pest, and weather information, as well as information about the soil, water levels, and fields. A number of sensors, including moisture sensors, electromagnetic sensors, and optical sensors, must be employed to gather all of this data. Large amounts of data are produced by sensor-enabled devices, which can be stored separately or in a cloud database (Vikranth, 2021).

In order to offer effective answers for several agriculture applications, the wireless sensor network (WSN) integrates sensor interpretation, mechanization management, digital network transmission, data storage, and data dispensation (Rajasekaran & Anandamurugan, 2019). By utilizing wireless sensor networks, precision agriculture would increasing efficiency, output, and revenue of diverse farming production methods (WSNs) (Sanjeevi, 2020). IoT aids farmers by providing real-time updates by informing the public about their crops using the wireless network. Various sensors, including temperature, moisture, and water level sensors, and flood gauge sensors, can be connected (Vikranth, 2021). With the help of these many sensors, the farmer can receive conscious notifications on what is going on in the field (Veena, 2018). As wireless communication technology develops, short-range communication is made possible by compact, multipurpose sensors with low power requirements and cheaper costs.

The management of farms utilising robotics, drones, artificial intelligence (AI), and the Internet of Things (IoT) is known as "smart farming," and it aims to boost product quantity and quality while lowering the need for human labour (Navarro, Costa, and Pereira, 2020). Utilizing information and technology, smart farming solutions optimize farming inputs and procedures along the food supply chain, including transportation, distribution, and retail, in order to increase crop and livestock production's economic yield. Cyber systems that enable monitoring, intelligent forecasts, decision assistance, automated control, and long-term planning are among the technologies that rely on big data analytics (Wolfert et al., 2017; Lokhorst, De Mol, and Kamphius, 2019). In order to increase the economic yield of crop and livestock production as well as to optimize farming inputs and procedures along the entire food supply chain, including transportation, distribution, and retail, smart farming solutions leverage information and technology. Cyber systems that enable monitoring, intelligent forecasts, decision support, automated

Figure 1. Smart Framing

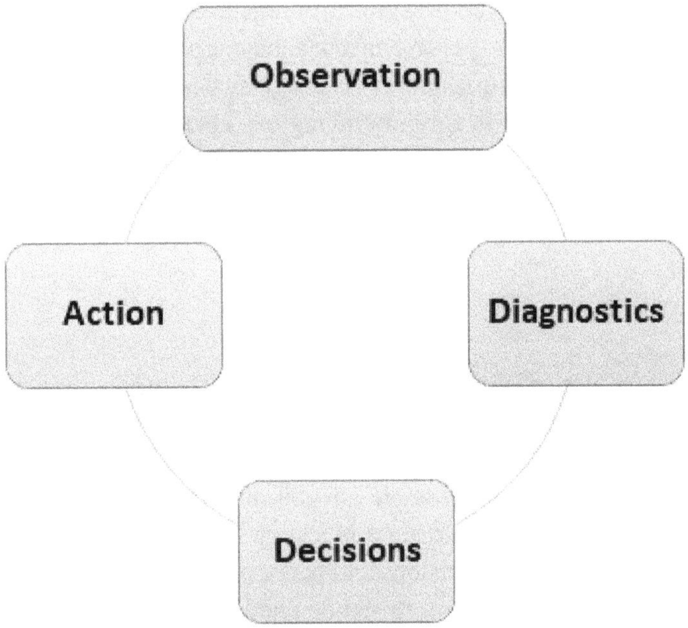

control, and long-term planning are among the technologies that depend on big data analytics (Sarker et al., 2020; Lokhorst, De Mol, and Kamphius, 2019).

The IoT-based smart farming lifecycle: The data you can extract from devices and send over the internet is the basis of the Internet of Things (IoT) (Ruan et al., 2019; Verdouw et al., 2021). IoT devices installed on farms should gather data and process it repeatedly so that farmers can react quickly to emerging problems and changes in the environment. This will optimize the agricultural process. This cycle is used in smart farming as figure 1.

1. **Observation**. Sensors capture observational data from the atmosphere, soil, livestock, and/or crops.
2. **Diagnostics.** The sensor data is sent to an IoT platform located in the cloud that has predefined decision rules and models, often known as "business logic," that determine the status of the object being studied and pinpoint any needs or inadequacies.
3. **Decisions**. After problems are identified, the user and/or machine learning-driven IoT platform components decide whether location-specific therapy is required and, if so, which treatments should be used.
4. **Action.** Once end users have evaluated and taken action, the cycle resumes (Sciforce, 2021).

Need of Smart farming

India's population is quickly increasing, and food security will be an issue unless inexpensive technology and farmers are provided with remedies that are utilised to decrease crop deficits (Nabi et al., 2022; Gupta et al., 2020). The low use of water, fertilizers, pesticides, insecticides, and herbicides, as well as the specialized equipment required, are all significant advances. The previous farming method based on a far-fetched estimate of hypothetical average conditions reality, necessitating precision in recognising field-specific changes and adjusting management plans accordingly (Regan, 2019). Variations in yield across a farmer's whole administrative region are frequently reported. These are the outcomes of soil features and environmental factors, as well as management practices. A farmer's duty of retaining information about field conditions and essential remedies, which he obtained via years of trial and error, was difficult. Furthermore, the change in agricultural regions made it even more difficult for the farmer to use the same techniques as in past years. Smart agriculture has tackled this challenge by automating and simplifying data collection, analysis, and delivery of results enabling better management decisions that may be implemented more quickly on individual sites within large fields.

Component of Farming

Right input, right time, right place, right amount, and right way are the five R's of precision agriculture (Khosla, 2008). These 5 R's must be met in order for PA to be implemented smoothly. The main goal of Precision agriculture is to collect and send the accurate data that various equipment used in precision agriculture require. However, PA is only possible now because of recent technological advancements. These technologies can work together to create a precision agricultural system that can help farmers make site-specific management (SSM) decisions and perform other tasks. Yield monitors, global positioning systems, and other technologies have raised demand for the agricultural production system as computing and data management skills have improved. By utilising crucial data to make site-specific crop production decisions, Additionally, precision farming can be viewed as a management strategy that enhances conventional approaches. It allows for economic savings, higher produce quantity and quality, and fewer negative environmental effects, as well as the oversight of geographical and a field's temporal variability. As a result, it's safe to say that PA consists of three key parts.

Information

The following discussion covers both crop characteristics and soil characteristics (such as tillage requirements, salt, waterlogging, topography, fertility levels, texture, moisture content/retention, etc.) are all factors that must be considered in order to obtain the greatest outcomes in PA. The frequency of pests (such as insects, diseases, and weeds), weather and climatic factors, additional plant responses to biotic and abiotic stresses, harvest and postharvest management, marketing and market information, and farmers' socioeconomic situations, among other factors (Ahmad & Mahdi, 2018). Almost all of the processes can benefit from the data, resulting in a significant improvement over traditional agriculture. You can use it to create thorough maps of farms, communities, and areas (for instance, varying soil properties, groundwater, insect and weed occurrence, geography, environmental contamination, etc.) (Ahmad & Nabi, 2021). Making judgments specific to a site is based on this knowledge (Belal et al., 2021).

1.3.2 Technologies:

Precision farming and emerging technologies work together to keep farmers informed and offer them with all of the benefits that come with it. The technologies have been proved to boost output, productivity, and profitability by a significant amount. At the farm/village/regional levels, remote sensing, geographic information systems Accurately identifying deficiency areas and quantifying the economic significance of soil water-fertilizer-pest-crop-related constraints can be done with the aid of (GIS), GPS, auto-analyzers, sensors, actuators, computers, suitable software, massive storage technologies, real-time computing devices, and other tools (Ahmad & Nabi, 2021). Implementing these can help with integrated management systems for health of the soil, nutrients, pests, energy, water, and other crop genetic resources. Sensors can be used to monitor and improve the postharvest quality of produce (Ahmad & Mahdi, 2018). Drones can be used for spraying as well as observation. Furthermore, the best illustration of PA technology is driverless tractors. Finally, IoT systems have unquestionably aided PA improvements (Belal et al., 2021).

Management

"Management is the key to success in Precision Agriculture," as the saying goes. Precision agriculture requires precise management to function properly. Management creates a comprehensive administration system by combining information obtained and available technology. To reap the greatest benefits, the user must be knowledgeable enough to apply the aforementioned information and technology. Precision and competitive management complement the 5 R's. Due to the complexity of agriculture and the Together, agronomists, plant breeders, soil scientists, agro-meteorologists, entomologists, plant pathologists, weed scientists, biotechnologists, economists, extension workers, and farmers are required to develop and deploy knowledge-intensive precision farming technology (Belal et al., 2021). Information management, a decision a particular precision agricultural service provider, and a decision support system (DSS) are all essential for the success of precision agriculture (Ahmad & Nabi, 2021).

SMART FARMING WITH IOT SENSORS

A brand-new and developing technology trend with applications in almost every industry is the Internet of Things (IoT) (Navarro, Costa, & Pereira, 2020). The entire system becomes intelligent when items are connected to the internet and to one another. We have used IoT in many different circumstances, like intelligent cities, smart homes, smart retail, and so forth. IoT applications in agriculture and farming practises is urgently needed, because 9.6 billion people are anticipated to live on the planet by the year 2050, and the agriculture business will need to provide at an even quicker rate to fulfil that need. This is made feasible by the use of modern technology, particularly the Internet of Things (IoT) (Doshi et al., 2019). Labor-free farms are now a possibility because to the Internet of Things. It can be employed not just in the principal farming methods, but also in the care of animals and greenhouse farming, farm management, and so on. The most important IoT tool is a sensor; sensors are gadgets that gather key data, which is subsequently processed to create the required analysis (Sishoda, Ray, & Singh, 2020). In agriculture, sensors are generally used to obtain readings for measuring soil moisture content, diagnosing illnesses, and analysing NPK concentrations (Ratnaparkhi et al., 2020).

A digital technology known as the Internet of Things (IoT) helps farmers increase output while lowering technological barriers. Because farmers have sufficient understanding of new technical developments and are now aware of them. IoT has the potential to alter agriculture businesses as well as motivate farmers to address technological challenges.Farmers can increase their output by utilising IoT to maintain a constant eye on soil acidity, temperature, weather, and a variety of other elements. Farmers can also use IoT-based gadgets to keep an eye on their cattle. Farmers may get key information from IoT-based sensors on things including rainfall, crop yields, pest infestation, and soil nutrition, all of which are vital to production and provide detailed information that can be used to improve agricultural operations over time. A crucial piece of using technology ensuring a seamless agriculture logistics flow is the Internet of Things (IoT), which brings precise, real-time, and shared characteristics to the agricultural supply chain (Jain, Krishna, and Saritha, 2012).

Primary Advantages of IoT

- Water management may be done effectively with IoT sensors.
- IoT allows for continuous land monitoring, allowing for any necessary precautionary steps to be taken at a later date.
- It assists farmers in reducing manual labour, making farming more efficient and time-consuming.
- IoT allows for easy soil management, such as pH and moisture levels, which aids farmers in seed sowing according to soil level.
- Diseases of plants and crops can be found with RFID chips and sensors. Data is read from RFID tags and sent to the appropriate reader over the internet using this technology. The concerned farmer or scientist can access this data and take the required precautions from a distance, protecting crops from disease outbreaks (Wang and Liu, 2014).
- Crop sales will expand in the worldwide market as farmers can now connect with the global market from any location.

IoT's Applicability in Agriculture

By removing the need for farmers to undertake physical labour, IoT adoption in agriculture has transformed it into smart farming, increasing production across the board (Khattab, Abdelgawad, & Yelmarthi, 2016). The use of sensors and their connectivity to monitor fields in real-time has completely changed how farming is done (Stočes et al., 2016). This includes keeping an eye on vital factors like soil moisture, temperature, and humidity. Additionally, fewer of these inefficient resources, such as water and electricity, are being consumed (Malavade and Akulwar, 2016).

Weather Conditions

Climate change significantly affects agriculture. Furthermore, both the amount and quality of agricultural produce are significantly impacted by a lack of awareness of climate. On the other side, IoT technologies let you monitor the weather in real-time. Both inside and outside of agricultural districts, sensors have been installed. They gather environmental data, which is then utilised to identify the best crops for sustaining growth in the particular climate (Garg, Khan, and Alam, 2019). Every sensor in the Internet of Things (IoT) network is capable of reliably detecting meteorological Real-time information on variables including humidity, rainfall, temperature, and more. All of these variables can be monitored and adjusted using a variety of sensors to fulfill your intelligent Farming requires. These sensors maintain an eye on the state of the crops and the climate in the area (Yadav, Garg, and Luthra, 2020). When unusual weather circumstances are identified, an alarm is sent out. Physical presence during inclement weather is no longer required, resulting in increased production and greater agricultural benefits for farmers.

Precision Farming

Precision farming is one of the most well-known IoT uses in agriculture. Applications for smart farming that are examples to improve farming efficiency and management include animal inventory monitoring, field observation, vehicle tracking, and monitoring (Suciu et al., 2019). Precision agriculture seeks to analyse sensor data and take appropriate action. Farmers can gather data from sensors using precision farming, then analyse that data to make quick, informed decisions. In order to improve agriculture's efficiency and efficacy, precision farming techniques such as vehicle tracking, livestock management, and irrigation management are all essential (Navarro, Costa, and Pereira, 2020). You can evaluate the state of the soil and other pertinent variables with the aid of precision farming to improve operational effectiveness. In addition, you can examine water and nutrient levels by keeping an eye on the connected devices' current operational status.

Greenhouse with Intelligence

The Internet of Things is helpful, weather stations can now autonomously Alternate climate scenarios in response to a set of guidelines, enabling we'll create greenhouses smarter (Sai Sravani et al., 2020). Human involvement in greenhouses is no longer necessary thanks to the deployment of IoT, improving the efficiency and accuracy of the entire process. For instance, sophisticated and affordable greenhouses can be built using solar-powered Internet of Things sensors (Donzia, Kim, and Hwang, 2018). These sensors enable accurate real-time greenhouse monitoring by gathering and transmitting real-time data. The sensors allow for the tracking of Email or SMS updates regarding greenhouse conditions and water usage. The Internet of Things allows for autonomous and intelligent irrigation. These sensors can measure pressure, vapour pressure, temperature, and light intensity.

Data Analytics

Device data from IoT requires more storage than a typical database system can provide offer. The smart agriculture system relies heavily on end-to-end IoT platforms with cloud-based data storage (Greasely and Edwards, 2021). These systems are anticipated to have a big impact on making it possible to carry out superior activities. In the IoT environment, sensors serve as the main data collection tool. Using analytics technologies, The data is evaluated and transformed into knowledge (Neethirajan and Kemp, 2021). Weather, livestock, and agricultural conditions can all be analysed using data analytics. The knowledge gained provides for better Using technological breakthroughs to make decisions. By collecting data from sensors using IoT devices, you may learn about the current condition of your crops. You can make better harvesting decisions with the aid of predictive analytics. By offering knowledge, 58 Digital Technologies for Smart Agriculture. Trend analysis can be used by farmers to predict the weather and agricultural harvesting for the future. IoT has helped farmers in the agriculture sector maintain soil fertility and crop quality, which has improved output in both quantity and quality (Tomar, Kaur, and Singh, 2018).

Agricultural Drones

Agriculture operations have changed nearly entirely thanks to technological improvements most recently upheaval agricultural drones being introduced. Plant health assessments, agricultural planting, crop spraying, crop monitoring, and field evaluation are all done with drones on the ground and in the air (Kamilaris et al., 2016). With Drone technology has provided agribusiness with real-time data for smart strategy and planning a lift and makeover. Drones equipped with multispectral or thermal sensors can identify regions that require irrigation. Sensors assess the plants health as they begin to grow as well as determine their vegetation index. shrewd drones helped to lessen the final effect on the environment. The amount of chemical that enters groundwater has dramatically lowered as a result (Tomar, Kaur, and Singh, 2018).

Livestock Monitoring

Farmers can utilise IoT applications to gather information on the whereabouts, wellbeing, and health of their livestock. This data aids them in determining the health of their cattle (Frost et al., 1997). For instance, finding sick animals and removing them from the herd would stop the disease from spreading throughout the entire herd. Ranchers' ability to find their animals with IoT-based sensors reduces labour costs significantly. JMB North America is an example of a corporation that uses an IoT system. Which business offers cow monitoring systems to cattle producers. Among the various solutions offered to support livestock owners maintain track of the animals that are expecting and will shortly give birth. A sensor-powered battery is released when the water inside them breaks. The then information given to the rancher cattle manager. As a result, the sensor enables farmers to focus more on their work. The above-mentioned IoT application has offered agriculture up-to-date technology. These have aided when connecting the output, quality, and gaps in quantity. Numerous sensors connected to the Internet of Things can collect real-time data that can be used to reduce crop loss. IoT has altered the economics by way of agriculture making it possible for products to be processed more quickly and delivered to supermarkets as soon as possible (Prathibha, Hongal, and Jyothi, 2017).

SMART FARMING WITH ROBOTS

These days, a variety of robots are used for smart farming. Robots like milkycows, pick weeds, plant seeds, harvest crops, etc. Such Robots will help to farm with increasing yields and revenue. Agriculture offers a lot of chances for automation because farming methods are repetitive and labor-intensive. Agricultural procedures are rapidly being automated by robots.

- **Self-Driving Robots:** These robots operate avoiding obstacles while navigating the field. Good quality cameras allow the robots to freely travel by creating a virtual 3D representation of the surface as per need. Sometimes it also depends on the constraints they are trained to respect or navigate around.
- **Seeding Robots:** Robots with GIS mapping helps in the automated seeding process. Combining GIS with an automated mechanism will seed more correctly by creating a map of a field and integrating information about soil characteristics. Such seeding robots will keep concernedabout-density and soil quality.
- **Weeding robots:** These types of robots will help to remove unplanted plants. Some devices such as weed-chipper are popular for automated welding processes. These robots work with identifying pre-defined unwanted plates in reference to existing plants.

- **Spraying robots:** these robots are linked with computer vision cameras to identify weeds for precise herbicide applications. Advanced systems may even recognize particular plants and only activate the appropriate spray nozzles. The tech-savvy devices protect from waste.
- **Picking Robots:** It is necessary to harvest these crops, which requires a lot of labour and time that are sometimes in short supply. If a farmer is unable to harvest the produce in time, it will be squandered. Farmers regularly run out of labour during the harvest due process is time intensive and laboros.

SMART FARMING WITH DRONES

The drones are used to control Crop scouting, irrigation, field soil analysis, crop health, crop treatment, and crop damage evaluations (Tripicchio et al., 2015). Drone surveying increases agricultural yields while cutting costs and time. Drones for agriculture may be used to analyze soil and fields to aid with field planning. Drones can be valuable to farmers in a variety of ways, but before purchasing pricey equipment, it is crucial to understand their capabilities and limits (Ram Kumar et al., 2018).

There are various types of drones as Micro Air Vehicles (MAV), Vertical take-off and landing (VTOL), Low Altitude- Short Endurance (LASE), and Low Altitude. Generally, Fixed-wing and multi-rotor drones can both be used for agricultural purposes. Some popular agriculture drones as DJI Agras MG-1 – Best Crop Spraying Drone, DJI Agras MG-SenseFlyeBee SQ, Sentera NDVI Upgrade, DJI Smart Farming Package, Precision Hawk Scouting Package, Parrot Bluegrass Fields Agricultural Quadcopter, etc (Bacco et al., 2018).

Spraying drones are useful for fertilizers, herbicides, fungicides, pesticides, Seeds, Desiccants, etc. Data collection is also carried out using RGB maps & NDVI maps (Wolfert et al., 2017).

SMART FARMING WITH GPS AND GIS

Global Positioning System (GPS)

The Global Positioning System (GPS) (Hakkim et al., 2016) In precision agriculture, the most crucial equipment is the GPS. It is characterized as a satellite-based navigation system that allows for the recognition of places within one meter of the field's actual site (100 m to 0.01 m) (Hakkim et al., 2016). In order to determine characteristics regarding crops, soil, water, impediments, pest occurrence, and weed invasion at that spot, this location includes information that is the the exact field location's latitude, longitude, and elevation, among other things (Yousefi and Razdari, 2015). Precision agriculture can be handled by a user with such an advanced and precise location system (Lu et al., 1997). The The following are a few advantages of GPS in agriculture:

- Application of inputs (e.g. seeds, fertilisers, insecticides, herbicides, and irrigation water) to target locations
- Mapping of soil and crop measurements
- Farm management
- Yield monitoring

In order to optimise operating costs and reduce expected completion times, Advanced and precise location data is frequently required for monitoring and carrying out agricultural practises across a large geographic area. The use of the Global Positioning System (GPS) for sample location acquisition utilising the Global Navigation Satellite System is necessary because agricultural techniques with extensive farms difficult operating off on remote sensing until precise positional data is available unavailable (GNSSs). As a result, the usage of GPS-based sensors in precision farming has expanded due to the growth of positional accuracies using GNSS signals (PA). It has paved the way for the deployment of GPS in the PA, according to the demands and expectations of the users. Additionally, GPS tools have improved with time, and today's market offers strong GPS tools that work with iOS, Android, and Windows, among other operating systems. With growing trends in PA leading to innovations in route optimization techniques for the most efficient use of fertiliser spraying and associated agricultural practises are taking on a new appearance known as smart farming, the availability of GPS signals in a variety of A market opportunity has been identified for instruments or equipment (Pandey, Tripathi, and Sharma, 2021).

The technology is put to good use. Additionally, GPS technologies have become stronger over time and are available on the market now for use with a variety of operating systems, including iOS, Android, and Windows. With growing trends in PA leading to innovations in route optimization techniques for the most efficient use of fertiliser spray and related activities in agricultural practises, transforming into a new look as Smart farming, the availability of GPS signals in a variety of instruments or equipment has been seen as a market opportunity (Neményy et al., 2003). GPS gadgets have progressed from traditional hand-held systems that have ground accuracy of about 10 meters, whereas differential GPS (DGPS) devices have centimeter-level accuracy. The limited field areas make scenarios in locations with steep topography, like Himachal, much more challenging (Milner, 2016). Due to the cropping pattern, farmers already have fewer options in this terrain, therefore PF implementation is challenging.

The global positioning system (GPS) allows in-field variations to be recorded as regionally encoded data. The correct position can be followed and continuously determined. Because this system more carefully considers agricultural areas and locations than previous methods, the user has access to a larger database. However, only in locations where GPS position tracking has occurred can precise yield data be supplied. The spatial coordinates for yield monitor data are provided by GPS receivers in conjunction with yield monitors. This can be turned into field-by-field yield maps. Information obtained from various To create field management strategies for chemical application, growing, and harvesting, satellite data and GPS references can be employed (Sood et al., 2015).

Geographic information systems (GIS) and the Global Positioning System (GPS) were combined to create precision agriculture, often known as site-specific farming (GIS). With the use of these technologies, enormous amounts of geographic data may be processed and analysed quickly by fusing real-time data collecting with accurate position data. GPS-based Farm planning, field mapping, soil sampling, tractor navigation, crop scouting, variable rate applications, and yield mapping are all done with the help of precision farming apps (Sood et al., 2015).

Geographic Information System (GIS)

The Geographic Information System (GIS) is a ground-breaking using using technology allows users to interact with data linked to a geographically mapped location on the planet. It's a database created specifically for working with map data. GIS is a platform for compiling, storing, retrieving, and analyzing data connected to the characteristics of a certain place (i.e. a geographic context) (Usery, Pocknee, and Boydell, 1995). Utilizing GIS allows data from a range of sources, including site-specific and geographical data, to be stored and organized in several layers. Each "coverage" layer in a GIS is made up of topologically relevant geographic components (such as Along with related information, geography, soil types, surface drainage, subsurface drainage, etc (Usery, Pocknee, and Boydell, 1995). GIS is undeniably more than a simple map capable of providing relative solutions. Instead, it includes calculation and analytic capabilities, allowing it to provide comprehensive field views and make appropriate agricultural technology judgments. Many sectors use GIS systems for data synthesis and decision assistance, however GIS shouldn't be mistaken with decision-making systems (Belal et al., 2021).

GIS software allows data from a range of sources, including site-specific and geographical data, to be stored and organised in several layers. Each "coverage" layer in a GIS is made up of topologically relevant geographic components (such as topography, soil types, subsurface drainage, and other factors) as well as related information. GIS is undeniably more than a simple map capable of providing relative solutions (Aguilar-Manjarrez and Ross, 1995). Instead, it includes calculation and analytic capabilities, allowing it to provide comprehensive field views and make appropriate agricultural technology judgments. GIS systems are used for data synthesis and decision support in many industries; however, GIS should not be confused with decision-making systems (Belal et al., 2021).

GIS can be used to create pictures, such as drawings, animations, and other cartographic products, in addition to maps. Globally, GIS is becoming more crucial to agriculture production, from mobile GIS in the field to analytical science of production data at the farm manager's office, by assisting farmers in increasing production, lowering expenses, and managingthey use their land more wisely. GIS applications including crop yield estimates, soil amendment assessments, and erosion detection and remediation can help farmers better understand and manage natural inputs in farming even when they cannot be controlled.. The In this study, a spatial crop model is initially created by fusing an environmental policy integrated climate (EPIC) model with a geographic information system (GIS) to simulate regional crop productivity (Usery, Pocknee, and Boydell, 1995). I worked on a Crop Growth Monitor System that combined data from the AVHRR and the VGT. GIS allows different 'layers' of data to be overlaid, such as ecological circumstances, genuine physiognomy, and pressure indexes related to people. A thematic and layer-based system called GIS enables you to overlay and examine indices for various site modifications. In planning and management, technology is used to its best potential (Usery, Pocknee, and Boydell, 1995).

SMART FARMING WITH SOFTWARE

In Smart Farming, a variety of software is required for a number of jobs. Below are some examples of software with specific functionality (Hakkim et al., 2016):

- Mapping software
- Software for map producers and variable-rate applications
- software to combine many maps
- Software with contemporary geostatistical characteristics

Software with instruments for statistical analysis Controller tools are common in precision agriculture technology, and precision farming software contains them (Grogan, 2012). IoT enhances software maintenance – for example, by automatically updating equipment – and brings new options for farm management, like using a controller to remotely operate a tractor that drives safely. The possibilities of modern IoT and precision agriculture allow for simultaneous control of dozens of equipment components.

Farm management software includes features for managing contracts, crops, customers, finances, greenhouses, inventories, labour forces, livestock, pricing, suppliers, Bar-coding/RFID, order processing, financial computations and analysis, and traceability, to name a few (FMS) (Paunova-Hubenova and Trichkova-Kashamova, 2021). FMS is a cutting-edge technology that enables farms to utilise their resources more effectively. It includes, among other things, software delivery strategies, data analytics tools, and best agricultural practises based on farm software. Helpful software of this type to increase farm production and decrease input waste by utilising GPS, sensing, together with communication technology. to boost agriculture's productivity, it is also integrated with hardware devices (Paunova-Hubenova and Trichkova-Kashamova, 2021).

Farm Management Software on a Global Scale: Current Status and Trends

The importance of agriculture in today's world economy, as well as the effects of ongoing globalisation processes, have already sparked the creation of specialised software instruments that are anticipated to facilitate, accelerate, and simplify management procedures pertaining to various facets and/or stages of the food production chain. Improved process management increases productivity and profitability, two factors that are prerequisites for growing competitiveness. As a result, there has been an enormous rise in the number of "virtual" "farm management software solutions" on a global scale in recent years. To improve farm productivity by providing instruments for technology-enabled/-enhanced farming, they, but only make up a small portion of a much larger assortment of modern ICT technologies. These can be broadly categorised into the key groups listed below (Hristoski et al., 2017):

- **Farm management software (FMS):** helps farmers manage their resources more effectively, such as agricultural produce, farm animals, and other resources.
- **Smart/Precision Farming Software (PFS) and Predictive Data Analytics:** systems that prioritise the use of big data and internet of things (IoT) paradigms, including predictive analytics, in addressing farm-related challenges for better farm-related decision-making, in order to conserve energy, These systems heavily use on big data and IoT paradigms, including predictive analytics, to boost efficiency, optimise herbicide and pesticide application, and minimise risk.
- **Sensors:** Smart sensors collect data to assist farmers in monitoring Crop health, climate, soil moisture, and soil quality.
- **Robotic:** Farmers can use smart sensors to collect data on crop health, the environment, soil moisture, and the weather.
- **Drones:** agriculture demands are served by unmanned aerial vehicles (UAVs) and related services Autonomous vehicles (for instance, crop field monitoring).
- **Animal Data:** From breeding patterns to genomics, there is software and hardware dedicated to better understanding livestock.
- **Smart Irrigation:** Water usage monitoring and automation systems for farms
- **Next Gen Farms:** Alternative farming practises that are technologically advanced and can be used in places and situations where conventional farming is impractical.
- **Marketplaces:** systems that connect farmers and suppliers directly or customers in online markets relevant to agriculture without the involvement of an intermediary. Some are standard e-

commerce platforms, while others make use of cutting-edge technologies improve the efficiency of physical marketplaces.

Primary Applications/Functionalities are Supported by FMS Solutions

- Maintaining records
- farming economics
- management of resources and inventory
- **Other:** Market notifications, client management, accounts payable and receivable, profit centre analysis, tax management, and so forth.

From 2016 to 2022, keeping records is predicted to have the biggest market share for an on-premise FMS delivery type (Hristoski et al., 2017). On the global market today, there are over 150 cutting-edge software systems dedicated solely to farm management. However,The needs of smallholder farmers' extremely basic daily operations cannot be met by those software solutions, which are too strong, sophisticated, and appropriate. This concept emphasises the requirement for the development of small-scale FMS goods aimed only at smallholder farmers.

SMART FARMING WITH DATA ANALYTICS

Data analytics necessitates the storage, consolidation, processing, and extraction of data. We must utilisedata mining methods toretrieve data and details about databases. It is crucial for the process of making decisions on a variety of agricultural challenges. The ultimate goal using data mining to obtain knowledge from Data transformation into a unique human-comprehensible format for advanced application. The function of big data in agriculture offers the possibility of increasing Farmers' financial gain from this sector's digital revolution, which we carefully explore. The types of data needed for a smart farming system are examined, as well as the design and conceptual layout of a proposed smart agricultural system. Additionally, the smart agricultural system must incorporate IoT and data analytics, as well as implement various components of the smart farming system (Vikranth and Krishna Prasad, 2021).

Technologies like Map Reduce, STORM, Hadoop Distributed File System (HDFS), and others are used in agricultural big data solutions. So, if all of these technologies are implemented in agriculture, crop failure will be reduced, and farmers' fundamental objective of making a profit will be met. By boosting productivity and reducing environmental impact, these technologies enhance the farming cycle. Additionally, this data aids farmers in determining how much water and fertiliser to use, and information on herbicides and insecticides (Vikranth, 2021).

The term "big data" refers to the vast quantity of available data sets. Big data is generated by sensors, phones, and other gadgets. The major purpose of gathering such a massive data set is to hunt for relevant patterns using efficient processing. Big data is collected using various sensors and then organised, disseminated, and organised into meaningful information that can be analysed later to make the right decisions. A crucial part of Big Data is sensors (Hung et al., 2005). It's a technique for looking into undiscovered agreements, finding an unknown connection, and gathering beneficial information for decision-making analysis. Supply chain management, machine learning techniques, failure forecasting, predictive preservation, industrial forecasting, and recurrent pattern mining were all areas where IoT and data analytics were applied. In smart agriculture, the sensor monitors soil nutrients, humidity levels, watching the weather, and identifying pests. We can farm sustainably without wasting water and using less fertiliser if we follow these ideas. Farmers are adamant about utilising the smartphone application, which allows them to access the study's findings and determine whether or not to pursue profitable agriculture based on the findings.

By creating a platform for agricultural machinery data analysis, the problem can be solved. Tractors, cultivators, harvesters, and dryers are among the agricultural machinery studied in this study. This programme takes data about equipment conditions at first, then analyses those conditions using multiple indicators utilising a big data platform, and finally provides warning messages by recommending maintenance needs (Vikranth, 2021). In this study, the data inspection inter-system runs alongside the data analysis procedures are carried out by the data process platform, a system application, which then presents the results as analytical outcomes. This study makes use of both the Spark engine, which is well-suited to Hadoop technology and the MapReduce data analysis engine, which is built on Hadoop technology (Jan et al., 2017). describes weather data analytics, which captures meteorological data remotely and puts it in a cloud database that can be accessed from anywhere. Novel wireless sensor network applications arose from the proposed new approach of efficiently gathering data. The idea is to put real-time sensors into use in a subject's area of vision to capture data in a real-world setting and make it available for future machine learning.

Big Data might be the foundation for a number of new skills, such as identifying relationships for better irrigation, reproduction, animal feeding, and marketing, between field, weather, and commodity data. Demand for seeds can be predicted using prognostic analytics, food demand, and animal feed demand, allowing agribusiness suppliers to take taking the proper steps to satisfy claim deadlines. To assist in balancing the market in light of the quantity that is available, new pricing strategies may be implemented. He thinks one solution for sustainable farming and lowering its environmental impact is big data analytics (Vikranth, 2021).

Examples of Weather forecasting, crop yield prediction, yield collecting, irrigation supervision, crop virus and pest supervision, agricultural selling, and agricultural pest management are some examples of big data-based digital agricultural technology that can assist farmers from field preparation through harvest (Vikranth, 2021).

SMART FARMING WITH CONNECTIVITY

The development of connection is crucial to the IoT's actual success in agriculture. From the standpoint of the telecom industry, offering connectivity including other value-added services has tremendous potential and might have a significant impact on the chain as a whole (Ayaz et al., 2019). The majority of telecom companies worldwide provide connectivity services, however, this segment of the market for smart agriculture is minuscule. The cellular operators must provide a new variety of services aimed at the needs of growers in light of their value, particularly in rural areas. Accepting because the majority of those who work in this business are not well educated and are frequently uninformed about As a result of emerging technology, the operator should provide more than just connectivity (Bacco et al., 2018). If true, it will undoubtedly aid in boosting the market share of telecom and mobile operators. Additionally, in order,these operators must collaborate with the investors to deliver end-to-end solutions, which requires a larger expenditure even before benefits can be observed (Patil et al., 2016).

The type of collaboration and the participants, including System integrators, non-cellular connectivity service providers, device makers, solution suppliers, etc., determine the outcomes of attracting investors. On the one hand, the result of this alliance would aid operators in breaking into the market more deeply, ultimately increasing their market share. The chance can also help to forge strong connections between organizations and farmers in order to inform them of the advantages of smart agriculture (Usery, Pocknee, and Boydell, 1995).

Only when service providers take advantages of cellular technology's true advantages, such as the convenience of two-way communication, and the portability and flexibility can they create affordable but tailored solutions. They are required to fulfill the farmer's needs in the location of their choice. Additionally, legislative adjustments are necessary to enable access to dependable and high-quality inputs in order to ensure fast penetration in the agricultural industry. The study, which looked at 23 studies, the majority of which were from developing nations, concluded that smallholder farmers have a bright future thanks to cellular services and smartphone technology since they will be able to increase their output.

Low-power wide-area (LPWA) technology is also anticipated to revolutionize smart agriculture. Future smart farming is anticipated to be significantly influenced by its characteristics and support services, which are ideallylow power consumption, effective coverage, and suited to the topography and economics of agriculture. As a result, narrowband Internet of Things (NB-IoT) gained substantial industry backing and became a successful international standard for LPWA connectivity (Ayaz et al., 2019). By altering people's conceptions of what the Internet is capable of, it has the potential to bring about significant connection improvements in the agriculture sector. Leading cellular operators with ambitious IoT plans are predicted to be able to make considerable profits by partnering with LPWA technology vendors to offer smart agriculture services. The actions that must be taken to build the necessary infrastructure to accomplish technology-based agriculture are vital for short-, medium-, and long-range communication systems' long-term success (Ayaz et al., 2019).

CONCLUSION

In light of the ever-decreasing amount of arable land, focus must be directed to developing smarter, better, and more efficient crop-growing techniques to meet the rising food demand of the expanding world population technologies. The advent of Smart farming signaled the start of a new agricultural period, a technologically advanced takes advantage of technology to maximize benefits. Smart Agriculture focuses on the field's site-specific variations to precisely address all problems.Robots, drones, GPS, IoT sensors, software, connectivity, data analytics and other technology have all been used to help Smart Farming fulfill its goals. In comparison to traditional agriculture, production, productivity, and profitability have improved. Furthermore, Smart Farming has resulted in improved management. In other words, SF employs information and technology to carry out effective management techniques. It's critical to realise that SF is a method that makes agriculture more sustainable. In essence, SF is entirely focused on optimizing inputs in order to achieve the best possible result, taking into account the field's diversity.

REFERENCES

Aguilar-Manjarrez, J., & Ross, L. G. (1995). Geographical information system (GIS) environmental models for aquaculture development in Sinaloa State, Mexico. *Aquaculture International, 3*(2), 103–115.

Ahmad, L., & Mahdi, S. S. (2018). Components of Precision Agriculture. In *Satellite Farming* (pp. 19–30). Springer.

Ahmad, L., & Nabi, F. (2021). *Agriculture 5.0: Artificial Intelligence, IoT and Machine Learning*. CRC Press.

Ayaz, M., Ammad-Uddin, M., Sharif, Z., Mansour, A., & Aggoune, E. H. M. (2019). Internet-of-Things (IoT)-based smart agriculture: Toward making the fields talk. *IEEE Access: Practical Innovations, Open Solutions*, *7*, 129551–129583.

Bacco, M., Berton, A., Ferro, E., Gennaro, C., Gotta, A., Matteoli, S., & Zanella, A. (2018). Smart farming: Opportunities, challenges and technology enablers. *IoT Vertical and Topical Summit on Agriculture-Tuscany (IOT Tuscany)*, 1-6.

Belal, A. A. EL-Ramady, H., Jalhoum, M., Gad, A., & Mohamed, E. S. (2021). Precision Farming Technologies to Increase Soil and Crop Productivity. In Agro-Environmental Sustainability in MENA Regions (pp. 117-154). Springer, Cham.

Donzia, S. K. Y., Kim, H. K., & Hwang, H. J. (2018, July). A software model for precision agriculture framework based on smart farming system and application of iot gateway. In *International Conference on Computational Science/Intelligence & Applied Informatics* (pp. 49-58). Springer, Cham.

Doshi, J., & Patel, T., & kumar Bharti, S. (2019). Smart Farming using IoT, a solution for optimally monitoring farming conditions. *Procedia Computer Science*, *160*, 746–751.

Frost, A. R., Schofield, C. P., Beaulah, S. A., Mottram, T. T., Lines, J. A., & Wathes, C. M. (1997). A review of livestock monitoring and the need for integrated systems. *Computers and Electronics in Agriculture*, *17*(2), 139–159.

Garg, D., Khan, S., & Alam, M. (2020). Integrative use of IoT and deep learning for agricultural applications. *Proceedings of ICETIT*, 521–531.

Greasley, A., & Edwards, J. S. (2021). Enhancing discrete-event simulation with big data analytics: A review. *The Journal of the Operational Research Society*, *72*(2), 247–267.

Grogan, A. (2012). Smart farming. *Engineering & Technology*, *7*(6), 38–40.

Gupta, M., Abdelsalam, M., Khorsandroo, S., & Mittal, S. (2020). Security and privacy in smart farming: Challenges and opportunities. *IEEE Access: Practical Innovations, Open Solutions*, *8*, 34564–34584.

Hakkim, V. A., Joseph, E. A., Gokul, A. A., & Mufeedha, K. (2016). Precision farming: the future of Indian agriculture. *Journal of Applied Biology and Biotechnology, 4*(6), 0-7.

Hristoski, I., Kostoska, O., Dimovski, T., & Kotevski, Z. (2017). Farm management software for increased productivity and competitiveness.

Hung, M. C., Wu, J., Chang, J. H., & Yang, D. L. (2005). An Efficient k-Means Clustering Algorithm Using Simple Partitioning. *Journal of Information Science and Engineering*, *21*(6), 1157–1177.

Jain, D., Krishna, P. V., & Saritha, V. (2012). A study on Internet of Things based applications.

Jan, B., Farman, H., Javed, H., Montrucchio, B., Khan, M., & Ali, S. (2017). Energy efficient hierarchical clustering approaches in wireless sensor networks: A survey. *Wireless Communications and Mobile Computing*, 2017.

Kamilaris, A., Gao, F., Prenafeta-Boldu, F. X., & Ali, M. I. (2016, December). Agri-IoT: A semantic framework for Internet of Things-enabled smart farming applications. In *3rd world forum on internet of things (WF-IoT)* (pp. 442-447). IEEE.

Khattab, A., Abdelgawad, A., & Yelmarthi, K. (2016, December). Design and implementation of a cloud-based IoT scheme for precision agriculture. In *2016 28th International Conference on Microelectronics (ICM)* (pp. 201-204). IEEE.

Khosla, R. (2008). Myths of Precision Farming. *From the ground up, agronomy news.*

Lokhorst, C., De Mol, R. M., & Kamphuis, C. (2019). Invited review: Big Data in precision dairy farming. *Animal*, *13*(7), 1519–1528. doi:10.1017/S1751731118003439 PMID:30630546

Lu, Y. C., Daughtry, C., Hart, G., & Watkins, B. (1997). The current state of precision farming. *Food Reviews International*, *13*(2), 141–162.

Malavade, V. N., & Akulwar, P. K. (2016). Role of IoT in agriculture. *IOSR Journal of Computer Engineering*, *2016*, 2278–0661.

Milner, G. (2016). What is GPS? *Journal of Technology in Human Services*, *34*(1), 9–12.

Mohamed, E. S., Belal, A. A., Abd-Elmabod, S. K., El-Shirbeny, M. A., Gad, A., & Zahran, M. B. (2021). Smart farming for improving agricultural management. *The Egyptian Journal of Remote Sensing and Space Sciences.*

Nabi, F., Jamwal, S., & Padmanbh, K. (2020). Wireless sensor network in precision farming for forecasting and monitoring of apple disease: a survey. *International Journal of Information Technology*, 1-12.

Navarro, E., Costa, N., & Pereira, A. (2020). A systematic review of IoT solutions for smart farming. *Sensors (Basel)*, *20*(15), 4231. doi:10.339020154231 PMID:32751366

Neethirajan, S., & Kemp, B. (2021). Digital livestock farming. *Sensing and Bio-Sensing Research*, *32*, 100408.

Neményi, M., Mesterházi, P. Á., Pecze, Z., & Stépán, Z. (2003). The role of GIS and GPS in precision farming. *Computers and Electronics in Agriculture*, *40*(1-3), 45–55.

Ouafiq, E. M., Elrharras, A., Mehdary, A., Chehri, A., Saadane, R., & Wahbi, M. (2021). IoT in smart farming analytics, big data based architecture. In *Human Centred Intelligent Systems* (pp. 269–279). Springer. doi:10.1007/978-981-15-5784-2_22

Pandey, P. C., Tripathi, A. K., & Sharma, J. K. (2021). An evaluation of GPS opportunity in market for precision agriculture. In *GPS and GNSS Technology in Geosciences* (pp. 337–349). Elsevier.

Patil, A., Beldar, M., Naik, A., & Deshpande, S. (2016, March). Smart farming using Arduino and data mining. In *3rd International Conference on Computing for Sustainable Global Development (INDIA-Com)* (pp. 1913-1917). IEEE.

Patil, V. C., Al-Gaadi, K. A., Biradar, D. P., & Rangaswamy, M. (2012). Internet of things (Iot) and cloud computing for agriculture: An overview. *Proceedings of agro-informatics and precision agriculture (AIPA 2012)*, 292-296. .

Paunova-Hubenova, E., & Trichkova-Kashamova, E. (2021, October). Smart Solutions for Control and Management in Livestock Farms. In Big Data, Knowledge and Control Systems Engineering (BdKCSE) (pp. 1-7). IEEE.

Prathibha, S. R., Hongal, A., & Jyothi, M. P. (2017, March). IoT based monitoring system in smart agriculture. In 2017 international conference on recent advances in electronics and communication technology (ICRAECT) (pp. 81-84). IEEE.

Rajasekaran, T., & Anandamurugan, S. (2019). Challenges and applications of wireless sensor networks in smart farming—a survey. *Advances in big data and cloud computing*, 353-361.

Ram Kumar, R. P., Sanjeeva, P., & Vijay Kumar, B. (2018). Transforming the traditional farming into smart farming using drones. In *Proceedings of the Second International Conference on Computational Intelligence and Informatics* (pp. 589-598). Springer, Singapore.

Ratnaparkhi, S., Khan, S., Arya, C., Khapre, S., Singh, P., Diwakar, M., & Shankar, A. (2020). Smart agriculture sensors in IOT: A review. *Materials Today: Proceedings*.

Regan, Á. (2019). 'Smart farming'in Ireland: A risk perception study with key governance actors. *NJAS Wageningen Journal of Life Sciences*, *90*, 100292.

Ruan, J., Wang, Y., Chan, F. T. S., Hu, X., Zhao, M., Zhu, F., Shi, B., Shi, Y., & Lin, F. (2019). A life cycle framework of green IoT-based agriculture and its finance, operation, and management issues. *IEEE Communications Magazine*, *57*(3), 90–96. doi:10.1109/MCOM.2019.1800332

Sai Sravani, E., Sreehitha, A. V., Konda Babu, A., & Nandan, D. (2020). Evaluation and Study of IoT Entrances. In *Micro-Electronics and Telecommunication Engineering* (pp. 431–438). Springer.

Sanjeevi, P., Prasanna, S., Siva Kumar, B., Gunasekaran, G., Alagiri, I., & Vijay Anand, R. (2020). Precision agriculture and farming using Internet of Things based on wireless sensor network. *Transactions on Emerging Telecommunications Technologies*, *31*(12), e3978. doi:10.1002/ett.3978

Sarker, M. N. I., Islam, M. S., Murmu, H., & Rozario, E. (2020). Role of big data on digital farming. *Int J Sci Technol Res*, *9*(4), 1222–1225.

Sciforce. (2020, June 22). Smart Farming: The Future of Agriculture. *IOT Fort All*. https://www.iotforall.com/smart-farming-future-of-agriculture

Sishodia, R. P., Ray, R. L., & Singh, S. K. (2020). Applications of remote sensing in precision agriculture: A review. *Remote Sensing*, *12*(19), 3136.

Sood, K., Singh, S., Rana, R. S., Rana, A., Kalia, V., & Kaushal, A. (2015, October). Application of GIS in precision agriculture. In Paper presented as lead lecture in national seminar on "Precision farming technologies for high Himalayas (pp. 04-05).

Stočes, M., Vaněk, J., Masner, J., & Pavlík, J. (2016). Internet of things (iot) in agriculture-selected aspects. *Agris on-line Papers in Economics and Informatics, 8*(665-2016-45107), 83-88.

Suciu, G., Marcu, I., Balaceanu, C., Dobrea, M., & Botezat, E. (2019, June). Efficient IoT system for precision agriculture. In *15th International Conference on Engineering of Modern Electric Systems (EMES)* (pp. 173-176). IEEE.

Tomar, P., Kaur, G., & Singh, P. (2018). A prototype of IoT-based real time smart street parking system for smart cities. In *Internet of Things and Big Data Analytics Toward Next-Generation Intelligence* (pp. 243–263). Springer.

Tripicchio, P., Satler, M., Dabisias, G., Ruffaldi, E., & Avizzano, C. A. (2015, July). *Towards smart farming and sustainable agriculture with drones. In international conference on intelligent environments*. IEEE.

Usery, E. L., Pocknee, S., & Boydell, B. (1995). Precision farming data management using geographic information systems. *Photogrammetric Engineering and Remote Sensing*, *61*(11), 1383–1392.

Veena, S., Mahesh, K., Rajesh, M., & Salmon, S. (2018). The survey on smart agriculture using IOT. [IJRIREM]. *Int J Innov Res Eng Manag*, *5*(2), 63–66.

Verdouw, C., Tekinerdogan, B., Beulens, A., & Wolfert, S. (2021). Digital twins in smart farming. *Agricultural Systems*, *189*, 103046. doi:10.1016/j.agsy.2020.103046

Vikranth, K. (2021). An Implementation of IoT and Data Analytics in Smart Agricultural System–A Systematic Literature Review. *International Journal of Management* [IJMTS]. *Technology and Social Sciences*, *6*(1), 41–70.

Wang, X., & Liu, N. (2014). The application of internet of things in agricultural means of production supply chain management. *Journal of Chemical and Pharmaceutical Research*, *6*(7), 2304–2310.

Wolfert, S., Ge, L., Verdouw, C., & Bogaardt, M. J. (2017). Big data in smart farming–a review. *Agricultural Systems*, *153*, 69–80. doi:10.1016/j.agsy.2017.01.023

Yadav, S., Garg, D., & Luthra, S. (2020). Development of IoT based data-driven agriculture supply chain performance measurement framework. *Journal of Enterprise Information Management*.

Yousefi, M. R., & Razdari, A. M. (2015). Application of GIS and GPS in precision agriculture (a review). *International Journal of Advanced Biological and Biomedical Research*, *3*(1), 7–9.

Chapter 4
Application of Drone Technology in Agriculture:
A Predictive Forecasting of Pest and Disease Incidence

Ganeshkumar C.
https://orcid.org/0000-0002-0913-2849
Indian Institute of Plantation Management, Bangalore, India

Arokiaraj David
https://orcid.org/0000-0002-9591-2410
St. Francis Institute of Management and Research, India

Jeganathan Gomathi Sankar
https://orcid.org/0000-0002-5077-5109
BSSS Institute of Advanced Studies, India

Manjunath Saginala
Indian Institute of Plantation Management, Bangalore, India

ABSTRACT

The study examines the use of drone technology in agriculture to predict the occurrence of pest and disease outbreaks. The purpose of the study is to comprehend competition analyses and determine the best entry tactics for the aerial imaging and advice service market. They can determine the current state of drone technology and create strategies based on the different companies providing drone services to farms by studying how farmers perceive the drone employed in predictive pest and disease analyses. These observations are highly valuable for comprehending how drone technology can affect Indian agriculture. A virtual survey gives a useful and accurate image; whether or not they utilise it for predictive forecasting and identification, and if they depend on the advice given by various sources or not, and most importantly, how best the drone service available in the market, whether farmers can afford for it.

DOI: 10.4018/978-1-6684-6413-7.ch004

Copyright © 2023, IGI Global. Copying or distributing in print or electronic forms without written permission of IGI Global is prohibited.

INTRODUCTION

Growing crops is a lovely pastime that makes a farmer delighted for producing a new life. Despite its attractiveness, crop cultivation necessitates a variety of agricultural tasks and ongoing upkeep in order to provide better and healthy output. (López-Morales et al., 2020). It is estimated that the world's population will reach nine billion by 2050, and we have to increase our production to 50% to feed the population. Currently, more than 815 million people worldwide are starving, with Asia accounting for more than half of them. Consequently, in the coming years, food production requires a boost of nearly 70% to 80% (Joshi et al., 2020; Vandana et al., 2020).

The drone is the latest technology which has been introduced and can help in achieving the aim. Unmanned Aerial Vehicles (UAVs) are another term for drones (UAV) are generally remote-controlled aircraft with no pilots on board. They fly over the fields at low altitudes and capture whole data (Petkovic et al., 2017). They are rapidly evolving, replacing satellites and aircraft. They can capture high-quality images at cheap prices that satellites cannot do. They generally run on battery or fuel (Prasanna et al., 2020). They are being used in various fields including military, agriculture, film industry, humanitarian relief etc. Drones have initiated a technological revolution in the agriculture sector (Negash et al., 2019; Sharma, 2021). They can gather large volumes of spatial data. The usage of drones in the area of agriculture is rising especially crop production, early warning, forestry, fisheries etc. They are so efficient that they can gather whole soil data without even touching the soil (Arokiaraj et al., 2020b; Ren et al., 2020; Reger et al., 2018). Drones these days are equipped with modern technologies like navigation, controls, imaging, sensing etc. They are also being used for predictive forecasts of pests and diseases, spraying fertilizers and pesticides (Barasa et al., 2021; David, 2020b). However, with the advancement of technology, they are also being used in the whole production process from land preparation to harvesting (Muraru et al., 2019). Early insect identification can minimise output loss by up to 40%, based on crop species and growth stage (Ahirwar et al., 2019). As a result, farmers must devote their whole attention to crop monitoring (Srivastava, 2020). Rather than trekking around the field, farmers may now utilise technology such as drones to inspect their crops more quickly and correctly (Banumathi & Arokiaraj, 2011; Bhardwaj et al., 2021; Rejeb et al., 2022).

Drone technology is an invention that is having a far-reaching impact on today's society, affecting our lives as well as business (Nayyar et al., 2020). Farmers may now boost their efficiency in some elements of the agricultural development, such as planting, crop monitoring, crop spraying, irrigation, mapping, and livestock management, thanks to high-tech drones (Pathak et al., 2020; Ben Ayed et al., 2021; Rachmawati et al., 2021). This farming method is built on seeing, measuring, and acting on real-time crop data. It eliminates a contemporary agriculture and gives farmers the option to maximise harvests and increase crop production (Klauser & Pauschinger, 2021; Ganeshkumar et al., 2022). The cost of agricultural drones has drastically decreased in recent years, resulting in an explosion of drones and their application in agriculture (Veroustraete, 2015). In the recent years, the agricultural drone industry is forecasted that 38% will be increased with the use of drones. Agricultural drones have a variation in the applications, including:

- Monitoring overall crop fitness
- Investigation of land and crop
- Inspection for weeds and spot treating plants
- Supervision of livestock and checking the healthiness

Drones are outfitted with technologies such as propulsion technology, infrared cameras, GPS and navigation systems, programmable controllers, and automated flight planning (Ravi et al., 2018; Călina, J., et al., 2020; Ganesh Kumar et al., 2017). Furthermore, with custom-made data processing tools, any obtained data may be immediately used to improve strategic decisions (Iost Filho et al., 2020; Wójtowicz et al., 2016; Daponte, P., et al., 2019).

The best agricultural drones in the global markets:

1. **Honeycorp Agdrone:** The wing of the 2017 Honeycorp AgDrone is made of Kevlar fibre. This material will not split like carbon fibre or fall apart like foam, making it an extremely robust, adaptable, and strong choice for agricultural drones (Feleen & David, 2021; Bogue, 2021).
2. **DJI Matrice 100:** It is among the greatest quadcopters for smart agriculture on the marketplace currently: it features a twin battery that allows for up to 40 minutes of flying time and a plethora of settings (Ayamga et al., 2021$_{a}$). DJI systems' basic features, including GPS, flight controllers, and DJI Lightbridge, will aid in navigating even the most challenging jobs, giving the pilot more time to focus on flying (Ganeshkumar et al., 2021; Kumar et al., 2022).
3. **DJI T600 Inspire1:** The DJI T600 Inspire Quadcopter is another drone appropriate for agricultural applications, although its batteries do not last long. You get 4k recorded video, independent flight and camera control, and a carbon fibre frame. The propellers are readily installed; simply pressing down and twisting to secure them (Ayaz et al., 2019; Dileep et al., 2020; David, 2020a; Esposito et al., 2021).
4. **Agras MG-1 DJI:** DJI, a Chinese drone company, has released the Agras MG-1. The Agras is an octocopter engineered to spray pesticides or fertilizers over large areas of farmland. It can cover large areas in a short period, covering 4,000-6,000 m2 in about 10 minutes (Ganeshkumar et al., 2020; Mogili et al., 2020; Madaan et al., 2021).

The market for agricultural drones is rising vastly and many companies have entered this business. By 2050 the market for UAVs will rise to $32.4 billion. Development of advanced agricultural technologies, unmanned aerial vehicles (UAVs), or drones has dramatically improved jeopardy, and serious harm and revolutionized how we formulate for the potential catastrophes that affect helpless farmers' and fishers' livelihood opportunities, as well as the country's food security. But the question that needs to consider is whether the farmers can afford it, and the availability of drone technology at the grass-root level for predictive forecasts of pests and diseases. To substitute the questions requires a study to obtain the market analysis on available solutions in the market for aerial imaging (drones) and predictive pest and disease advisory to understand competitiveness in the market (India) and formulate some apt strategies to enter the aerial imaging and advisory service segment (Pratheepkumar et al., 2017; Eli-Chukwu et al., 2019; David, 2020c).

REVIEW OF LITERATURE

Unmanned Aerial Vehicles (UAVs), often known as drones, have been a game-changer in the agriculture industry since 2011 because they are more economical and easier to deploy (Frankelius et al., 2017). In the agriculture sector, practical uses for UAVs are increasing faster than ever before (Arokiaraj et al.,

2020$_a$). Contrary to retrieving satellite imagery or traditional airborne data, recollecting aerial photographs with a UAV integrated with GPS and digital cameras mode is less expensive and quicker. UAVs are among the most significant emerging technical instruments, owing to their versatility and expanding usefulness (Ndiaye et al., 2020; Wynn, 2019; David et al., 2020; Srivastava et al., 2022).

Nowadays, the most prevalent uses of UAVs include: a) agricultural data collecting (using various types of cameras); b) crop growth monitoring; c) sprayers (fertilizers, herbicides, and insecticides); and d) picture capturing to construct 3D models. According to worldwide research produced by Price Waterhouse Coopers (PwC), the agricultural UAV industry is valued at $32.4 billion US dollars (Mazur et al., 2016). A key consideration is that UAV technology will enable individual farms to be extremely data-driven, resulting in higher production and yields (Thalluri et al., 2021; Stehr, 2015; Puri et al., 2017; David et al., 2022; David & Ravi, 2017).

Despite the fact that India is heavily reliant on agriculture, it falls far short of using cutting-edge techniques in order to achieve a productive farm. The use of UAVs in precision agriculture, photogrammetry, and remote sensing has already begun in developed nations [UMRao Mogili BBVL Deepk]. It is highly rapid and might lessen a farmer's burden. UAVs are naturally outfitted with the cameras and sensors for the maintenance of crop, as well as the application of pesticide sprayers. Previously, a variety of UAV types were used for military and civilian purposes. Yamaha created the first UAV model in agriculture. Yamaha RMAX unmanned helicopter was introduced for the crop monitoring and pest control applications in the area of agriculture (Ayamga et al., 2021$_b$). Yamaha, on the other hand, ceased manufacture in 2007. The purpose of technical analysis of UAVs in precision farming is to determine their application in agricultural activities such as crop height and crop monitoring (Rani, et al., 2019; Ganeshkumar & Khan 2021; Pobkrut et al., 2016; Pachayappan et al., 2020).

OBJECTIVES OF THE STUDY

1. To understand the solutions available in the market, provided by various players, for predictive pest and disease advisory and spraying services.
2. To obtain market analysis on available solutions in the market.
3. To study the competition in the market based on the market share, brand value, and quality of advisory.

Table 1. Indicators and variables

Sl.n.	Objective	Indicators	Variables
1	To understand the solutions available in the market, provided by various players, predictive pest and disease advisory, and spraying services.	Different companies in the market, provide the solution.	• Satisfaction level. • Premium charge on service. • Advisory services. • The willingness of the farmer.
2	To obtain market analysis on available solutions in the market.	Market analysis.	• Promotions. • Advertisements. • Business strategies.
3	To study the competition in the market based on the market share, brand value, and quality of advisory.	Competitors.	• Services gave by competitors. • Brand value. • Quality of the advisory services. • Tie-ups with FPOs and Farmers.

RESEARCH METHODOLOGY

Our major goal in this research is to examine the field in order to uncover viable options for predicting diseases and pests' occurrence, with the support of drone technology. For that, we are surveying the farmers who are using such services (Siddhartha et al., 2021; Arokiaraj, 2015). After collecting the data, we will analyze the data in the medium of excel and finally produce an appropriate strategy, based on the data collected, to enter the company into this new segment. The drone company name, their drone service, UAVs companies were given in the table 2, table 3 and table 4.

Research Design

Exploratory research is to be conducted to study the current market phenomenon and to acquire new market insight. The fundamental goal of this exploratory study, as the names suggest, is to investigate the competitiveness and apt the strategies to enter the aerial imaging and advisory service segment through a thorough market analysis of available solutions (Panpatte et al., 2021). It emphasizes idea generation and inquisitive cognition. The approach is suitable for inquiries that are able to permit for consideration of all facets of the area. The required evidence is only weakly detailed at this part, and the scientific method is fluid and unstructured. This overall research design is flexible with an unstructured process are,

The following non-probability sampling approaches are available:

1. Convenient sampling.
2. Snowball sampling.
3. Quota sampling.
4. Purposive or Judgmental sampling.

Research Methods

1. Primary research is done to collect the information from the farmers with the help of telephonic interviews.
2. Secondary research is done for collecting secondary information via websites, blocks and journals.

Sampling Technique

Convenient Sampling Technique

It is a sampling method of first reachable to data that appoints for the research data without much constraints (Siddhartha et al., 2019). This sampling method needs collecting the individual information or data wherever they can be found that can be used for the research purpose which is known as convenient sampling. Reasons to choose this sampling technique are:

- We can choose the samples based on our convenience wherever we can find them.
- Samples are chosen purposively because of the few.

Snowball Sampling Technique

It is primarily a nonprobability sampling strategy in which current research respondents acquire prospective study volunteers from their friends. It also depends on references from responses given by the respondents to produce more subjects (android users), ensuring that the samples were representative of the public. Reasons to choose this sampling technique are:

- Secretive groups can be identified easily.
- A referral system helps to find samples quickly.

Required Sample Size

Considering our population as finite, with a sample size of 100 farmers using drone technology from various states in India like Uttar Pradesh, Andhra Pradesh, Telangana, and Tamil Nadu.

Method of Collecting the Data

The collection of the data in the questionnaire is done using a telephonic survey with the farmers. The questionnaire is designed to collect farmers' data in four sections. The first section is to collect the basic information about the farmers and the agriculture (Ganeshkumar et al., 2019; Victer Paul et al., 2020). The second section includes the questions based on the information about the drones and their usefulness, that are available to the farmers. The third section includes the information of those farmers who are using the same services. The last section is to collect the reference data of the other farmers, using similar services. During the survey, we will come across the farmers who are also not using such services. So, those data will be removed from the source and data of the farmers who's all the four sections are filled up, are the only person of our interest.

DATA ANALYSIS AND INTERPRETATION

Research Sample Size

Considering our population as finite, with a sample size of 100 farmers using drone technology from various states in India like Uttar Pradesh, Andhra Pradesh, Telangana, and Tamil Nadu.

- From the overall sample, UP state has the highest farmers' distribution with 68%.
- The south Indian states (AP, TS, TN) has a farmers' distribution of 32% in the sample.
- The sample is categorized into four age groups, based on the availability of knowledge and ability to work. The categories are: <25, 26 – 35, 36 – 50 and >50.

Due to the non-availability of farmer's leads, we initially collected the primary farmer leads through social media like YouTube and other similar sources as mentioned in figure 1, figure 2, and figure 3.

Figure 1. Sample size of 100 farmers using drone technology from various states in India

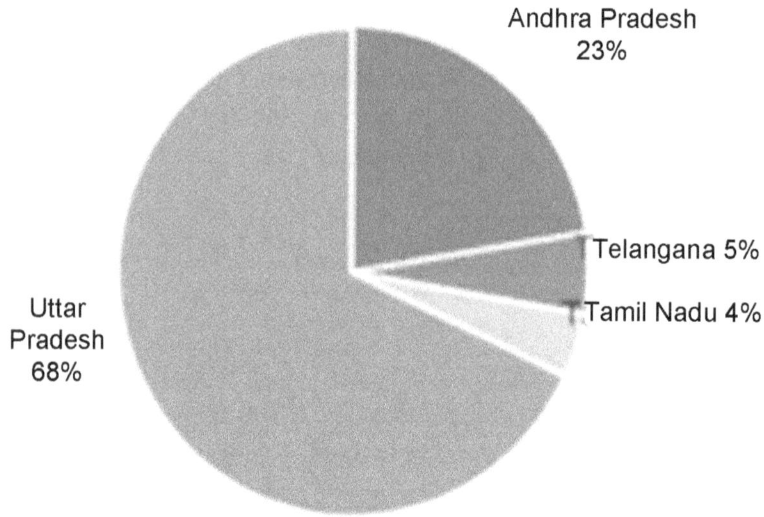

Figure 2. Crops grown by farmers in Uttar Pradesh

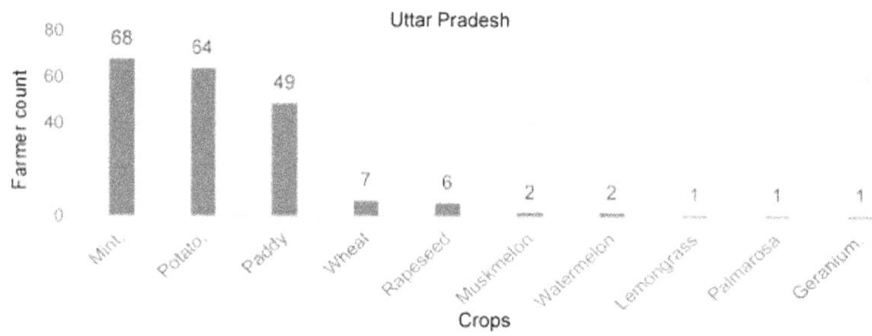

Figure 3. Crops grown by farmers in AP, TS and TN states

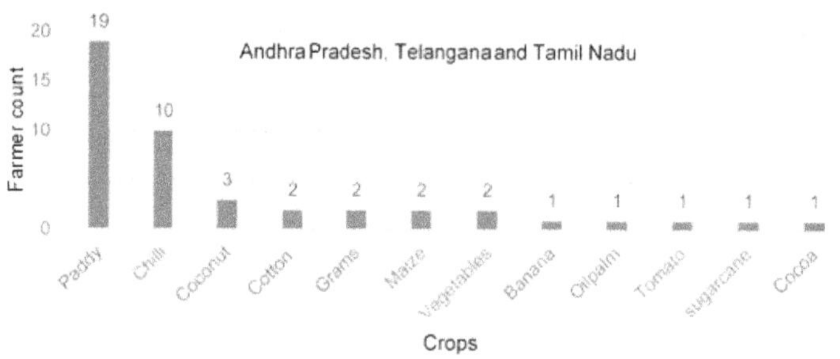

Inference:

- The UP farmers grow major crops like mint (68), potato (64), paddy (49) including wheat, rapeseed, muskmelon, and watermelon.
- Very few (1) farmers also grow lemongrass, palmarosa and geranium in UP.

- Cumin is the widely grown crop in the Bara-Banki region of UP.
- Paddy (19), chilli (10), coconut (3), cotton (2) etc are been grown in southern states.

CROPS WISE PEST AND DISEASES

Table 2. List of major pest and diseases in crops grown by farmers

Crops	Major pests	Major diseases
Mint	Cutworm, Mites	Wilt
Paddy	Hoppers	Brown Leaf spot, Khaira (ZN deficiency)
Wheat	Armyworm	Karnal bunt, Powdery Mildew, Rust, False smut
Cotton	Bollworm, Aphids, Thrips, Jassids, Whitefly	Blight, Leaf spot
Rapeseed	Aphids, Bollworm	Wilt
Muskmelon	Cutworm	Powdery mildew, Downy mildew
Lemongrass	Bugs	Leaf spot, Blight
Palma Rosa	Aphids, Thrips	Leaf spot
Pulses	Bugs	Wilt
Coconut	Beetles, weevils	Blight, Bud rot
Cocoa	Bugs, caterpillars	Anthracnose, Black pod
Banana	Aphids	Wilt, Anthracnose
Sugarcane	Shoot borers(top, internal)	Red rot
Chilli	Whitefly, Thrips	Anthracnose, Powdery mildew

Figure 4. Major pest found in UP

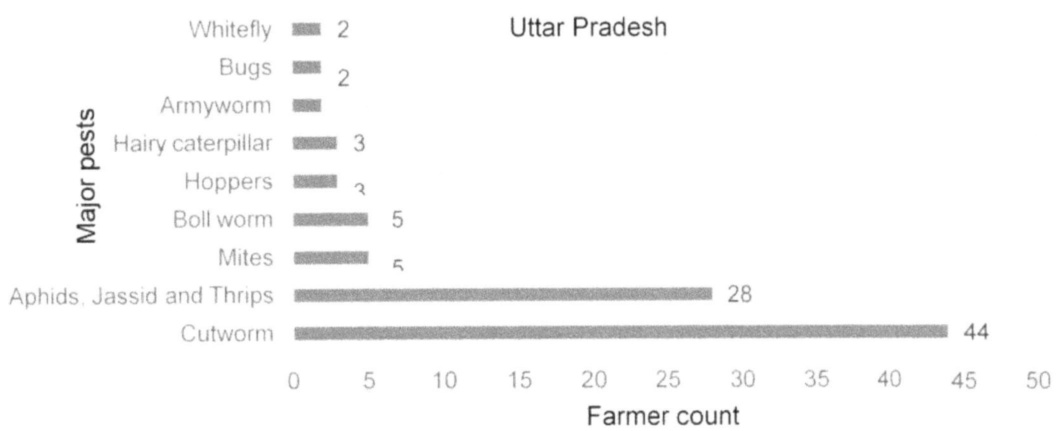

Figure 5. Major pests found in AP, TS, and TN

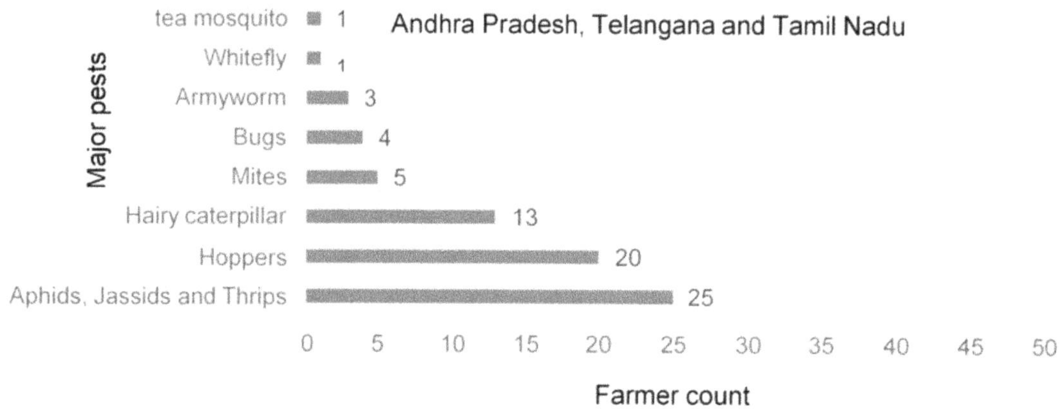

Inference:

- Based on the figure 4, in UP the major pests observed by the farmers are cutworms, aphids, jassids, thrips, mites, bollworms, hoppers, hairy caterpillars, armyworms, bugs and whitefly.
- Out of these insects, cutworm and aphids are the most destructive pests, that occurs in the cumin crop in UP.
- Based on the figure 5, pests observed in fields by farmers are aphids, jassids, thrips, hoppers, hairy caterpillars, mites etc, in southern states (AP, TS, and TN).

Figure 6. Major diseases found in the UP

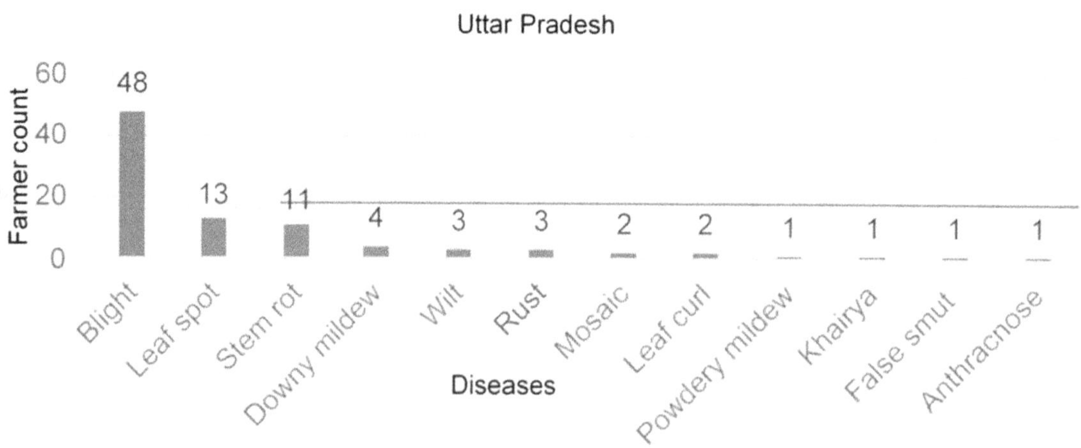

Figure 7. Major diseases found in AP, TS, TN

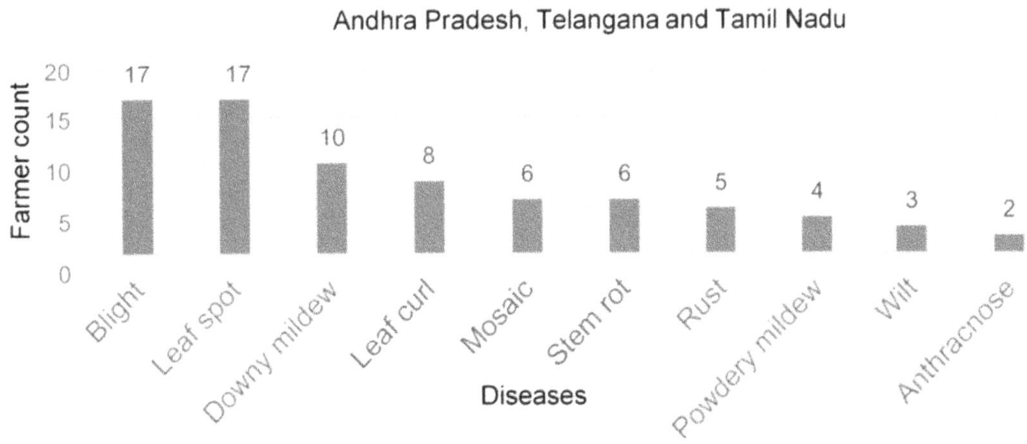

Inference:

- Based on the figure 6, and figure 7, blight is the major disease observed by farmers in UP, AP, TS, and TN.
- Leafspot, stem rot, downy mildew, wilt, rust and other diseases were observed in various crops grown by farmers as mentioned.
- Potato crop cultivated in UP state is more affected due to potato blight. Figure 8 and figure 9 mentioned the land cultivation.

Figure 8. Percentage distribution of farmers according to land cultivated in UP

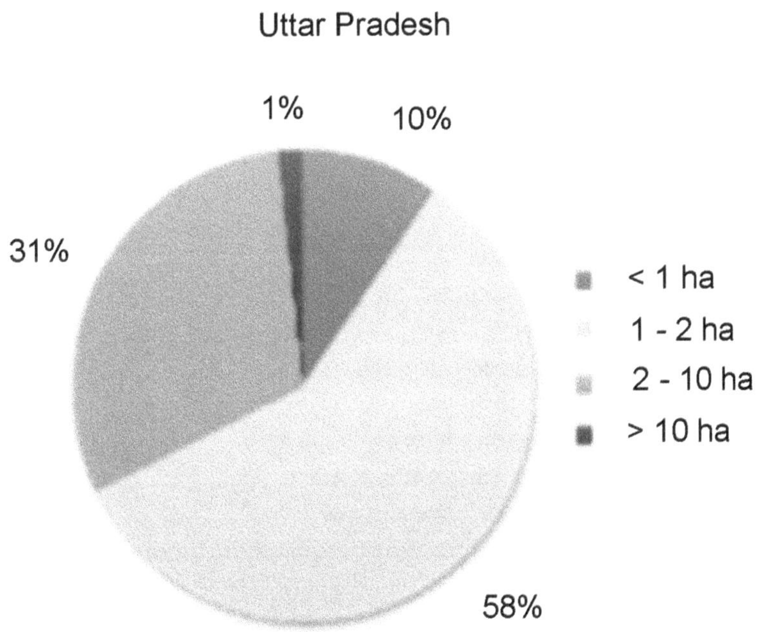

Figure 9. Percentage distribution of farmers according to land cultivated in AP, TN, and TS

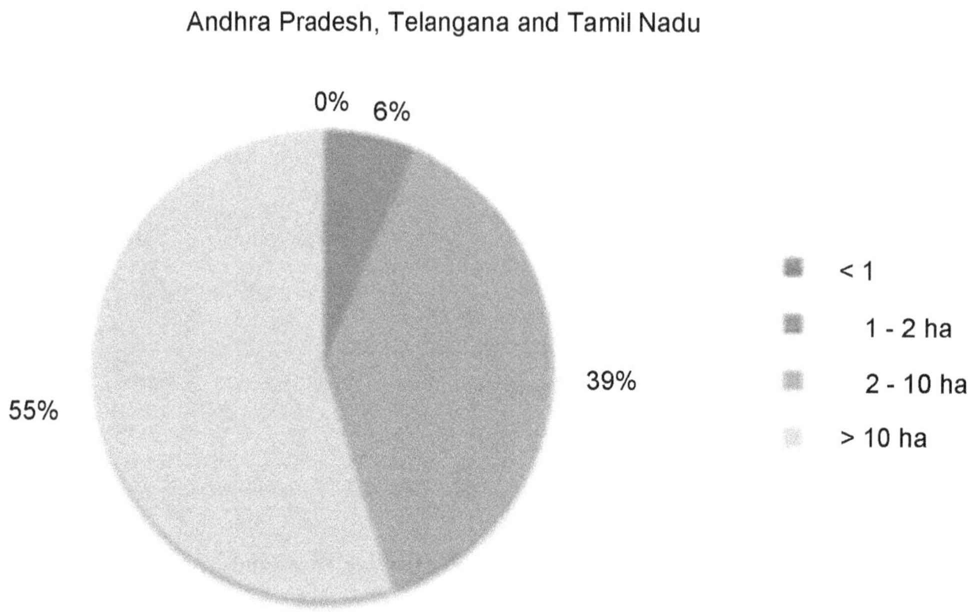

Inference:

- 58% of farmers in UP have 1-2 hectares and followed by 31% of 2-10 ha landholdings.
- In southern states, 55% of farmers having >10ha and 39% with 2-10 ha landholders use this service.
- Only very few farmers < 2 ha lands use this service due to various reasons. Figure 10 and figure 11 explained about the advantages of drones in agriculture.

Figure 10. Farmer distribution based on knowledge about advantages of drones in AP, TS, and TN

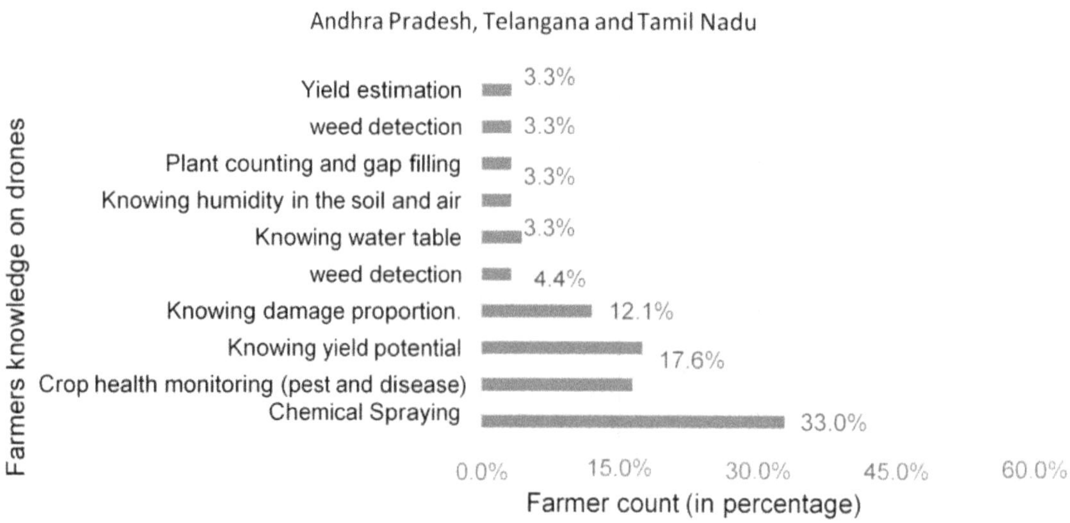

Figure 11. Farmer distribution based on knowledge about the advantages of drones in UP

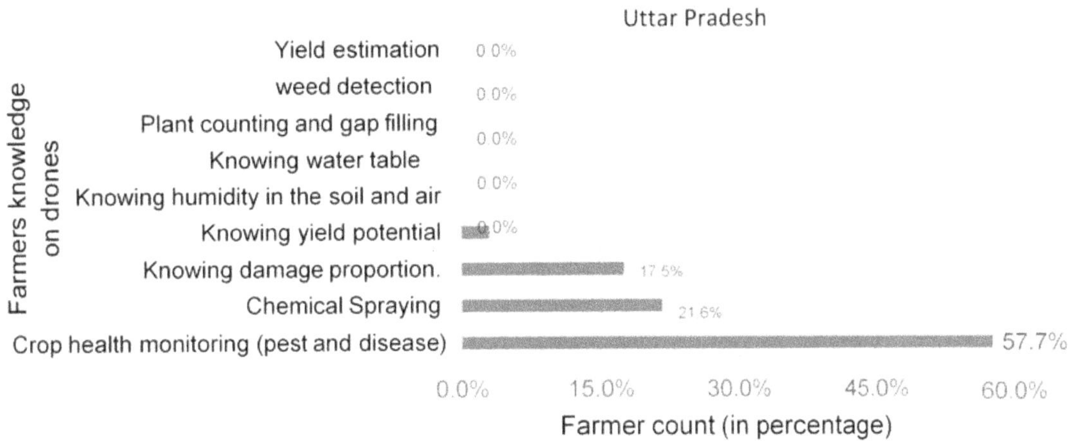

Inference:

- The technology used by organizations results in multiple benefits for UP farmers in monitoring, spraying majorly and yield prediction, water table etc.
- The majority of farmers in south India use drones majorly for chemical spraying but are aware of other potential benefits also.

Figure 12. Farmer distribution based on the frequency of requirement of the service in UP

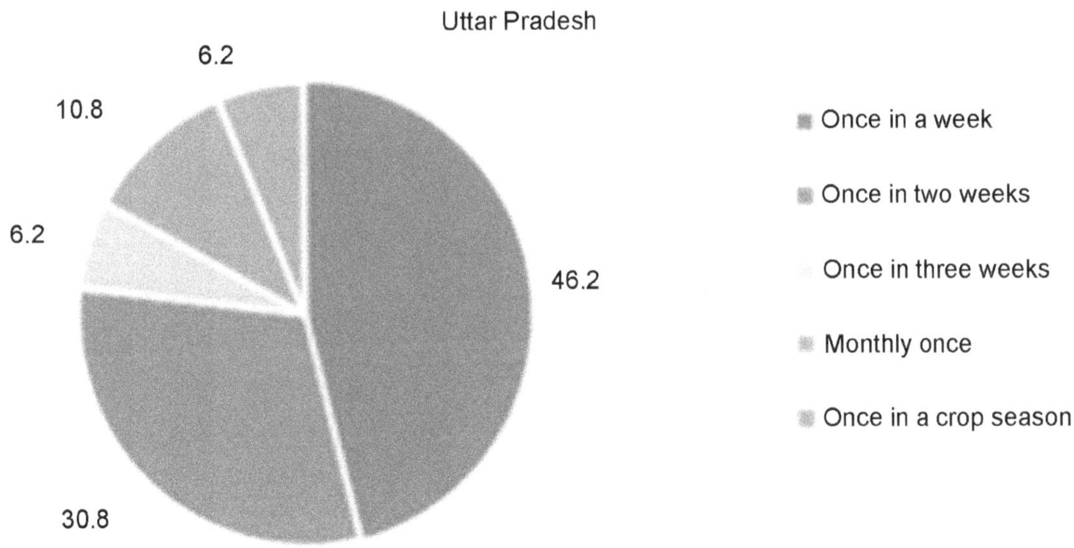

Figure 13. Farmer distribution based on the frequency of requirement of the service in AP, TS, and TN

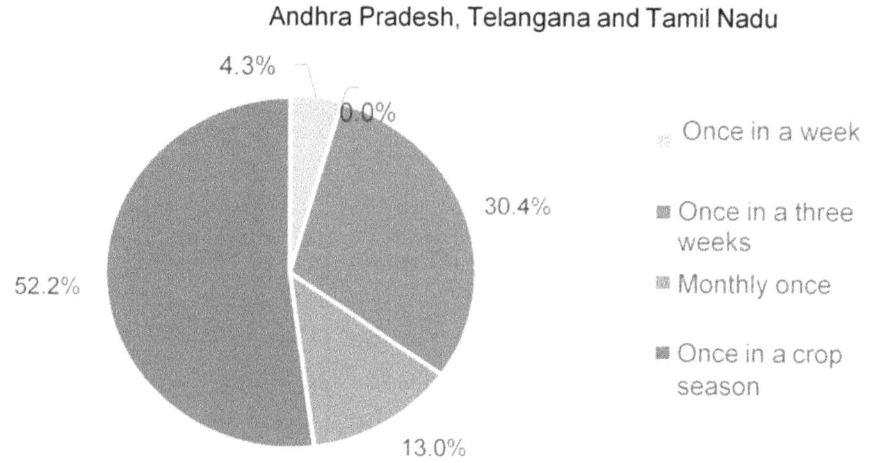

Inference:

- Figure 12 and figure 13 portray the farmers' requirement of the services.
- Almost 46.2% of farmers are willing to use the service once a week, and 30.8% of farmers are willing to use it once in two weeks and rest monthly or for a crop season in UP.
- In southern states, the farmer uses drones mostly once in crop season (52.2%) or twice in crop season (30.4%).
- The alerts are informed either in the form of messages or personally by the company employees available in the specific region like Masauli village of Bara Banki, UP.
- Farmers are willing to get result-oriented services that help to produce timely predictions of pests and diseases.

Figure 14. Distribution of farmers on price levels that are chosen for service in UP

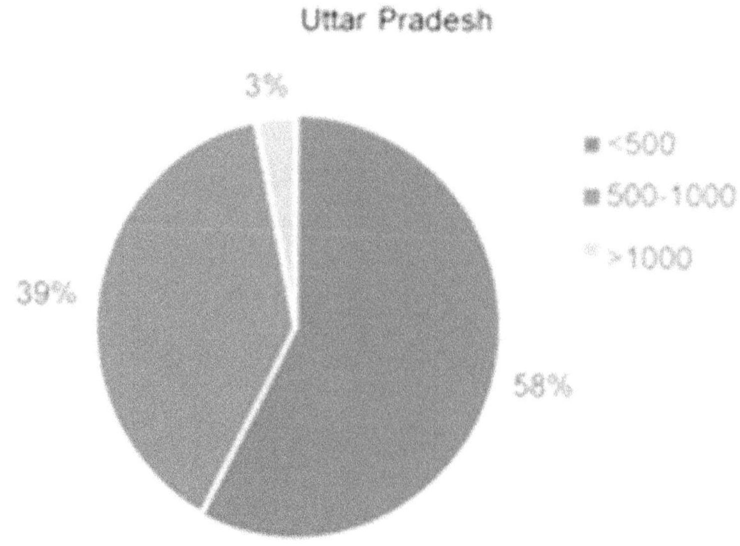

Figure 15. Distribution of farmers on price levels that are chosen for service in AP, TS, and TN

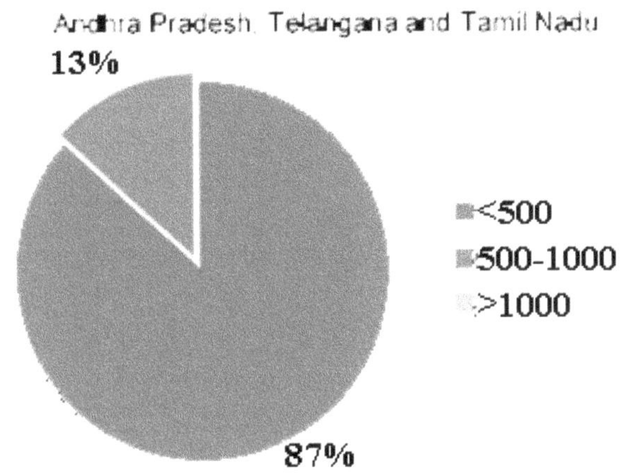

Inference:

- Figure 14 and figure 15 explains the farmers' perception about the price levels.
- In UP Maximum farmers (58%) are willing to choose services below Rs. 500/acre and 39% of farmers choose service at Rs. 500-1000. Very few farmers were also told to have such service (similar to Bharat Rohan's service) above 1000.

Figure 16. Distribution of farmers using different measures for prediction in UP.

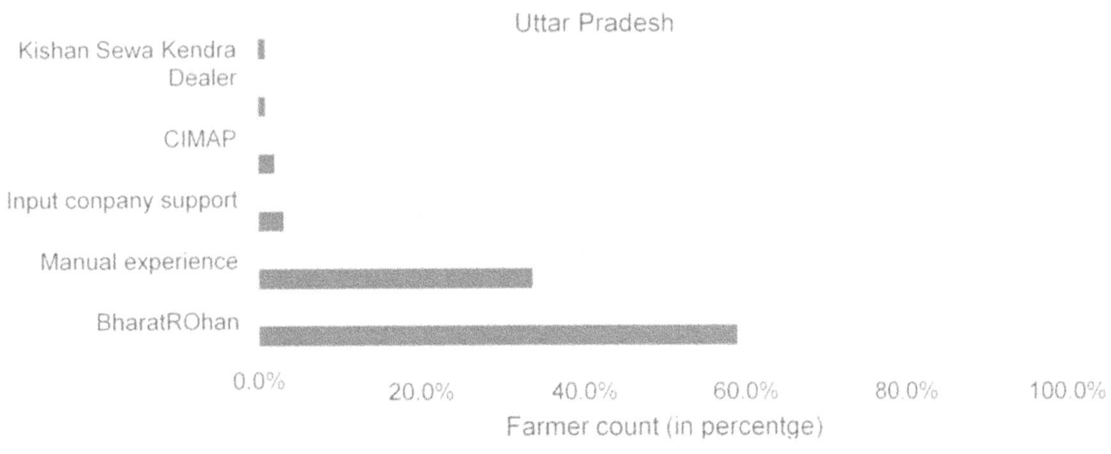

Figure 17. Distribution of farmers using different measures for prediction in AP, TS, and TN

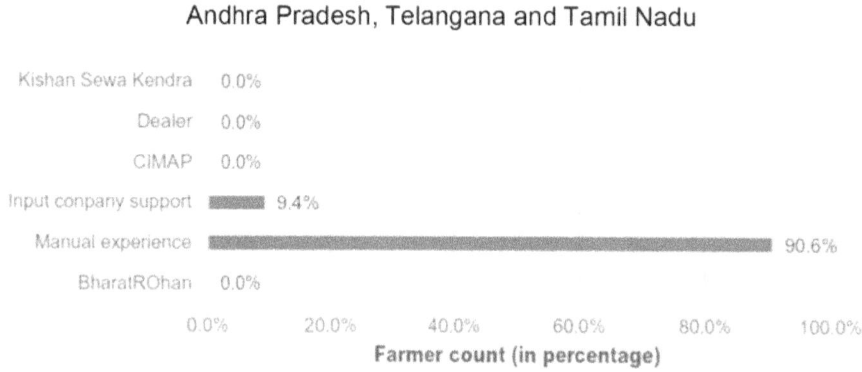

- Most of the farmers' perception was: "If the service helps me in increasing my income by Rs. 10,000 more, then there is no issue if the services charges even to more than Rs. 1,000.
- The majority of farmers use the service if the price is <Rs500 in southern states.
- Very few farmers (3%) even afford >Rs1000.

Inference:

- Figure 16 and figure 17 explains the distribution of farmers using different measures for prediction.
- The majority of farmers use manual experience in southern states and UP is the dominant service

Figure 18. Distribution by the ways the company reach the farmer in UP

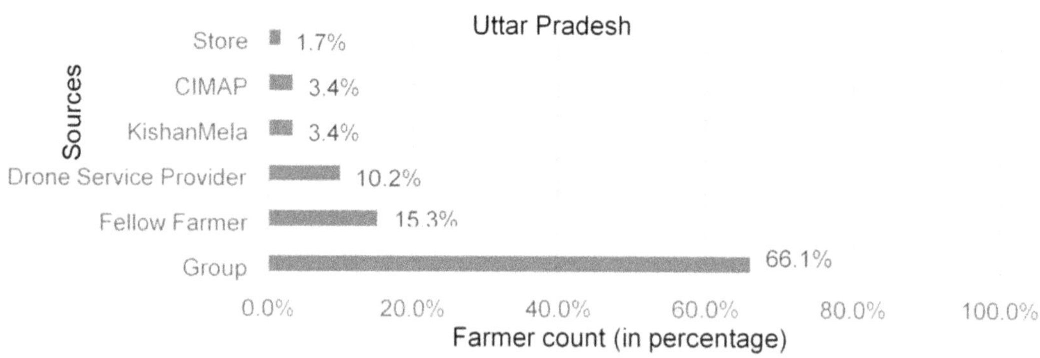

provider, BharathRohan gives advisory on prediction by using unique UAV/drone-based hyper-spectral remote sensing and artificial intelligence.

- Other corporates like UPL, Syngenta, Bayer and government organizations' services are been used by farmers.

Inference:

- Figure 18 and Figure 19 shows the distribution by the ways the company reach the farmer
- The majority of the service providers approached the farmer using informal groups and advertisements.

Figure 19. Distribution by the ways the company reach the farmer in AP, TS, and TN

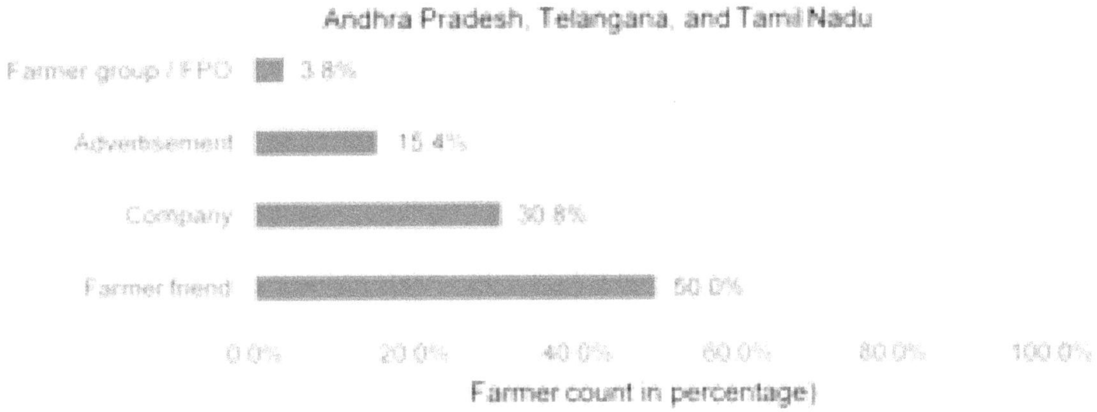

Figure 20. Farmers' distribution based on the channels that influence them to use the service in UP

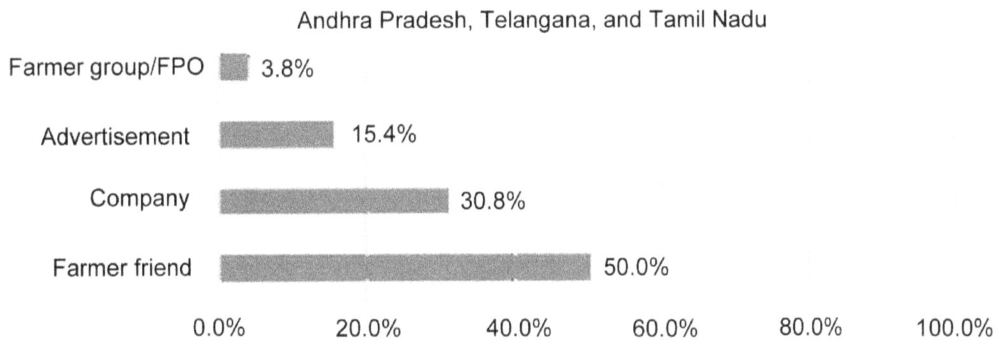

Figure 21. Farmers' distribution based on the channels that influence them to use the service in AP, TS, and TN

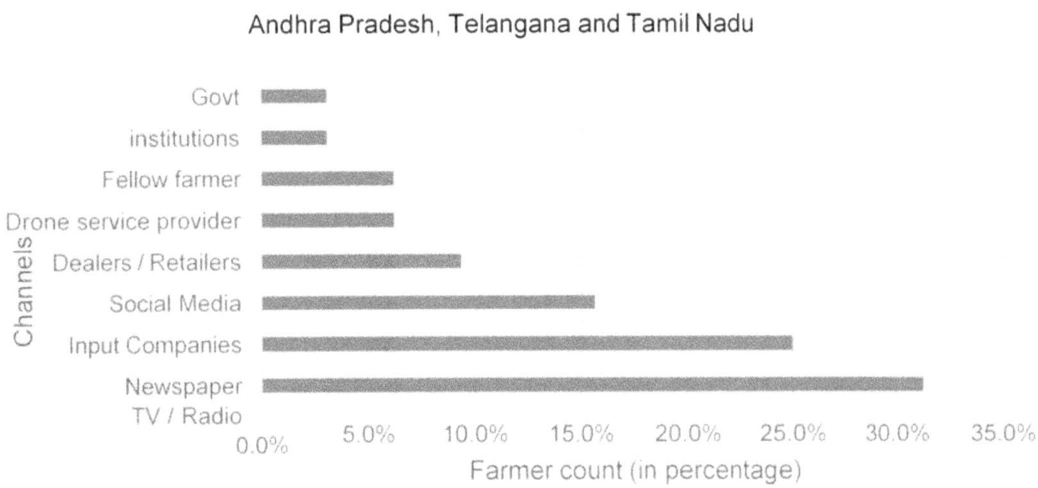

- Exhibitions and advertisements and marketing activities, social media (YouTube) act as a good influencer to use the technology more for the farmers of South India (AP, TS, and TN).
- We can conclude that the UP farmers are less inclined to social platforms.
- Whereas south Indian farmers are more inclined to social platforms and posed good knowledge about the technologies available in their areas

Inference:

- Figure 20 and Figure 21 portrays the farmers' distribution based on the channels that influence them to use the service.
- In UP due to being a dominant player in market the word of mouth, advertisements, and ease of technology worked out well and other corporate and government institutions worked to penetrate this technology.

Figure 22. Farmer distribution based on experience with the service of the company in UP

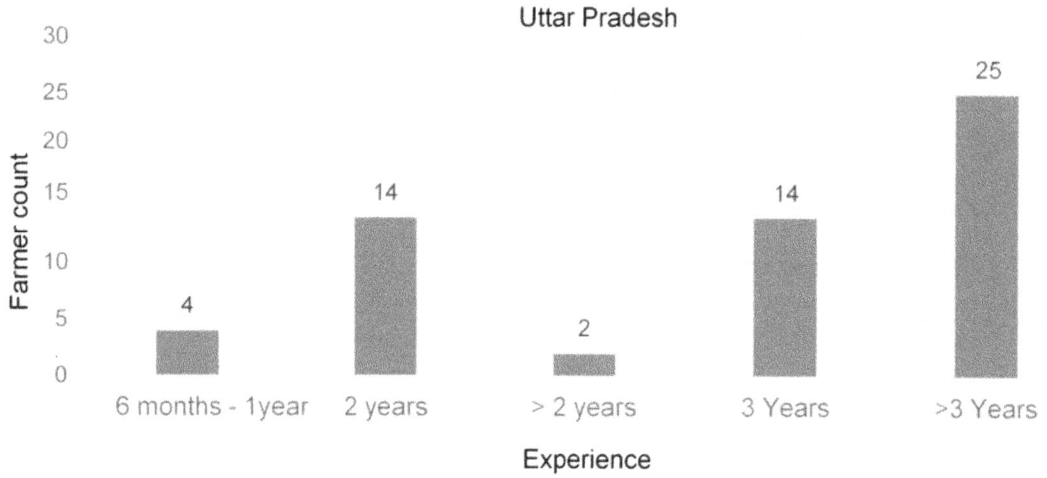

Figure 23. Farmer distribution based on experience with the service of the company in AP, TS, and TN

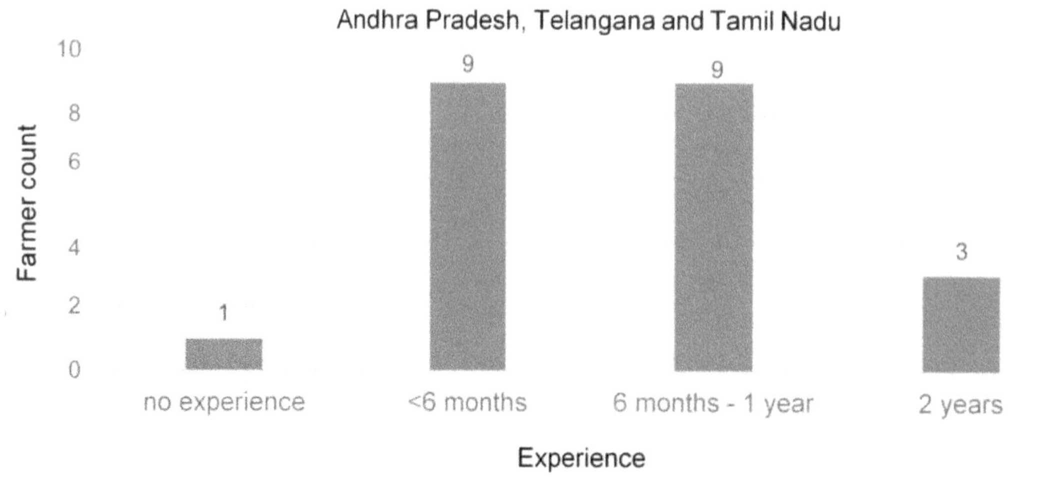

- Farmers joined company service also through a government organization like CIMAP- Central Institute of Medicinal and Aromatic Plants, Lucknow.
- The majority of the farmers in south India, get influenced by the advertisements in mass media like TV/Radio, and newspapers.
- Some Important agrochemical representatives and a few drone service providers and dealers also hold a source of influence in using UAV technologies in south India.

Inference:

- Figure 22 and Figure 23 depicts the farmer distribution based on experience with the service of the company.
- Majority of the farmers using this service for the past 3 years and even before but at least have 6 months' experience with the service.
- BharatRohan has good fame in the Bara Banki, UP state. Farmers have a very good experience of its services.
- Moreover, farmers are concerned with the service. Accordingly, their act of thought is that ADAMA is a better Brand and has better product provision. Thus, if it provides better services, farmers are readily willing to join.
- Due to no strong service providers,' the majority of the farmers have an experience of 6 months- 1 year with the service.

Inference:

- Figure 24 shows the farmer distribution on the increment of yield after joining the company's service.
- In UP, nearly 87% of farmers are conveyed that up to 30% of yield has been increased. And 13% of farmers conveyed to increase yield above 30%.
- Analysing the data showed that, taking consideration of the data recorder, it is found that the majority of farmers with land less than 2 ha have confessed to an increase in yield of more than 20%. Whereas many farmers holding land more than 2 ha have told that there is a minor increment of yield (I.e., 0-10%).
- Due to the non-availability of advanced service providers, the increment in the yield cannot be estimated exactly in south Indian states (AP, TS, TN).

Figure 24. Farmer distribution on the increment of yield after joining the company's service in UP

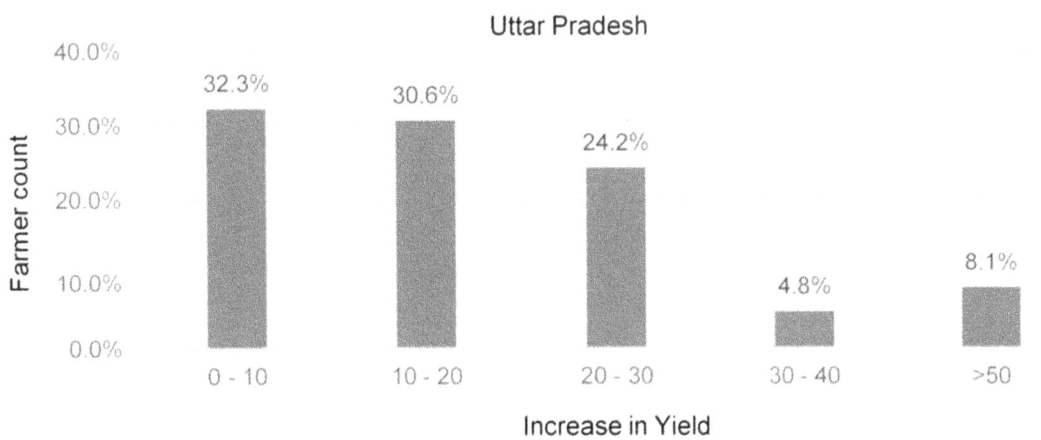

Figure 25. Farmer count based on cost reduction in different commodities in UP

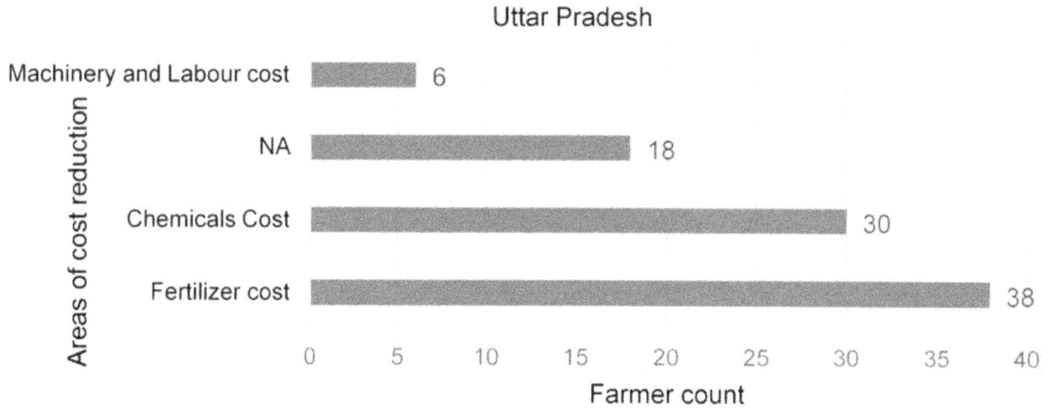

Figure 26. Farmer count based on cost reduction in different commodities in AP, TS, and TN

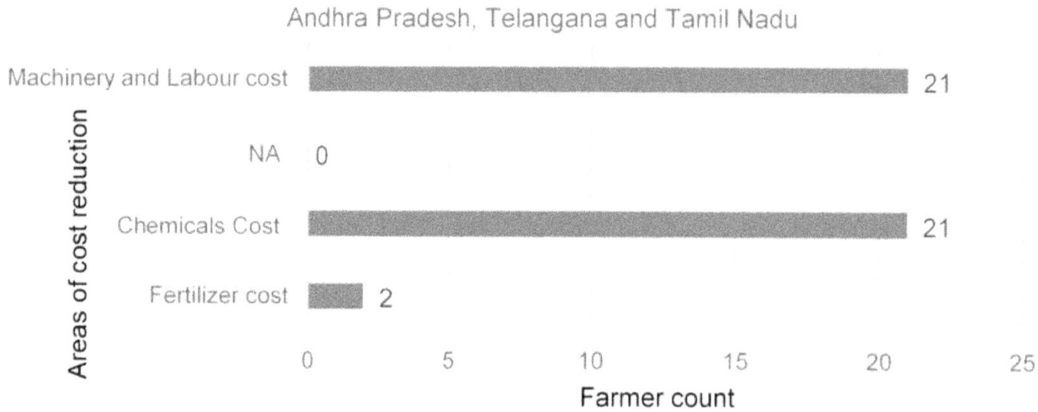

Inference:

- Figure 25 and figure 26 shows the farmer count based on cost reduction in different commodities
- The site-specific management in the field is well addressed by the service providers as there are huge decreases in fertilizer cost, chemical cost and machinery and labour costs respectively in UP.
- 18 farmers out of 68 farmers conveyed that they haven't noticed any reduction in cost in UP.
- Due to no such advanced technology used by the service providers the responses support mostly the decrease in machinery and labour cost and chemical costs in south India (AP, TS, TN).
- Thus, it is noticed that after using the services, there is a reduction in the frequency and quantity of the doses of chemicals and fertilizers, but the cost is an independent factor.

Figure 27. Farmer's rating based on the quality of service provided by respective company in UP

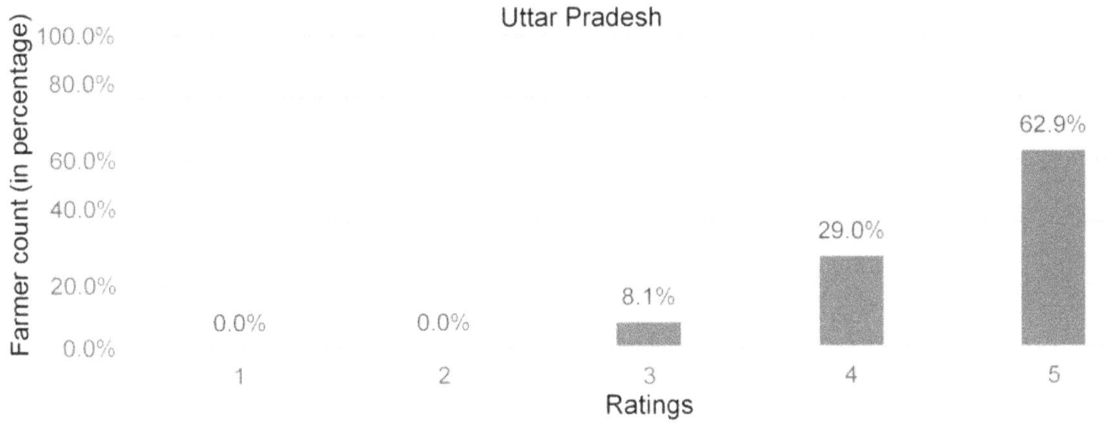

Figure 28. Farmer's rating based on the quality of service provided by respective companies in AP, TS, and TN

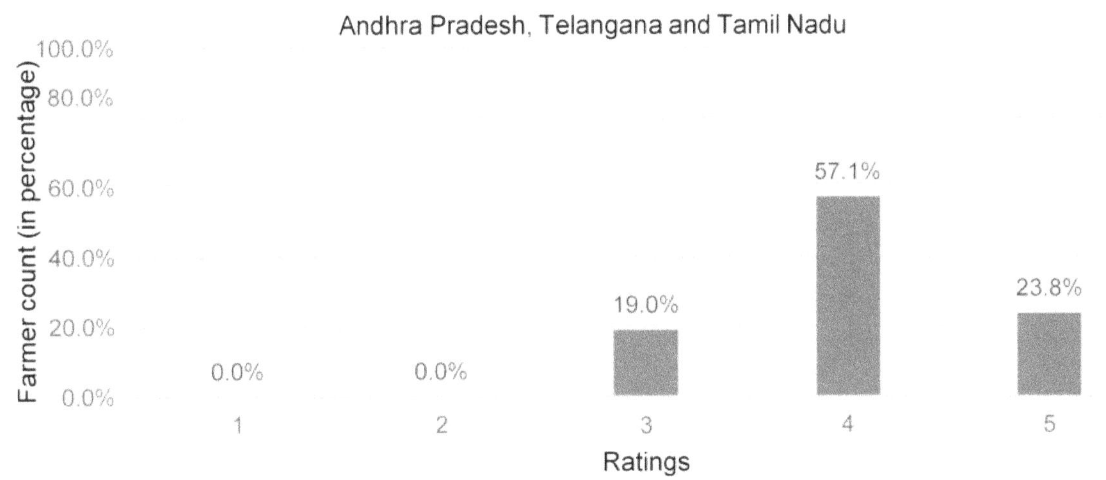

Inference:

- Figure 27 and figure 28 shows the farmer's rating based on the quality of service provided by respective company.
- Almost 63% of farmers reported 5 /5 rating for the quality of service provided followed by 4/5 with 29%.
- 57% of farmers gave 4/5 rating for the service used followed by 5/5 rating from 23.8% of farmers.
- The quality of the service is good from the service provider companies, but few farmers also gave low ratings.
- One of the reasons was farmers do not agree fully with the information provided.
- Based on the research, our thought for the issue might be a communication gap between the provider and the farmer, or maybe due to technical issues like unavailability of information on time, or farmer does not follow.

Figure 29. Farmer's rating based on the quality of service provided by respective company in UP

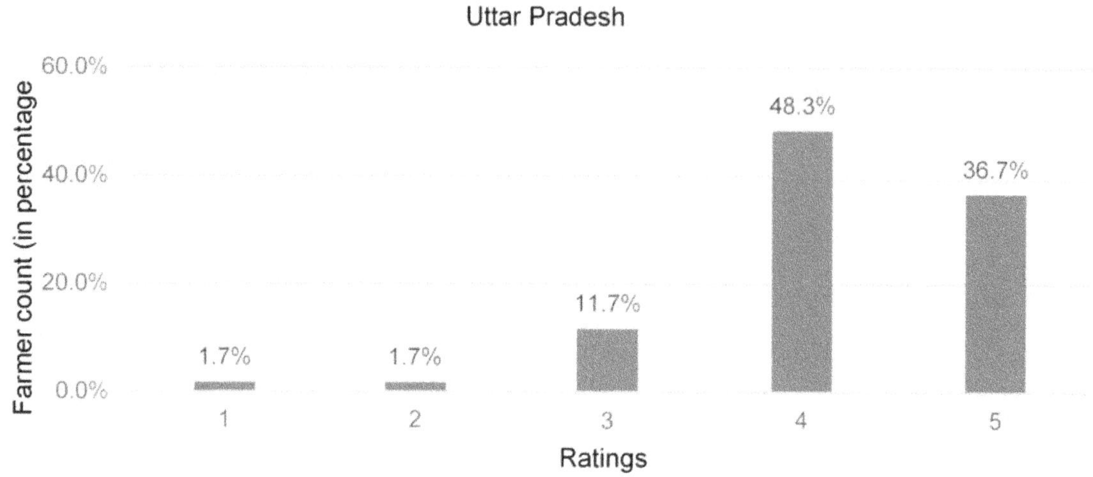

Figure 30. Farmer's rating based on the quality of service provided by respective companies in AP, TS, and TN

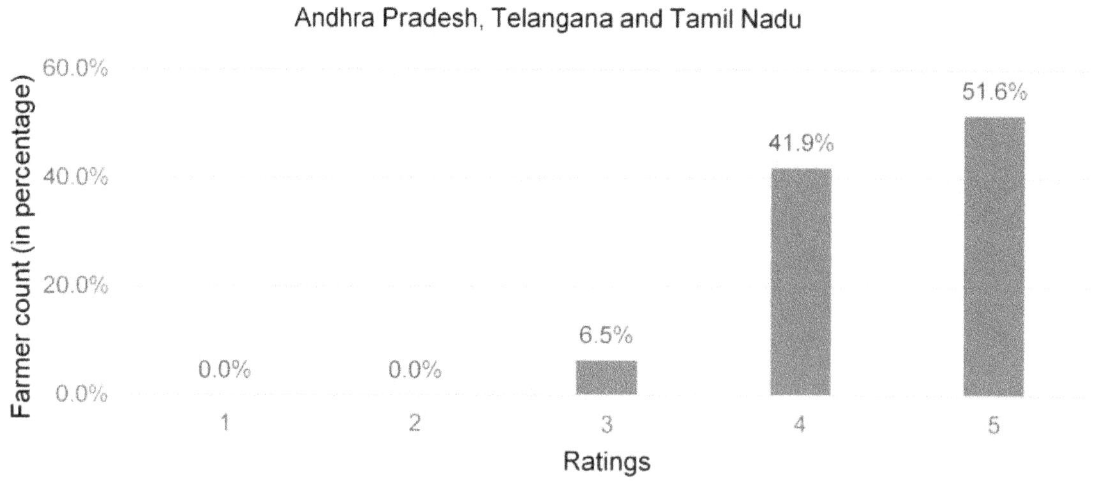

Inference:

- Figure 29 and figure 30 the farmer's rating based on the quality of service provided by respective companies.
- Every farmer in the sample strongly believes in the helpfulness of UAV's service as multiple benefits of drones are known to them.
- Few farmers gave low ratings due to uncertainties found in agriculture, few farmers are dissatisfied with the quality of service. Very few farmers in UP gave low ratings due to the ineffectiveness of the products provided by the company.
- In UP 48.3% of farmers have rated 4/5 and 36.7% rated 5/5 because the cost of chemicals and fertilizers had decreased. It also provides seeds and chemicals to their doorstep.

- In AP, TS, and TN almost 93% of the farmers expressed helpfulness of service (aerial imaging), if provided, as it helps in predicting and forecasting pests and diseases, and site-specific management.

Figure 31. Status of tabled-4 companies work on drone predictive analysis

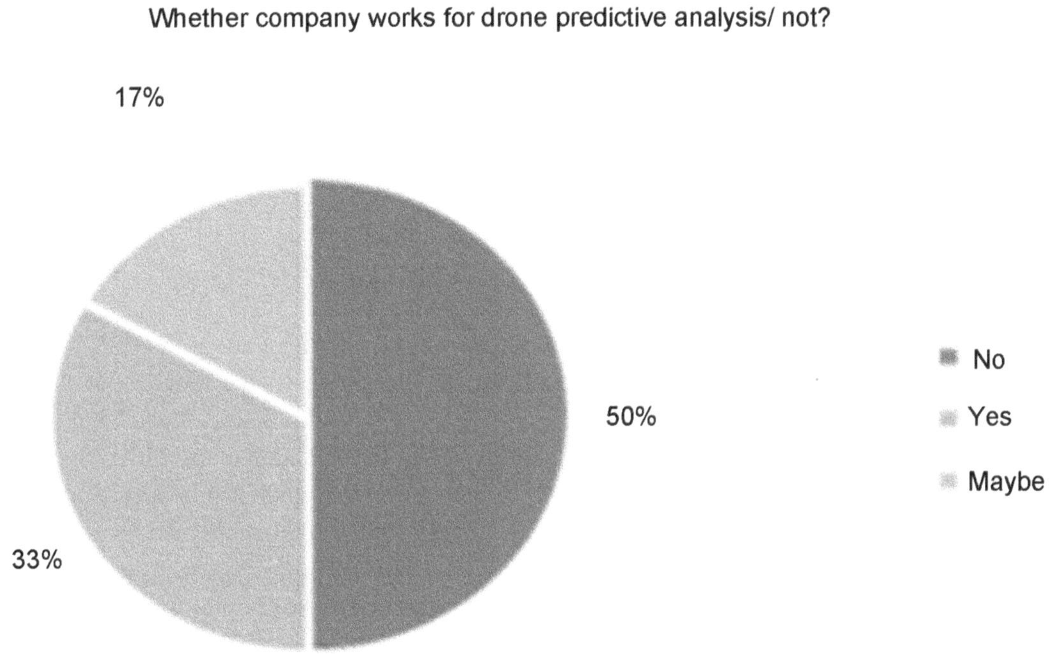

Inference:

Figure 31 shows the Status of tabled-4 companies work on drone predictive analysis. Among the sample companies, only 50% do predictive analysis.

Figure 32. Service charges for drone spraying

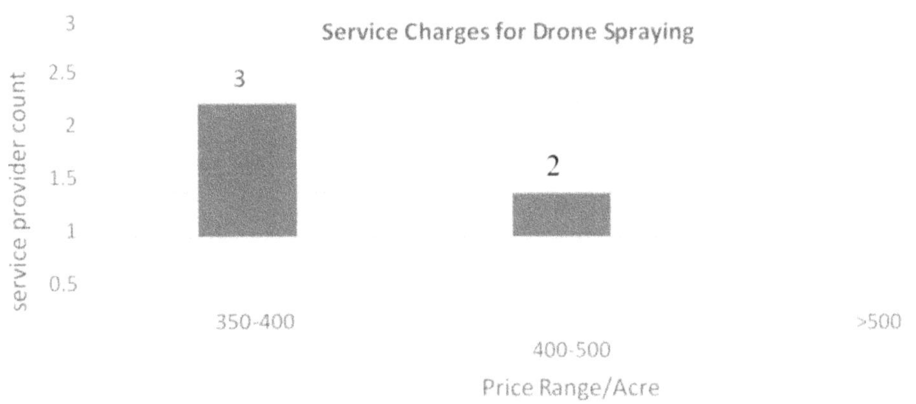

Inference:

Figure 32 depicts the price range

The price range of Rs350-500/acre is charged by various companies

DISCUSSION OF THE STUDY

- There is huge scope in the penetration of agricultural technology all over India at affordable prices by the farmers.
- Farmers are even ready to try drones in their fields for spraying, crop health monitoring etc.
- The majority of farmers use drones for low height crops like paddy, potato, chilli, mint etc. and even in Tamil Nadu drones are been used for coconut crops. This can be applicable even for the oil palm which is largely grown in Telangana and Andhra Pradesh.
- Leaf eating caterpillars, and sucking pests are major pests in the crops for which input company chemicals can be the solution for them.
- Farmers need service and also personalized product service if given attentionally.
- The majority of farmers require service at a low cost <Rs.500/acre, the range is Rs.400-500/acre in southern states.
- Through thorough research, we found that companies like BharatRohan, UPL, Syngenta, and Bayer are trying the service even at a low cost for cross-selling of their products.
- Through word of mouth, social media advertisements, local drone dealers and service providers are easily tapping the market.
- Southern states have no consistent service provider with advanced technology to huge customers.
- Multiple benefits of this UAV's service are not exactly known to the farmers which drags the interest of farmers towards the usage of drone service.
- Farm input cost is decreased with the usage of drone service which can a marketing lead through advertisements.
- Site-specific management of crops is being done in a UP and not in southern states by local service providers due to a lack of advanced technology and a huge customer base.
- Farmers with landholdings of 2-10 hectares and above are interested to use the service at reasonable prices and with better customer management strategies.
- In southern states especially in AP and Telangana, awareness is increasing among farmers about the benefits of drones because of various government institutions, corporates, and local service providers.
- Customers (farmers) satisfaction level is scaling with additional features of UAVs.
- Product services.
- Some potential farmers are willing to get personal owned drones for their field management and are ready to train themselves to start the service to fellow farmers.
- One of the major problems quoted by farmers is the problem with the propellers and batteries of the drone.

RECOMMENDATIONS

- Strong advisory services with a consultancy team can benefit the penetration of the service in the market.
- Though in UP, BarathRohan is dominant but can enter a niche market with multiple benefits for bulk farmers, FPOs with lesser chemical price, strong advisory, and better customer relationship management.
- It is easy to enter in southern states' market as they are not either strong players or using advanced technology.
- Finding out the potential customers for drone service and maintaining the real-time data can help to maintain the customer and also increase the frequency of usage of the service.
- Drone specified product diversification can be done.
- Social media channel- YouTube channels especially on agro-technology can benefit customer reach.
- Technology using startups like Crystal ball, Heliosprayers, Precision hawk, Equinox's drones etc are beneficial if collaborated.
- Product with fewer hardware errors and high battery backup is recommended based on the problems faced by farmers.
- Segmentation based on crop, land, and age will help to find the lead and potential customers.
- Finding premium customers- community building.
- For commercial-scale farmers- estates, private farms.
- Out reachable store, toll-free number, convenient booking, easy app user interface.

DRONE FLYING RESTRICTIONS IN INDIA

The most rules and regulations for using drones in India are,

- Except the Nano group, all drones must be disclosed with a Unique Identification Number (UIN).
- The actions of commercial drone require a permit, with the exception of Nano category should fly with feet of 50 below and also Microdrones should fly 200 feet below.
- No Micro drone should fly higher than 60 metres above ground level (AGL) or 25 metres per second (m/s) should be faster.
- No Small drone shall rise higher than 120 metres AGL or faster than 25 metres per second.
- Medium and large drones must operate under the circumstances indicated in the DGCA Operator Permit.
- Always the Drone pilots must keep a real and clearly visual line of sight whenever they fly.
- Drones are not permitted to fly higher than 400 feet upwards.
- It is not permitted to be flown in "No Fly Zones," which include regions nearby the airport, international boundaries, Vijay Chowk in Delhi, military facilities, the Government Administrative Complex in state capitals, and important places.
- To operate in regulated airspace, submit a flight plan and get a unique Flight Information Center (FIC) number / Air Defense Clearance (ADC).

No Takeoff/ No Permission Policy in India

The drone pilots must seek proper permission to fly their drones with the help of smartphone applications that can be grant or deny it.

Drone Pilots and Data Security

It is responsible for safeguarding any data collected throughout a drone operation. To securely keep or discard such data, appropriate processes and hardware must be followed. Furthermore, the data obtained during operations must not be transmitted to any third party without the explicit approval from the data possessors.

CONCLUSION

It is forecast that the world population will rise more than nine billion in 2050. Thence we need to feed the entire population, so there is a necessity to fulfil the people's hungry. At present, there are around 815 million of people who are hungry in the world among that more than half of them are in the Asia continent. Food production requires a boost upto 70% to 80%. We required a system and new technology to accomplish the future requirement, with help of blockchain technology the farmers can improve the production and health of crops. The use of drones is the latest technology which has been introduced that can be helpful to fulfil the future scarcity of food and also it can able to protect the crops from the various diseases. This technology is more useful for the farmers in order to monitor the growth of crops and health conditions will be more accurately and more efficiently. A virtual survey provides a practical and real picture of the farmers (for both drone users & non-users) for predictive forecast and identification of pest and disease incidence.

REFERENCES

Ahirwar, S., Swarnkar, R., Bhukya, S., & Namwade, G. (2019). Application of drones in agriculture. *International Journal of Current Microbiology and Applied Sciences*, *8*(01), 2500–2505. doi:10.20546/ijcmas.2019.801.264

Arokiaraj, D. (2015). A Study on Environmental Responsibility of the Stakeholders of Auto Industry in Chennai. International *Journal of Innovative Technology and Exploring Engineering, 9*(1).

Arokiaraj, D., Ganeshkumar, C., & Paul, P. V. (2020). Innovative management system for environmental sustainability practices among Indian auto-component manufacturers. *International Journal of Business Innovation and Research*, *23*(2), 168–182. doi:10.1504/IJBIR.2020.110095

Arokiaraj, D., Ramyar, R. A., Ganeshkumar, C., & Gomathi Sankar, J. (2020b). An empirical analysis of consumer behaviour towards organic food products purchase in India. *Calitatea Qual Access Success, 21*.

Ayamga, M., Tekinerdogan, B., & Kassahun, A. (2021). Exploring the challenges posed by regulations for the use of drones in agriculture in the African context. *Land (Basel)*, *10*(2), 164. doi:10.3390/land10020164

Ayamga, M., Tekinerdogan, B., Kassahun, A., & Rambaldi, G. (2021b). Developing a policy framework for the adoption and management of drones for agriculture in Africa. *Technology Analysis and Strategic Management*, *33*(8), 970–987. doi:10.1080/09537325.2020.1858047

Ayaz, M., Ammad-Uddin, M., Sharif, Z., Mansour, A., & Aggoune, E. H. M. (2019). Internet-of-Things (IoT)-based smart agriculture: Toward making the fields talk. *IEEE Access: Practical Innovations, Open Solutions*, *7*, 129551–129583. doi:10.1109/ACCESS.2019.2932609

Banumathi, M., & Arokiaraj, D. (2011). Eco-labeling–The Need for Sustainable Marketing. In *National Conference in the era of Global Recovery,*(pp. 511-515).

Barasa, P. M., Botai, C. M., Botai, J. O., & Mabhaudhi, T. (2021). A Review of Climate-Smart Agriculture Research and Applications in Africa. *Agronomy (Basel)*, *11*(6), 1255. doi:10.3390/agronomy11061255

Ben Ayed, R., & Hanana, M. (2021). Artificial intelligence to improve the food and agriculture sector. *Journal of Food Quality*. doi:10.1155/2021/5584754

Bhardwaj, H., Tomar, P., Sakalle, A., & Sharma, U. (2021). Artificial Intelligence and Its Applications in Agriculture with the Future of Smart Agriculture Techniques. In *Artificial Intelligence and IoT-Based Technologies for Sustainable Farming and Smart Agriculture* (pp. 25–39). IGI Global. doi:10.4018/978-1-7998-1722-2.ch002

Bogue, R. (2021). Robots are poised to transform agriculture. *Industrial Robot: the international journal of robotics research and application*.

Călina, J., Călina, A., Miluţ, M., Croitoru, A., Stan, I., & Buzatu, C. (2020). Use of drones in cadastral works and precision works in silviculture and agriculture. *Publisher NARDI Fundulea, România, 37*, 273–284.

Daponte, P., De Vito, L., Glielmo, L., Iannelli, L., Liuzza, D., Picariello, F., & Silano, G. (2019, May). A review on the use of drones for precision agriculture. *IOP Conference Series. Earth and Environmental Science, 275*(1), 012022. doi:10.1088/1755-1315/275/1/012022

David, A. (2020$_a$). Air Pollution Control Monitoring & Delivery Rate Escalated by Efficient use of Markov Process in Manet Networks: to Measure Quality of Service Parameters. *Test Engineering & Management, The Mattingley Publishing Co., Inc.*

David, A. (2020b). The consumer purchasing process of organic food product: An empirical analysis. [QAS]. *Journal of Management System-Quality Access to Success, 21*(177), 128–132.

David, A. (2020c). *Corporate and individual environmental responsibility towards the automobile.* Book Rivers.

David, A., Kumar, C. G., & Paul, P. V. (2022). Blockchain technology in the food supply chain: Empirical analysis. [IJISSCM]. *International Journal of Information Systems and Supply Chain Management, 15*(3), 1–12. doi:10.4018/IJISSCM.290014

David, A., Muthukumar, A., Ayyappan, D., & Ravi, S. (2020). Accident Avoidance System by Using Sensors Module. *Test Engineering & Management, The Mattingley Publishing Co., Inc.*

David, A., & Ravi, S. (2017). The direness of cultivable land spotted on agricultural: A special reference to rice production in South India. *Abhinav National Monthly Refereed Journal of Research in Commerce & Management, 6*(09), 55-59.

Dileep, M. R., Navaneeth, A. V., Ullagaddi, S., & Danti, A. (2020, November). A study and analysis of various types of agricultural drones and their applications. In *Fifth International Conference on Research in Computational Intelligence and Communication Networks (ICRCICN)* (pp. 181-185). IEEE.

Eli-Chukwu, N. C. (2019). Applications of artificial intelligence in agriculture: A review. *Engineering, Technology &. Applied Scientific Research, 9*(4), 4377–4383.

Esposito, M., Crimaldi, M., Cirillo, V., Sarghini, F., & Maggio, A. (2021). Drone and sensor technology for sustainable weed management: A review. *Chemical and Biological Technologies in Agriculture, 8*(1), 1–11. doi:10.118640538-021-00217-8

Feleen, F., & David, A. (2021). A Comparative Study of Work from Home Vs Work from Office: Preference of Women Employees in IT Industry. *Design Engineering (London), 7*(1), 5763–5775.

Ganesh Kumar, C., Murugaiyan, P., & Madanmohan, G. (2017). Agri-food supply chain management: A literature review. *Intelligent Information Management, 9*(02), 68–96. doi:10.4236/iim.2017.92004

Ganeshkumar, C., David, A., & Jebasingh, D. R. (2022). Digital Transformation: Artificial Intelligence Based Product Benefits and Problems of Agritech Industry. In *Agri-Food 4.0*. Emerald Publishing Limited. doi:10.1108/S1877-636120220000027010

Ganeshkumar, C., Jena, S. K., Sivakumar, A., & Nambirajan, T. (2021). Artificial intelligence in the agricultural value chain: Review and future directions. *Journal of Agribusiness in Developing and Emerging Economies*. doi:10.1108/JADEE-07-2020-0140

Ganeshkumar, C., & Khan, A. (2021). Mapping of Agritech Companies in Indian Agricultural Value Chain. In *Proceedings of the Second International Conference on Information Management and Machine Intelligence* (pp. 155-161). Springer, Singapore. 10.1007/978-981-15-9689-6_18

Ganeshkumar, C., Prabhu, M., & Abdullah, N. N. (2019). Business analytics and supply chain performance: Partial least squares-structural equation modeling (PLS-SEM) Approach. *International Journal of Management and Business Research*, *9*(1), 91–96.

Ganeshkumar, C., Prabhu, M., Reddy, P. S., & David, A. (2020). Value Chain Analysis of Indian Edible Mushrooms. *International Journal of Technology*, *11*(3), 599–607. doi:10.14716/ijtech.v11i3.3979

Iost Filho, F. H., Heldens, W. B., Kong, Z., & de Lange, E. S. (2020). Drones: Innovative technology for use in precision pest management. *Journal of Economic Entomology*, *113*(1), 1–25. doi:10.1093/jee/toz268 PMID:31811713

Joshi, E., Sasode, D. S., Singh, N., & Chouhan, N. (2020). Revolution of Indian Agriculture through Drone Technology. *Biotica Research Today, 2*(5 Spl.), 174-176.

Klauser, F., & Pauschinger, D. (2021). Entrepreneurs of the air: Sprayer drones as mediators of volumetric agriculture. *Journal of Rural Studies*, *84*, 55–62. doi:10.1016/j.jrurstud.2021.02.016

Kumar, A., Rani, M., & Kumar, P. (2022). Drone Technology in Sustainable Agriculture: The Future of Farming Is Precision Agriculture and Mapping. In *Agriculture, Livestock Production and Aquaculture* (pp. 3–12). Springer. doi:10.1007/978-3-030-93262-6_1

López-Morales, J. A., Martínez, J. A., & Skarmeta, A. F. (2020). Digital transformation of agriculture through the use of an interoperable platform. *Sensors (Basel)*, *20*(4), 1153. doi:10.339020041153 PMID:32093147

Madaan, G., Swapna, H. R., Kumar, A., Singh, A., & David, A. (2021). Enactment of sustainable technovations on healthcare sectors. *Asia Pacific Journal of Health Management*, *16*(3), 184–192. doi:10.24083/apjhm.v16i3.989

Mogili, U. R., & Deepak, B. B. V. L. (2020). An intelligent drone for agriculture applications with the aid of the mavlink protocol. In *Innovative Product Design and Intelligent Manufacturing Systems* (pp. 195–205). Springer. doi:10.1007/978-981-15-2696-1_19

Muraru, S. L., Cardei, P., Muraru, V., Sfîru, R., & Condruz, P. (2019). Researches regarding the use of drones in agriculture. *International Multidisciplinary Scientific GeoConference: SGEM, 19*(6.2), 683-690.

Nayyar, A., Nguyen, B. L., & Nguyen, N. G. (2020). The internet of drone things (IoDT): future envisions of smart drones. In *First international conference on sustainable technologies for computational intelligence* (pp. 563-580). Springer, Singapore.

Ndiaye, M., Salhi, S., & Madani, B. (2020). When Green Technology Meets Optimization Modeling: The Case of Routing Drones in Logistics, Agriculture, and Healthcare. In *Modeling and Optimization in Green Logistics* (pp. 127–145). Springer. doi:10.1007/978-3-030-45308-4_7

Negash, L., Kim, H. Y., & Choi, H. L. (2019, November). Emerging UAV applications in agriculture. In *2019 7th International Conference on Robot Intelligence Technology and Applications (RiTA)* (pp. 254-257). IEEE. 10.1109/RITAPP.2019.8932853

Pachayappan, M., Ganeshkumar, C., & Sugundan, N. (2020). Technological implication and its impact in the agricultural sector: An IoT Based Collaboration framework. *Procedia Computer Science, 171*, 1166–1173. doi:10.1016/j.procs.2020.04.125

Panpatte, S., & Ganeshkumar, C. (2021). Artificial intelligence in agriculture sector: case study of blue river technology. In *Proceedings of the Second International Conference on Information Management and Machine Intelligence* (pp. 147-153). Springer, Singapore. 10.1007/978-981-15-9689-6_17

Pathak, H., Kumar, G. A. K., Mohapatra, S. D., Gaikwad, B. B., & Rane, J. (2020). *Use of drones in agriculture: Potentials, Problems and Policy Needs*. ICAR-National Institute of Abiotic Stress Management.

Paul, V., Ganeshkumar, C., & Jayakumar, L. (2019). Performance evaluation of population seeding techniques of permutation-coded GA traveling salesman problems based assessment: Performance evaluation of population seeding techniques of permutation-coded GA. [IJAMC]. *International Journal of Applied Metaheuristic Computing, 10*(2), 55–92. doi:10.4018/IJAMC.2019040103

Petkovic, S., Petkovic, D., & Petkovic, A. (2017). IoT devices VS. drones for data collection in agriculture. *DAAAM International Scientific Book, 16*, 63–80. doi:10.2507/daaam.scibook.2017.06

Pobkrut, T., Eamsa-Ard, T., & Kerdcharoen, T. (2016, June). Sensor drone for aerial odour mapping for agriculture and security services. In *13th International Conference on Electrical Engineering/Electronics, Computer, Telecommunications and Information Technology (ECTI-CON)* (pp. 1-5). IEEE.

Prasanna, S., & Jebapriya, J. (2020). IoT based agriculture monitoring and smart farming using drones. *Mukt Shabd, 9*.

Pratheepkumar, P., Sharmila, J. J., & Arokiaraj, D. (2017). Towards mobile opportunistic in cloud computing. *Indian Journal of Scientific Research, 17*(02), 2250–0138.

Puri, V., Nayyar, A., & Raja, L. (2017). Agriculture drones: A modern breakthrough in precision agriculture. *Journal of Statistics and Management Systems, 20*(4), 507–518. doi:10.1080/09720510.2017.1395171

Rachmawati, S., Putra, A. S., Priyatama, A., Parulian, D., Katarina, D., Habibie, M. T., & Valentino, V. H. (2021, August). Application of Drone Technology for Mapping and Monitoring of Corn Agricultural Land. In *International Conference on ICT for Smart Society (ICISS)* (pp. 1-5). IEEE. 10.1109/ICISS53185.2021.9533231

Rani, A., Chaudhary, A., Sinha, N., Mohanty, M., & Chaudhary, R. (2019). Drone: The green technology for future agriculture. *Harit Dhara*, *2*(1), 3–6.

Ravi, S., David, A., & Imaduddin, M. (2018). Controlling & calibrating vehicle-related issues using RFID technology. *International Journal of Mechanical and Production Engineering Research and Development*, *8*(2), 1125–1132. doi:10.24247/ijmperdapr2018130

Reger, M., Bauerdick, J., & Bernhardt, H. (2018). Drones in Agriculture: Current and future legal status in Germany, the EU, the USA and Japan. *Landtechnik*, *73*(3), 62–79.

Rejeb, A., Abdollahi, A., Rejeb, K., & Treiblmaier, H. (2022). Drones in agriculture: A review and bibliometric analysis. *Computers and Electronics in Agriculture*, *198*, 107017. doi:10.1016/j.compag.2022.107017

Ren, Q., Zhang, R., Cai, W., Sun, X., & Cao, L. (2020, February). Application and development of new drones in agriculture. *IOP Conference Series. Earth and Environmental Science*, *440*(5), 052041. doi:10.1088/1755-1315/440/5/052041

Sharma, R. (2021). Review on Application of Drone Systems in Precision Agriculture. *Journal of Advanced Research in Electronics Engineering and Technology*, *7*(2), 5–7.

Siddhartha, T., Nambirajan, T., & Ganeshkumar, C. (2019). Production and retailing of self-help group products. *Global Business and Economics Review*, *21*(6), 814–835. doi:10.1504/GBER.2019.102590

Siddhartha, T., Nambirajan, T., & Ganeshkumar, C. (2021). Self-help group (SHG) production methods: insights from the union territory of Puducherry community. *Journal of Enterprising Communities: People and Places in the Global Economy*.

Srivastava, K., Pandey, P. C., & Sharma, J. K. (2020). An approach for route optimization in applications of precision agriculture using UAVs. *Drones (Basel)*, *4*(3), 58. doi:10.3390/drones4030058

Srivastava, V., Singh, A. K., David, A., & Rai, N. (2022). Modelling Student Employability on an Academic Basis: A Supervised Machine Learning Approach with R. In *Handbook of Research on Innovative Management Using AI in Industry 5.0* (pp. 179–191). IGI Global. doi:10.4018/978-1-7998-8497-2.ch012

Stehr, N. J. (2015). Drones: The newest technology for precision agriculture. *Natural Sciences Education*, *44*(1), 89–91. doi:10.4195/nse2015.04.0772

Thalluri, L. N., Adapa, S. D., Priyanka, D., Sarma, A. V. N., & Venkat, S. N. (2021, October). Drone Technology Enabled Leaf Disease Detection and Analysis system for Agriculture Applications. In *2nd International Conference on Smart Electronics and Communication (ICOSEC)* (pp. 1079-1085). IEEE. 10.1109/ICOSEC51865.2021.9591837

Vandana, C. P., Prakash, A., Bhowal, A., & Taj, D. M. (2020). Drone Assisted Effective Pesticide Sprayer. *International Journal of Scientific Research in Computer Science. Engineering and Information Technology*, *3*(6), 363–370.

Veroustraete, F. (2015). The rise of drones in agriculture. *EC agriculture, 2*(2), 325-327.

Victer Paul, P., Ganeshkumar, C., Dhavachelvan, P., & Baskaran, R. (2020). A novel ODV crossover operator-based genetic algorithms for traveling salesman problem. *Soft Computing*, *24*(17), 12855–12885. doi:10.100700500-020-04712-2

Wójtowicz, M., Wójtowicz, A., & Piekarczyk, J. (2016). Application of remote sensing methods in agriculture. *Communications in Biometry and Crop Science*, *11*(1), 31–50.

Wynn, C. R. (2019). *Drone Technology: Is It Worth the Investment in Agriculture* [Doctoral dissertation, The University of Nebraska-Lincoln, USA].

APPENDIX

Table 3. Companies Providing Service on Predictive Forecast

Sr. No.	Company Name	Contact Details
1	BharatRohan Airborne Innovations Private Limited.	B2/85, Basement, Vishesh Khand, Gomti Nagar, Lucknow, Uttar Pradesh – 226010 Amandeep Panwar (CEO) – 96505 88865 Vimal Kumar Verma (BharatRohan employee) – 70683 64100. Email ID - info@bharatrohan.in
2	CropIn	www.cropin.com
3	Fasal	www.fasal.co
4	Equinox's Drones	https://www.equinoxsdrones.com/agriculture
5	Precision Hawk	https://www.precisionhawk.com/agriculture

Table 4. Lead companies providing service on the predictive forecast (UAVs Companies)

Sr. No.	Company Name	Contact Details
1	Sky Lark	IKP Eden, 16, Tavarekere Main Rd, Bengaluru, Karnataka 560029. PH NO. (080) 5005 8600 MAIL:- info@skylarkdrones.com
2	Aerial Photo	BVM 8, 1ST FLOOR, DLF PHASE 2, GURGAON Phone: +91-9910888251, 9910941170. Email: get@aerialphoto.in
3	IG Drones	iHub Gujarat, KCG Campus, Navrangpura, Ahmedabad, Gujarat, Pin 380015. PHONE: +91 6370868100 +91 9776982070 EMAIL:- info@igdrones.com
4	Indrones	Registered Office: H-103, Nerul Station Complex, Above Dwarka Hotel, Nerul (W), Navi Mumbai – 400706. R&D Lab: Plot 90, Sector 16A, Nerul (W), Navi Mumbai - 400706 Email: info@indrones.com General Inquiries: +91-9594-160-160 Urgent Inquiries: +91-813-097-7119.

Table 5. Companies providing UAVs (Drone Service)

Gopi Raja (Drone raja)	9989838337	Vijayawada - Kankipadu
Sudheer	9704040591	Guntur - Kollipora
Crystal ball- Madhusudhan Reddy	8555087826	Hyderabad
Axpect Atomizer	9494503535	Vijayawada
Aerospace Technologies	7989059604	Vijayawada
Jagadeesh Chowdary	8074892715	Nellore

Chapter 5
Prediction of Grape Fruit Disease Using Shuffled Frog Leaping Algorithm-Based Convolutional Neural Network

Gokula Krishnan V.
https://orcid.org/0000-0002-3913-3156
Saveetha School of Engineering, Saveetha Institute of Medical and Technical Sciences, India

Deepa J.
Easwari Engineering College, India

Pinagadi Venkateswara Rao
ACE Engineering College, India

K. Sreerama Murthy
Sreenidhi Institute of Science and Technology, India

Hemamalini S.
Panimalar Engineering College, India

Avula Chitty
CVR College of Engineering, India

ABSTRACT

A healthy plant has a significant impact on the environment and human health. To put it another way, early discovery of plant diseases results in a momentous increase in the quality of the finished product. Using IoT-based agriculture, deep learning, and image processing techniques are essential to developing an effective way for detecting and controlling the spread of disease in agricultural products. Leaf drop is a common symptom of many diseases, as is the disappearance of the plant or plants nearby. This means that early detection and prevention are critical to ensuring healthy grape growth and yield. A method for detecting grape disease utilizing a CNN methodology and real-time data from environmental characteristics collected to predict illness likelihood. The image is subjected to multiple layers of CNN processing. The shuffled frog-leaping technique, a meta-heuristic algorithm, is used to optimize the CNN's learning rate (SFLA). The results suggest that the proposed SFLA-CNN outperforms existing approaches in terms of performance.

DOI: 10.4018/978-1-6684-6413-7.ch005

Copyright © 2023, IGI Global. Copying or distributing in print or electronic forms without written permission of IGI Global is prohibited.

INTRODUCTION

The internet of things aims to reduce human-to-human interaction by automating routine tasks. Data from sensors is collected by a controller, which then processes it and sends it out to actuators, which in turn complete the automation process (Sangeetha et al., 2022). IoT in agriculture aims to automate every aspect of the process, from seed to fork, in order to make it as efficient as possible. Traditional methods of yield management, livestock management, weed control, soil management, and water management, plant management, and animal tracking involve a lot of human touch and high labor costs and power consumption. There has been a lot of interest in agricultural production and efficiency in the development of intelligent systems that monitor and control a variety of field characteristics. Automated measuring systems collect data to produce precise readings, allowing them to make appropriate decisions (Babbar et al., 2022).

Soil moisture, humidity, temperature, and pH levels are some of the environmental characteristics now being tracked by smart farming systems. In machine learning, a new method of calculating intelligence is employed. Smart farming and agriculture also has a wide variety of ML applications. A few examples of how machine learning can be used in this industry include yield prediction, water management, and livestock management. There has been a lot of study done on IoT and ML for smart agriculture, but it has been discovered that there are a number of separate studies. ML and IoT are combined in this study to create better and more precise farming solutions (Popa, Prostean, and Popa, 2019; Medela et al., 2013; Mahdavinejad, 2018).

It is possible, however, that the rapid spread of undesired identifiers in plants may cause major damage to their leaves and fruits producing a loss in economic benefits (Barbedo, 2016). The appearance of leaves and petioles can be used to diagnose the majority of associated diseases, which are characterized by rots and abnormalities (McKinley and Talboys, 1979; Ebihara, Uematsu, & Nomiya, 2010). To prevent the spread of disease, the affected leaves must be removed or treated as soon as possible. Diagnosing plant diseases is therefore critical for both economic gain in agriculture and the development of high-quality crops.

When it comes to plant disease identification, previous literature has mostly concentrated on visual traits and disease characteristics (Harris, 1990). In recent years, plant disease diagnosis has been improved using machine learning approaches (Fujita et al., 2016). Although these methods are computationally expensive, they fail to detect all diseases. Some complex categorization issues can't be generalized by these procedures, which has a restriction. There was a substantial impact on agricultural production in recent years due to the quick development of deep learning techniques in the field. In addition, deep learning methods do not rely on specific features and are independent of real-world variables, such as changing lighting conditions, different-sized objects, and varying backgrounds. By training on a huge number of photos, they have the ability to detect and classify a wide range of objects.

Agricultural IoT

Research papers on IoT-based farm structures have been published (Kamilaris et al., 2016; Jayaraman et al., 2016). A microprocessor, a variety of sensors, actuators, and wireless interfaces are typically included in an IoT solution. In addition to WiFi and LoRaWAN, Zigbee, Bluetooth Low Energy, and other wireless technologies may be used. The network layer is formed by a local WSN gateway sending data over an Internet gateway. Data processing, such as data visualization, data analysis, data storage, and data security, are all necessary in order to make sense of the data acquired by the service tier. It is the application layer that is most critical here, allowing end users to monitor and control various farm processes and make critical decisions based on forecasts and market trends. As a result of IoT-based agriculture, a wide range of data has been collected from many sources on and around agricultural farms, ranging from voltage values to photographs, actuator states to robot positions. Data that is of high-quality leads to information that is of high quality and accuracy. You can't use ML algorithms to build predictive models unless you have high-quality data to work with. When these updated datasets are fed into ML algorithms, more accurate analysis and predictions can be made. It is possible to collect enormous amounts of data from sensors and incorporate cloud computing applications such as agricultural maps and cloud storage using the Internet of Things. Monitoring and communication between all stakeholders are made possible with the availability of real-time data from any location at any time. (Castelli, 2018).

The main aim of the work is to develop a deep learning model for the prediction of leaf disease in grapefruits. The learning rate of the CNN model is optimized by meta-heuristic algorithm, which will effectively predict the disease. The experiments are carried out on the publicly available dataset in terms of various metrics. The remaining paper is constructed as follows: Section 2 presents the related works of the existing techniques; the explanation of the proposed deep learning technique is given in Section 3. The validation of optimized CNN with existing techniques in terms of performance metrics is provided in Section 4. Finally, the conclusion of the research work is described in Section 5.

RELATED WORKS

Wang et al. (2017) used noise truncation and masking to obtain attention characteristics. Stacked CNN with multiple attentions greatly enhanced image categorization in this case. According to Chen et al. (2017), an attention model based on multi-layer feature maps employs both channel-wise and spatial attention. Due to the method's ability to combine data from various feature maps, it performs at its best. Multi-path recurrent connections in a spatial attention network were proposed by Zhao et al. (2018) to parse scenes in a pointwise manner. Using numerous convolutional layers, the method generates pixel-by-pixel global attention maps for every location in the input image. An attention-guided module that uses multi-level attention features was proposed in 2018 by Zhang et al. (2018a). Use of multi-level attention enhances the textual data needed for classification. Hierarchical holistic care was merged into the stacked CNN for pose estimation presented by Chu et al. (2017).

Reverse Reweighting Networks (RRNs) are used to apply different markers to the concatenated multilayer features that Zhang et al. (2018b) suggested. An approach based on the combination of RPN and Multi-Scale Detection Network (MSDN) for pedestrian detection in low light circumstances was developed by Xu et al. (2017). To detect several illnesses in maize leaves, Zhang et al. (2018c) used GoogleNet and Cifar 10 networks. It's a time-consuming process, so there's a downside. New deep learning architectures by Yu et al. (2020) include the FS-SubNet and SAC-SubNet models. An application of FS-SubNet and SAC-SubNet on a segmented feature map yielded semantic spot information that can be used to improve classification accuracy. Using SLIC superpixels, Tetila et al. (2020) first segmented the soybean leaves. When it came to unhealthy leaf detection, they used four different deep learning models: Inceptionv3, Resnet-50, VGG-19 and Xception.

Multitask prediction and distillation (PADNet) employing deep features across multi-modal information was proposed by Xu et al. (2018). Scene parsing and depth estimation using PADNet make use of the

Figure 1. Proposed Architecture.

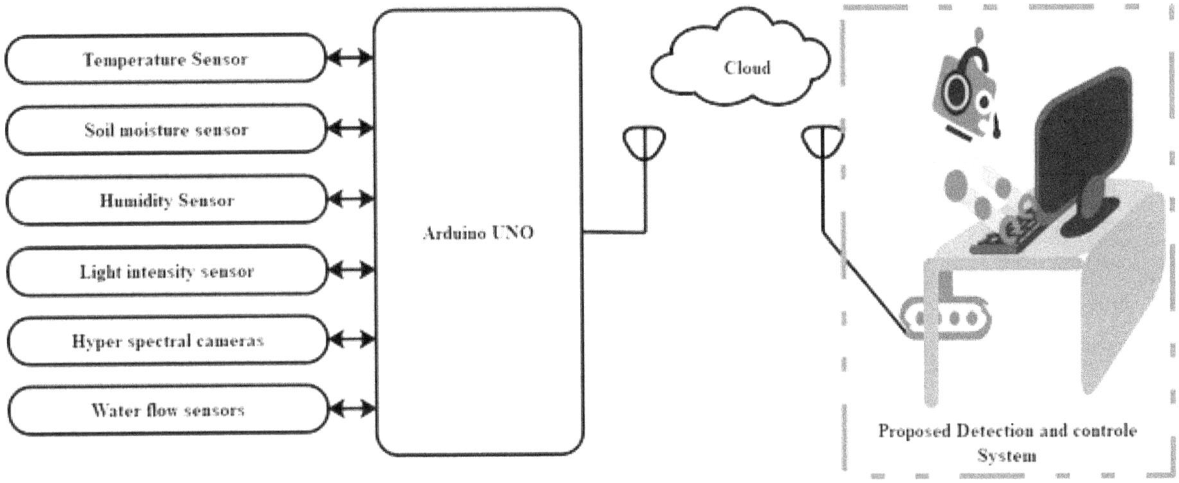

auxiliary intermediate information. Deep CNN was used by Ma et al. (2018) to obtain real-time photos of cucumber leaves (LeNet5). The deployed model is less expensive to run, but the results it produces are less precise. AlexNet and SqueezeNet Deep architectures have been applied to tomato leaves by Durmus, Gunes, and Kirci (2017). Additionally, they claim that a robotic arm can detect and remove sick leaves.

Many pre-trained models are utilized for disease prediction, although most of the strategies are based on the leaves of various plants such as tomato, rice, soybean, maize, cucumber, and so on. This study isn't motivated by grapefruit, however.

PROPOSED METHODOLOGY

IoT is integrated with deep learning technique for disease prediction is provided in this research work. The various components are used to collect information such as temperature, humidity, moisture of soil, flow of water and light intensity for effective prediction. After the data is collected, it is transferred to cloud using Arduino UNO and deep learning technique takes this information for final classification. The early stage of prediction will helps to improve the crop quality and leads to huge profit. Figure 1 shows the working flow of proposed research work.

Sensors for Measuring the Plants

Temperature, soil moisture, and light sensors are all used in this suggested system to measure the plant's temperature value, moisture level, and light intensity. The sensors' data is processed by the Arduino UNO microprocessor. The suggested model is utilized to create a training sample of leaf images and suitable threshold values for leaf attributes in order to conduct this disease analysis. This study makes use of three different types of modules: hardware, analysis, and decision-making. Deep learning algorithms are activated by a hardware module that includes an Arduino controller unit, sensors, and cameras. In this subject, you'll learn about leaves and leaf patterns from the perspective of training and testing. This

Table 1. Description of sensors

Components of IoT	Model
Temperature Sensor	LM 35 TO-92-3
Soil Moisture Sensor	LM358
Humidity Sensor	DHT11
Light Intensity Sensor	BH1750
Hyper Spectral Cameras	Hyspex
Water Flow Sensors	YF-S201
Microcontroller	Arduino UNO
Communication Technologies	GSM module and WiFi

module's sensors are employed in the fields of agriculture. They are utilized to keep tabs on every aspect of the production process. A wide variety of sensors are also in use to keep tabs on the surrounding environment. Relays connect these parts to the various terminal devices, such as water pumps and other computer systems. The components necessary to fulfill the aim have an impact on the IoT architecture design. Table 1 depicts the IoT components that are used to collect data.

Dataset Description

In this experiment, the PlantVillage dataset (Figure 2) (Hughes, & Salathé, 2015). was utilized to collect data on the grape crop. There are 3251 training images and 811 validation images with four different classes in the plant village dataset. One of the four seminars is devoted solely to healthy images. There are between 339 and 1107 images in each class. To ensure that each class's training is free of bias, 1000 training photos are taken in each class, and 100 of these images are kept for each validation class. The general augmentation methods, such as zooming, inversion, flipping, and rotation, have been used to balance the photographs in each class (Bloice, Stocker, and Holzinger, 2017; Perez and Wang, 2017). The below Table 2 summarizes the dataset.

Hybrid Classification

In this section, hybrid classification includes learning rate optimization for CNN using SFLA optimization is described briefly. Initially, the CNN is explained and once the data is collected, it is directly given as an input to the deep learning technique for prediction process. The feature extraction process is carried out within the CNN model that is described as follows.

Convolutional Neural Network

Because it uses a back propagation technique to learn new information (Bakar, Tahir, and Yassin, 2010), CNN may be used to identify target classes by taking input from a vector of previously learned data. For this purpose, CNN output is compared to the desired aim to see how far off we are from our original expectation we are. Mathematically assuming the role of the next CNN,

Figure 2. Examples images of our grape dataset.

$$E(\omega)=\frac{1}{2}\sum_{A=1}^{p}\sum_{j=1}^{N_l}\left(o_{j,p}^{l}-y_{j,p}\right)^{2} \tag{1}$$

To minimize the cost function E(), we must find a minimizer =1,2,...,v Rv where v=LWeightNum (k) and denote that the weight space Rv is equal to the sum of weights in each k layer of the total L layers of the CNN network (WeigtNum(.)). The CNN network has a total of L layers.

Table 2. Images in exercise folders for grapes in each class.

Class Name	Original Images	Images After Growth
Grapes Black Rot	944	1000
Grapes Esca (Black Measles)	1107	1000
Grapes Leaf Blight	861	1000
Grapes Healthy	339	1000

$$\nabla E_i(\omega_i) = \left(\frac{\partial E_i}{\partial \omega_i^1}, \ldots, \frac{\partial E_i}{\partial \omega_i^v}\right) \tag{2}$$

$$\omega_{i+1} = \omega_i - n\nabla E_i(\omega_i) \tag{3}$$

Here, n denotes the rate of learning. N has been chosen using the Shuffled Froggy Jump Algorithm (SFLA), which is detailed in detail here:

Shuffled Frog-leaping Algorithm (SFLA)

Another organically oriented meta heuristic approach was initially industrialized by Eusuff and Lansey (2016) and is based on the second clustering approach named SFLA. The SFLA principally developed as a population-based meta-heuristic capable of performing a cognizant heuristic search through any mathematical function seeking solutions for combinatorial optimization problems. It can be employed to solve several complicated optimization problems, with the essence of the non-linearity, non-differentiability and multi-modality (Eusuff, 2006). It links the PSO algorithm's benefits to those of the genetically based memetic algorithm. Local searches for each memeplex are carried out in this method, which generates a virtual population of solutions based on frogs (solutions). Within any memeplex, the individual frogs each have their own set of thoughts that can be impacted by those of the other frogs. Memeplexes are formed after a predetermined number of cycles of memetic development. As long as the defined convergence requirements are met, the local search and mixing procedure will continue. (Khorsandi et al., 2011). The SFLA can be summarized in the following Table 3:

RESULTS AND DISCUSSION

The experiments and its discussion are provided in this section, where the proposed model is implemented with Python language. Initially, the performance metrics are discussed as follows:

Performance Metrics

The efficiency of the presented method is measured in terms of specificity, sensitivity, and accuracy as mentioned in Table 4. The confusion matrix for the terms TP, TN, FP, and FN derived from the predicted outcome different measures, where TP as the true positives, TN as the true negatives, FP as the false positives, and FN as the false negatives. Table 4 displays the formulas for different parametric metrics used in this study.

Proposed System Evaluation

Table 5 shows the comparison between the existing techniques with the proposed method with the selected parameters (Accuracy, Sensitivity, Specificity, F-Measure and FRP etc.) without optimization.

Table 3. SFLA procedure

Step	Details
1	**Initialization:** Choice m and n, where m represents the amount of memeplexes and n denotes the amount of frogs in every memeplex. Thus, the total sample size, *F* would be *F=m•n*.
2	**Population Generation:** An early population of F frogs is created random uniformly. For a S- dimensional problem, each frog *i* is signified by *S* variables, such as *Fi=(fi1,fi2,…,fiS)*.
3	**Sorting and Distribution:** The frogs are first classified according to their fitness values in descending order, after which the entire population is grouped into memeplex, each with n frogs. In this step, the first frog is shifted to the first memeplex, the second frog is shifted to the second, frogm shifts to memeplex.
4	**Memeplex Evolution:** This phase is based on a local quest, in which the FR and FW are respectively found in each memeplex. The frog is identified separately as Fg with global best fitness. The Eq. (4) and (5) are then implemented to increase frog health to the worst possible exercise. $D_i = rand() \times (F_g - F_w)$ ……… (4) $F_w^{i+1} = F_w^i + D_i, \left(D_{min} \le D_i \le D_{max}\right)$ ……… (5) Where *rand()* is random number among zero and 1, *Dmin* and *Dmax* are the minimum and the maximum displacement allowed in a frog's position respectively. If this process generates a better solution, it substitutes the worst frog; else, if Eq. (4) and (5) do not increase the worst solution, *Fb* of Eq. (4) is replaced with Fg adjusting to Eq. (6): $D_i = rand() \times (F_g - F_w)$ ……… (6)
5	**Shuffling:** The new population is classified according to their fitness in descending order after a predefining sum of steps to memeplex evolution all the memeplex frogs are collected.
6	**Stop Condition:** The algorithm stops when a global solution or a fixed iteration sum has reached. Move to stage (2) otherwise and repeat.

Table 6 shows the comparison between the existing techniques with the proposed method with the selected parameters (Accuracy, Sensitivity, Specificity, F-Measure and FRP etc.) with optimization.

The below Table 7 shows the various optimization evaluations with the proposed CNN technique.

In addition, various set of experiments are conducted, i.e. 50% of training data – 50% of testing data for accuracy, which is shown in Figure 3. Then, the ratio of 70-30 (i.e. 70% of training data and 30% of testing data) experiments are conducted and provided in Figure 4 in terms of accuracy.

From the above figures, it is clearly stated that the proposed Alex Net achieved better performance for various set of experiments. But the LesNet-5 provides poor performance even for the ratio of 50-50

Table 4. Formula for various parameters

Quality Parameter	Formula
Accuracy	$\frac{TP+TN}{TP+TN+FP+FN}$
Sensitivity	$\frac{TP}{TP+FN}$
Spccificity	$\frac{TN}{TN+FP}$

Table 5. Without optimization

Network Type	ACC	SEN	SPEC	F-MEASURE	FPR
DNN	0.78	0.46	0.99	0.63	0.62
RNN	0.85	0.56	0.99	0.70	0.58
LSTM	0.82	0.57	0.99	0.78	0.52
Auto-Encoder	0.95	0.89	0.99	0.90	0.46
GAN	0.94	0.89	0.99	0.90	0.46
CNN	**0.96**	**0.96**	**0.99**	**0.96**	**0.46**

Table 6. With optimization

Network Type	ACC	SEN	SPEC	F-MEASURE	FPR
DNN	0.84	0.55	0.99	0.70	0.56
RNN	0.88	0.62	0.99	0.76	0.53
LSTM	0.87	0.65	0.99	0.78	0.50
Auto-Encoder	0.93	0.75	0.995	0.85	0.49
GAN	0.92	0.79	0.99	0.89	0.48
CNN	**0.98**	**0.96**	**0.99**	**0.97**	**0.46**

experiments. Finally, the comparison of proposed classifier with existing recent technique proves that the disease prediction is effectively carried out by proposed ALexNet with SFLA.

CONCLUSION

CNNs have been used to classify illness in grape leaves in this study. To identify the sickness is the primary goal of the proposed method, according to the author. As a result, four illnesses that had a greater impact on grape production were used to try out the method proposed in this paper. Anthracnose, leaf blight, and black rot are all examples of these diseases. Classification and detection of sick regions in leaves are accomplished through the use of a feature map, which extracts the image's key features. It is the unified framework of feature map development, learning weight optimization, object recognition

Table 7. Various optimization evaluations with proposed CNN

Method	ACC (%)
GA	86
PSO	79
Whale	79
BFO	90
SFLA	**98**

and classification that is innovative in this work. Using a dataset with a 98% accuracy rate, the experiment shows that the proposed method is effective. An efficient, simple, and accurate model has been

Figure 3. Comparison of Accuracy (50:50).

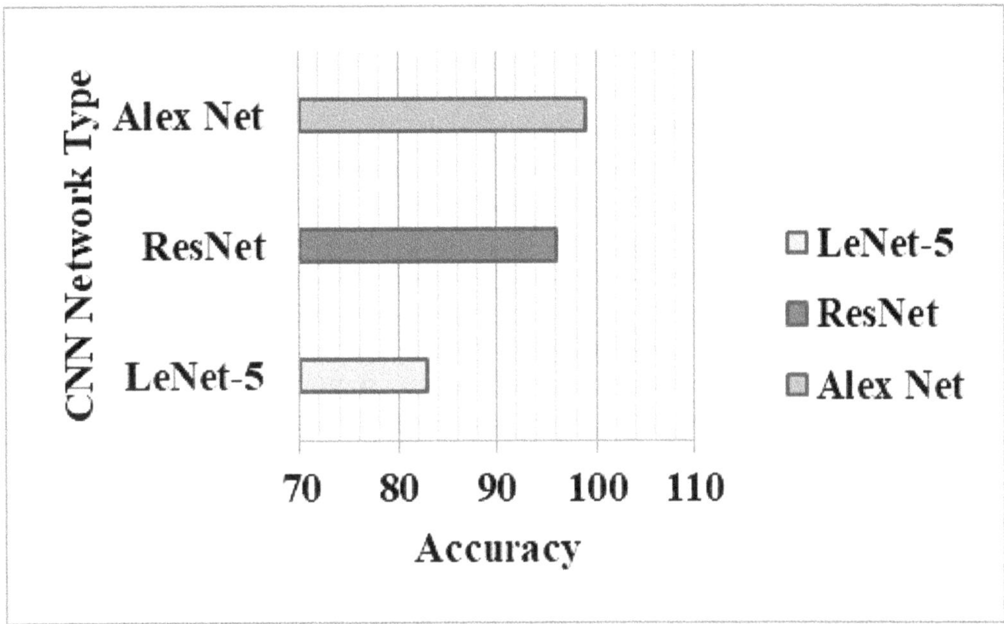

Figure 4. Comparison of Accuracy (70:30).

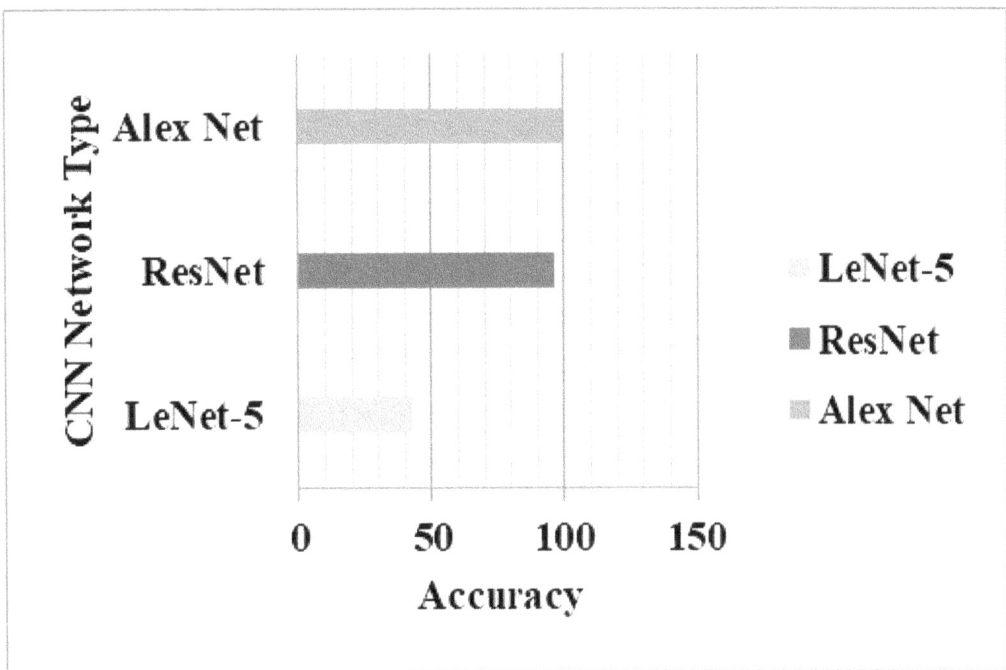

described that might be used in place of on-site identification by professionals. For future research, a model has been proposed to determine the farm's current fertilizer needs in order to avoid wasteful use of fertilizers and to lower the cost of fertilizers.

REFERENCES

Babbar, N., Kumar, A., & Verma, V. K. (2022). *Crop management: Wheat yield prediction and disease detection using an intelligent predictive algorithms and metrological parameters. Deep Learning for Sustainable Agriculture*. Academic Press.

Bakar, Z. A., Tahir, N. M., & Yassin, I. M. (2010, May). Classification of parkinson's disease based on multilayer perceptrons neural network. In *6th International Colloquium on Signal Processing & its Applications* (pp. 1-4). IEEE.

Barbedo, J. G. A. (2016). Expert systems applied to plant disease diagnosis: Survey and critical view. *IEEE Latin America Transactions*, *14*(4), 1910–1922. doi:10.1109/TLA.2016.7483534

Bloice, M. D., Stocker, C., & Holzinger, A. (2017). Augmentor: an image augmentation library for machine learning. *arXiv preprint arXiv:1708.04680.*

Castelli, M. (2018). Supervised Learning: Classification, Reference Module in Life Sciences. *Informatica, 31.*

Chen, L., Zhang, H., Xiao, J., Nie, L., Shao, J., Liu, W., & Chua, T. S. (2017). Sca-cnn: Spatial and channel-wise attention in convolutional networks for image captioning. In *Proceedings of the conference on computer vision and pattern recognition* (pp. 5659-5667). IEEE.

Chu, X., Yang, W., Ouyang, W., Ma, C., Yuille, A., & Wang, X. (2017). Multicontext attention for human pose estimation. In *Conference on Computer Vision and Pattern Recognition,* 5669–5678. IEEE .

Durmus¸, H., Gunes¸, E. O., & Kirci, M. (2017). Disease detection on the ¨ leaves of the tomato plants by using deep learning. In *6th International Conference on Agro-Geoinformatics*, 1–5.

Ebihara, Y., Uematsu, S., & Nomiya, S. (2010). Control of verticillium dahlia at a strawberry nursery by paddy-upland rotation. *Journal of General Plant Pathology*, *76*(1), 7–20. doi:10.100710327-009-0205-x

Eusuff, H., Lansey, M., & Pasha, F. (2006). Shuffled frog-leaping algorithm: A memetic meta-heuristic for discrete optimization. *Engineering Optimization*, *38*(2), 129–154.

Fujita, E., Kawasaki, Y., Uga, H., Kagiwada, S., & Iyatomi, H. (2016). Basic investigation on a robust and practical plant diagnostic system. In *15th International Conference on Machine Learning and Applications (ICMLA)*, 989–992. IEEE. 10.1109/ICMLA.2016.0178

Harris, D. C. (1990). Control of verticillium wilt and other soil-borne diseases of strawberry in britain by chemical soil disinfestation. *Journal of Horticultural Science*, *65*(4), 401–408. doi:10.1080/00221589.1990.11516072

Hughes, D., & Salathé, M. (2015). An open access repository of images on plant health to enable the development of mobile disease diagnostics. *arXiv preprint arXiv:1511.08060.*

Jayaraman, P. P., Yavari, A., Georgakopoulos, D., Morshed, A., & Zaslavsky, A. (2016). Internet of things platform for smart farming: Experiences and lessons learnt. *Sensors (Basel), 16*(11), 1884. doi:10.339016111884 PMID:27834862

Kamilaris, A., Gao, F., Prenafeta-Boldu, F. X., & Ali, M. I. (2016, December). Agri-IoT: A semantic framework for Internet of Things-enabled smart farming applications. In *2016 IEEE 3rd world forum on internet of things (WF-IoT)* (pp. 442-447). IEEE. 10.1109/WF-IoT.2016.7845467

Khorsandi, A., Alimardani, A., Vahidi, B., & Hosseinian, S. H. (2011). Hybrid shuffled frog leaping algorithm and nelder–mead simplex search for optimal reactive power dispatch. *IET Generation, Transmission & Distribution, 5*(2), 249–256.

Ma, J., Du, K., Zheng, F., Zhang, L., Gong, Z., & Sun, Z. (2018). A recognition method for cucumber diseases using leaf symptom images based on deep convolutional neural network. *Computers and Electronics in Agriculture, 154*, 18–24.

Mahdavinejad, M. S., Rezvan, M., Barekatain, M., Adibi, P., Barnaghi, P., & Sheth, A. P. (2018). Machine learning for Internet of Things data analysis: A survey. *Digital Communications and Networks, 4*(3), 161–175. doi:10.1016/j.dcan.2017.10.002

Mckinley, R., & Talboys, P. W. (1979). Effects of Pratylenchus penetrans on development of strawberry wilt caused by Verticillium dahliae. *Annals of Applied Biology, 92*(3), 347–357. doi:10.1111/j.1744-7348.1979.tb03884.x

Medela, A., Cendón, B., Gonzalez, L., Crespo, R., & Nevares, I. (2013, July). IoT multiplatform networking to monitor and control wineries and vineyards. In Future Network & Mobile Summit (pp. 1-10). IEEE.

Nagasubramanian, G., Sakthivel, R. K., Patan, R., Sankayya, M., Daneshmand, M., & Gandomi, A. H. (2021). Ensemble classification and IoT-based pattern recognition for crop disease monitoring system. *IEEE Internet of Things Journal, 8*(16), 12847–12854.

Perez, L., & Wang, J. (2017). The effectiveness of data augmentation in image classification using deep learning. *arXiv preprint arXiv:1712.04621.*

Popa, M., Prostean, O., & Popa, A.S. (2019). Machine Learning Approach for Agricultural IoT In Proc. *International Journal of Recent Technology and Engineering (IJRTE)*, 22-29.

Sangeetha, B. Parvathi, N. K., Ambresh P., Ambalgi, S. L., Abdul, H., Thilagam, K., & Vijayakumar. P. (2022). IOT based smart irrigation management system for environmental sustainability in India. Sustainable Energy Technologies and Assessments, 52, 101973.

Tetila, E. C., Machado, B. B., Menezes, G. K., Da Silva Oliveira, A., Alvarez, M., Amorim, W. P., De Souza Belete, N. A., Da Silva, G. G., & Pistori, H. (2020). Automatic recognition of soybean leaf diseases using uav images and deep convolutional neural networks. *IEEE Geoscience and Remote Sensing Letters, 17*(5), 903–907.

Wang, F., Jiang, M., Qian, C., Yang, S., Li, C., Zhang, H., & Tang, X. (2017). Residual attention network for image classification. In *Proceedings of the IEEE conference on computer vision and pattern recognition* (pp. 3156-3164).

Xu, D., Ouyang, W., Ricci, E., Wang, X., & Sebe, N. (2017). Learning crossmodal deep representations for robust pedestrian detection. In *Conference on Computer Vision and Pattern Recognition*, 4236–4244. IEEE.

Xu, D., Ouyang, W., Wang, X., & Sebe, N. (2018). PADNet: Multi-tasks guided prediction-and-distillation network for simultaneous depth estimation and scene parsing. In */CVF Conference on Computer Vision and Pattern Recognition*, 675–684. IEEE.

Yu, H., & Son, C. (2020). Leaf spot attention network for apple leaf disease identification In *Conference on Computer Vision and Pattern Recognition Workshops (CVPRW)*, 229–237. IEEE .

Zhang, X., Qiao, Y., Meng, F., Fan, C., & Zhang, M. (2018c). Identification of maize leaf diseases using improved deep convolutional neural networks. *IEEE Access: Practical Innovations, Open Solutions*, *6*, 30370–30377.

Zhang, X., Wang, T., Qi, J., Lu, H., & Wang, G. (2018a). Progressive attention guided recurrent network for salient object detection. In *CVF Conference on Computer Vision and Pattern Recognition*, 714–722. IEEE.

Zhang, X., Yuan, Y., & Wang, Q. (2018b). Roi-wise reverse reweighting network for road marking detection in *British Machine Vision Conference*, United kingdom, 1–12.

Zhao, H., Zhang, Y., Liu, S., Shi, J., Loy, C. C., Lin, D., & Jia, J. (2018). Psanet: Point-wise spatial attention network for scene parsing. In *Proceedings of the European conference on computer vision (ECCV)* (pp. 267-283).

Chapter 6
Automatic Seed Dispensing System Using Drone Technology

Jayapadmini Kanchan
Sahyadri College of Engineering and Management, India

Ganaraj K.
Sahyadri College of Engineering and Management, India

Madhura N. Hegde
Sahyadri College of Engineering and Management, India

Shwetha S. Shetty
Sahyadri College of Engineering and Management, India

ABSTRACT

The barren land detection management using manual methods has proven to be of great difficulty in terms of time and workforce. Also, several other methods such as vibration-based models, etc. have been costly to be implemented, as well as their questionable accuracy is along with their limitations. This system aims at designing and developing a drone that mainly focuses on addressing the issue of deforestation with the integration of digital intelligence and drone technology.

INTRODUCTION

Deforestation, the loss of wild woodland habitats due to human interest, has grown into a worldwide hassle as the call for timber climbs. Blended, these 4 problems affect not only the handiest wild flora and animals, but people as well. Shrinking forests can motive wide-accomplishing problems, which include soil erosion, water cycle disruption, greenhouse fuel emissions, and loss of biodiversity

There are various results of deforestation:

1. Modifications in soil:
    a. Breakdown of tree leaves results in loss of nutrients derived from the soil.

DOI: 10.4018/978-1-6684-6413-7.ch006

Copyright © 2023, IGI Global. Copying or distributing in print or electronic forms without written permission of IGI Global is prohibited.

   b. Elevated soil erosion through wind and rain.
   c. Elevated flooding due to a loss of water absorption with the aid of the usage of timber.
   d. Decreased capacity to help distinctive flowers because of nutrient loss.
2. Lack of biodiversity:
   a. Animals and vegetation that are not able to develop outside of a wooded location environment face extinction.
3. Climate trade:
   a. Stepped forward greenhouse gases (together with carbon dioxide) present inside the air, while trees aren't available to take in them.
   b. Accelerated carbon within the air from the burning of trees.
   c. Reduced cooling of land from water evaporation in forests.
4. Impacts on people:
   a. decreased availability of timber for cooking and heating. Losses.

The Deforestation May Have the Subsequent Consequences

It's going to result in scarcity of timber and other food merchandise: the wooden received from wooded area timber is used as gasoline, for making doorways, windows, and furnishings, making paper. It will cause a scarcity of timber for these types of cause. Wooded areas supply us useful products which include honey, gum, wax, catechu and many others.

It will cause a boom in temperature of earth's atmosphere main to worldwide warming: timber use carbon dioxide gasoline from the environment for the technique of meals making referred to as photosynthesis. While a variety of bushes are cut down for the duration of deforestation, then lesser number of bushes might be left. The lesser wide variety of timber will deplete much less carbon dioxide due to which carbon dioxide gasoline trap the solar warmness rays contemplated via the earth. Trapping of warmth rays by way of carbon dioxide increases the temperature of earth's surroundings. This will result in worldwide warming. The gradual boom inside the over-all temperature of earth's surroundings due to green residence impact due to the elevated degree of carbon dioxide inside the surroundings is referred to as worldwide warming.

Deforestation Will Reason Soil Erosion Making Soil Infertile and Result in Desertification

The roots of bushes bind the particles of topsoil collectively. Due to this binding of soil particles, the wind and water are not able to bring away the pinnacle soil due to which the topsoil does no longer come to be unfastened speedy.

Whilst the forests are reduce down at some stage in deforestation, there aren't any roots of timber that could bind the soil particle together and save you them from being carried away by robust winds or flowing rain water. Seeing that there's no tree cowl on the soil to melt the impact of heavy rain, the naked topsoil becomes unfastened speedy by means of the force of falling rainwater and erode hastily.

The topsoil of earth is most fertile. The removal of pinnacle layer of soil in the course of soil erosion, exposes the decrease, difficult and rocky layer of soil. This decrease layer of soil has much less humus and is less fertile. Plant life do not develop well on this less fertile soil. The fertile land gets converted into a desolate tract.

Deforestation Influences the Water Cycle Driving to Diminish in the Rainfall

The diminish in precipitation brings down the groundwater level and may cause droughts: The woodland trees put a parcel of groundwater, sucked through their roots, into the climate as water vapor by the method of transpiration. This water vapor makes a difference in bringing rain in that area. When the woodland trees are cut, then the lesser number of trees put less water vapor into climate through trough transpiration. Since less water vapor is put into the atmosphere, there is less rainfall in that area. When there's less precipitation in an area, then less water permeates into the ground. Due to this, groundwater level too gets lowered. The deficiency of surface water and groundwater due to diligent moo precipitation in an range can lead to droughts.

Deforestation Leads to Termination of Numerous Wild Creatures and Plants

Forests are the common territories of numerous animals, birds and plants. When timberland trees are cut down, the common territory of wild creatures.

Seed Balls: An Overview

Exceptionally plant you understand, begun some region, either as a reducing or as a seed or other frame. Seeds are the final item of a plant and are the starting of existence for every other. Their mode of dissemination shifts, with some being bodily planted, with others being relocated with the aid of humans or other fowls and creatures, and others being scattered off by way of air. Seed balls are exceptionally made to be scattered off in a bit of a controlled way and permit existence to unused vegetation. This newsletter talks about seed balls in element, specifically expertise what seed balls are, how to create them, how they work, what they are applied for, and multiple records about them. Seed balls or seed bombs are seeds that have been wrapped in soil matter and after that dried. The soil is often a mix of clay and compost. The seed is, consequently, pre-planted and may be sown by using storing the seed ball anywhere so long as the species can live to tell the tale there. The wrapping combination continues the seed comfy until it can legitimately germinate.

Seedballs are arranged with clay, supplements, and seeds of wanted plants. Damp clay obtained from riverbed is perfect which is dried in daylight, another the seed ball fabric is included and given a shape of a marble. Cocopeat, humus, compostor bovine fertilizer are utilized as supplements which work both as perforators and water retainers. It is best to utilize a blend of diverse seeds within the same ball, to boost the abdicate of the tree planting ophe little estimate of seed bombs makes them significantly flexible. You'll be able fair hurl them over wall, toss them on surrendered areas, or put them in your backyard. Be it nearby strolling zones or bicycle stopping; you'll be able toss seed bombs and disregard all about it.

Machine Learning and Image Processing

The Google Cloud Stage gives world course speed and exactness through frameworks that can utilized by essentially calling APIs. With these and a have of other APIs, GCP contains a device for close to any machine learning job. In this basic journey, you'll get hands-on hone with machine learning because it applies to picture preparing by taking labs that will empower you to name pictures, identify faccs and points of interest, as well as extricate, analyse and translate content from inside images. In conventional programming, a computer design composes a arrangement of bearings that taught a computer how to convert input information into a craved yield. Informational are generally based on an IF-THEN structure: Machine learning, on the other hand, is a mechanized handle that empowers machines to fathom issues with small or no human input, and take activities based on past observations.

Image Processing techniques leverage device mastering models like Deep Neural Networks to transform images on a ramification of tasks, like applying artistic filters, tuning an image for choicest excellent, or enhancing unique image info to maximise first-rate for pc imaginative and prescient obligations.

The intention of image processing is to beautify or compress the image records while in gadget gaining knowledge of, it is used to optimize differentiable parameters in order that a certain loss or fee characteristic is minimized. So, mixture of those has brought about a higher conception about popularity and processing of pictures. There are many fields and makes use of where, frameworks that examine photographs ought to have a good deal benefit. From high-tech uses to regions like agriculture, photo reputation etc., frameworks advantage the community and improve excellent of existence.

An Overview of Drone Technology

Drones can be organized with sensors, counting ultrasonic, laser, or lidar take-away sensors, time-of-flight sensors, chemical sensors, and stabilization and introduction sensors. Visual sensors provide still and video facts. Ruddy, green and blue sensors gather widespread visible ruddy, green and blue wavelengths, and multispectral sensors accumulate unmistakable and nonvisible wavelengths, which include infrared and vibrant. Accelerometers, whirligigs, magnetometers, indicators and GPS are furthermore commonplace ramble highlights.

Drones are referred to as "unmanned aerial motors" and to normal people they're a weapon or a device utilized by army human beings. However now even normal human beings can use them. For purposes like- videography, surveillance, traffic monitoring, climate monitoring, agriculture etc

From look and protect operations to overseeing catastrophe zones, ramble innovation is being used increasingly for open security operations. One range where rambles are making a critical affect is in look and protect missions. Rambles are cheaper and more adaptable than utilizing kept an eye on airplane and come with numerous benefits for look and protect missions such as being sent speedier, being able to move into contract regions, being prepared with proficient detecting gear and being simple to function.

RELATED WORK

Andrio (2019) proposed that UAV is a transportation platform that capable of convey substances from 1 vicinity to different area. With the aid of calculating the UAV payload and articulated that into the development of canister, a machine which can bring shipping and drop seeds has been evolved. This work includes following steps. First, a drone takes an aerial photograph of the demonstration plots to spotlight existing flowers and obstruction. The end result is the data this is used to broaden the pleasant seed dropping and planting sample. The map then is uploaded into the drone. Next step is growing mechanized canisters containing a germinated seed. The seeds then released from canisters set up in drone and destroy open upon impact with the soil to allow the germinated seeds to take root inside the predetermined positions.

The manner of applying the seeds is managed by way of the comments from the wi-fi sensors network deployed at seed box. The system of applying the seeds is controlled by using the comments from the wireless sensors community deployed at seed container. This Bot can perform fundamental basic capabilities like harvesting, planting and spray the pesticides. The utility of agricultural equipment in precision agriculture has skilled a growth in funding and research due to the use of robotics packages in the equipment layout and venture executions (Diwate, Nitnaware, and Argulwar, 2018)

Singh (2019) proposed that a sequence of elements that can be traversed (as many times as you want) and can be accessed by an indexed subscript.

Fortes (2017) proposed that the shelling out of seeds right into a soil surface is delegated through the GPS coordinates as soon as generated, previously generated, or by way of manual control of the operator. The seed dispersal mechanism is a box for storing seeds and a motor-controlled mechanism for releasing them. This motor device is designed to particularly adjust the flow at which the seeds are launched. The seed dispersing software strolling on the Arduino drives the motor proportionally to the velocity of the aircraft in order that the seeds are precisely launched at the favoured location. The seed container is built to in shape the quadcopter and be long-lasting in case of collisions.

Latha et al. (2014) proposed a method by means which we can detect weed by means of the use of photograph processing. Then, the input of the weed blocks to the automated sprayer which sprays the handiest in those blocks. In this paper, the important goal is to discover the weed within the crop by the usage of picture processing. Then we are able to supply the inputs of the weed regions to an automated spray pesticide only in those areas. For this, we need to take a picture of the sector with proper clarity to stumble on the weeds with greater accuracy. Taking an image can be completed with the aid of attaching a digicam to a tractor or taking them manually. Then we will apply photograph processing to that photograph using MATLAB to locate the weed.

Abdullahi, Sheriff, and Mahieddine (2017) proposed a new method for classifying and recognizing the fitness state of the plantation and immediately generating treatment solutions on the field RS statistics and sample recognition approach turned into used to estimate direct and independent crop location in the take a look at location [3]. In this overview, the authors reviewed the distinctive strategies for crop stock in the Indian situation. Optical and microwave data are used to categorize the crop. Chlorophyll and water had been represented by using optical facts, and crop geometry and dielectric properties had been characterized by microwave.

Saha et al. (2018) proposed that the Raspberry pi 3B model can be used to integrate the solutions and other related highlights into the drone, Unmanned Aerial cars are getting increasingly more famous to fulfil the demands of improved population and agriculture. Drones prepared with suitable cameras, sensors, and integrating modules will help in reaching smooth, efficient, precision agriculture. The proposed answers associated with these drones, if incorporated with numerous ML and internet of things ideas, can help in increasing the scope of additional development. In this paper, the related work in this area has been highlighted alongside proposed answers.

Dickson, Ni, and Han (2004) proposed that the vision guidance gadget uses an okay-means Clustering set of rules in photo processing to distinguish between crop and non-crop functions. An automatic vision steering gadget for an agricultural automobile is disclosed and defined. The vision steerage machine uses an ok-means clustering algorithm in picture processing to distinguish between crop and non-crop functions. The imaginative and prescient steering device utilizes moment algorithms to decide the area and orientation of crop rows, from which desired wheel angles are determined and steerage is commanded. The imaginative and prescient guidance device may additionally regulate the place and orientation of visually decided crop rows in line with a predetermined distance among crop rows. Similarly, the vision steerage gadget may also make use of a redundant wide variety of regions of hobby in figuring out crop row locations and orientations.

A paper by Cheng, Peng, and Liu (2013) proposes clustering evaluation technique is one of the most important analytical techniques in information mining; the technique of clustering a set of rules will affect the clustering outcomes immediately. This paper discusses the usual ok-approach clustering algorithm and analyzes the shortcomings of the usual k-manner set of rules, which includes the ok-way clustering set of rules has to calculate the gap between every information object and all cluster facilities in every new release, which makes the performance of clustering not excessive. This paper proposes a progressed ok-manner set of rules if you want to solve this question, requiring an easy statistics shape to keep some data in every generation, that's for use within the subsequent generation. The improved method avoids computing the gap of each data item to the cluster centers repeatedly, saving going for walks time. Experimental effects display that the advanced technology can efficiently enhance the rate of clustering and accuracy, decreasing the computational complexity of the k-approach.

Jawahar and Umarani (2019) proposed that a mixture of a hydrophilic polymer, growth hormone, pigment, and stabilizer has been proven to enhance seed germination. A series of experiments have conducted the usage of Thespesia populnea seeds to standardize the seed coating method, humid priming approach, and cube media composition. The effects revealed that seeds of T. populnea coated with 4 g kg1 Shidocoat (a gibberellic acid-based totally seed coating method developed with the aid of Tamil Nadu Agricultural college and marketed via Vriksha Agro Ventures, India) followed through humid priming (24 hours soaking + three days humid incubation) and embedded in cube media (seven-hundred g soil, one hundred thirty g sawdust, 80 g bone meal, forty g vermicompost, and 50 g arbuscular mycorrhizal fungi) recorded a growth of forty-six. Zero lengths, 23.1media. Due to its considerably superior overall performance, it is encouraged that the stepped-forward seed cubes of T.

Vibhute and Bodhe (2012) proposed that the picture processing method has been proved as a powerful device imaginative, and prescient machine for the agriculture area. Imaging techniques with extraordinary spectrums such as Infrared, hyperspectral imaging, and X-ray had been beneficial in determining the plant's indices.

Radhakrishna (n.d.) proposed the composition of seed balls makes itself sustainable and favourable for germination in maximum environments. The composition of seed balls makes them self-sustainable and favourable for germination in most environments. Making seed balls is a laugh and smooth. The concept of seed balls changed into began through Masanobu Fukuoka, a Japanese guy well-known for popularizing natural farming, the concept has been adopted internationally. Seed balls also can be used to "over-seed" current ecosystems, without detrimental to the soil structure — or to seed efficient flora into forested regions and steep hillsides where tillage isn't possible. Seed balls also can be used in combination with animals including pigs who will do the work of shuffling the mulch around imparting seed balls greater cover

SYSTEM DESIGN AND IMPLEMENTATION

The figure below represents the abstract architecture of Seed Ball Dispensing System.

Collect the database that is in the form of images obtained by smartphone. Then process the database, after which we train the model using YOLO Darknet. Then this trained model is tested with the test

Figure 1. Architecture diagram of Seed Ball Dispensing System

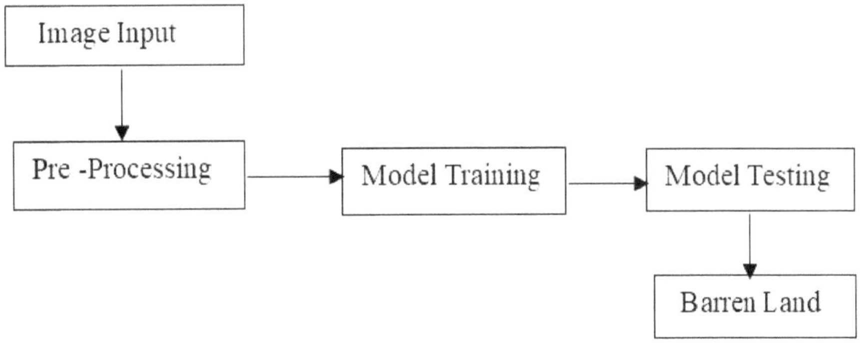

images where it detects the barren land and once the barren land is detected the seed balls are dispensed from seed dispensing system.

Implementation

Item detection is one of the conventional laptop vision troubles in which you attempt to decide what and wherein basically, what objects are within a given image, and wherein they may be inside the image. The item detection trouble is greater complex than labelling, which can also become aware of gadgets but does now not indicate in which the item is placed within the image. The set of rules applies the entire image to a single neural network, then divides the image into regions and predicts bounding boxes and possibilities for every area. Such bounding packing containers are weighted through the possibilities anticipated.

Data Collection

A deep learning algorithm detects the damages of the road. To make a model that detects cracks and potholes. We obtain the database from collecting smartphone images of barren land. It comprises pictures of barren land. These datasets are stored in a file.

Data Pre-processing

To begin with, the images from the data set is read and eventually converted into gray-scale images and then both the initial and the images converted to grey scale are displayed. We draw bounding boxes for all the images in the region of the barren land and also obtain its coordinate values. Then the data of these bounding box values is gathered in a .txt file for each of the image. This is known as annotation.

Data Augmentation

As the existing number of images weren't sufficient for our studies, we alleviated this problem by generating more input samples from the available images. This process is termed as data augmentation. This

Figure 2. Pseudocode for gray scaling

```
Pseudocode for grayscaling

    begin
    Read the image.
    convert to grey scale image.
    Display the initial and grey scale image.
    end

Psuedocode for bounding box values

    begin
    Retrieve the bounding box values which is classified based on the predefined classes.
    end
```

was achieved by altering multiple properties of the images such as resize, crop, flip using 'torchvisions. transformations' of PyTorch library. This aided the problem of lack of training samples.

Making Dataset Iterable

The training data is read from the database on demand by the model. That is, a pair of images is read for every instance passing through the model. For this, we need to keep the dataset iterable. This just means that the model can iterate though the dataset, fetching the required data for training. This is an advantage of using PyTorch where instead of reading all the data to the memory on start, we load them on fly during run time. This type of datasets is particularly suitable for cases where random reads are expensive, where the batch size depends on the fetched data. When using an iterable dataset with multi-process data loading.

Train Validation Split

The proposed model is trained using train data. But just by training on the entire dataset, we cannot confirm that the model is accurate and it will perform well on new examples. So, it is necessary to confirm the model performance on unseen set of data. To achieve this, keep a part of the data separate, specifically for testing purpose. In this application, we use 80 percent of data for training and the rest is used for testing. Once to achieve good accuracy over testing data, it can stop training and use the trained

Figure 3. Pseudocode for recursive training

```
Pseudocode for recursive training

begin
Collect the bounding box values and image
for each image do
    Recursive Training using bounding box values.
    Then the damage is detected.
end for
end
```

model for deployment. separate data into train, validation, and test splits to prevent your model from over fitting and to accurately evaluate your model.

System Instantiation

YOLO for Object Detection

The image is broken down into a SxS grid to perform the detection. Each of the cells must predict potential" bounding boxes" of N and the degree of certainty (or probability) of each of them, which means that boxes of SxSxN are determined. The vast majority of these boxes will have a very low probability, that's why the algorithm proceeds to delete the boxes that are below a certain threshold of minimum probability. A YOLO training collection consists of a series of images, each with a text file showing the coordinates and the class of each of the objects in the image. We need to label them and this is called as annotations. It can split the annotations into test and train sets. Until we reach a certain threshold, we train the images with the obtained weight files. After which we test the prediction model with test images. If barren land is found, it shows barren land detected, if not it shows no object.

Training the Model

Train the proposed model with the training dataset. For this, iterate through the dataset, obtain the weight values until a threshold is reached, also when lowest error is obtained.

Figure 4. Drawing Bounding Boxes using labelling

RESULTS AND DISCUSSIONS

Drone is controlled by an Arduino UNO model, for which we have programmed using Arduino IDE for setup, control & calibration. This program is written using both C# and Python language. Integrated with the Gyro model to provide maximum stability. Using labelling tool, bounding boxes were created on desired image and created XML file of our data set, which was further compressed and converted to a zip file consisting of the images and XML files.

Using Goolge Colab and Darknet we trained the data-set in order to generate weight files required to run our main Barren land Detection Program. This process takes 6-8hrs, this is done in order to achieve maximum accuracy. Once the required number of outcomes is achieved we will be able to get newly generated weight file.

Drone Assembly

We were able to Build a drone with all the necessary hardware setup and also tested for circuit stability.

CONCLUSION

The proposed project is developed with an objective to take a snapshot of land as input, process the image and train the system to recognize the pattern between vegetative land and barren land. The output is expected to trigger a mechanism to dispense seeds when barren land detected. Based on the literature survey and related research, Our solution uses YOLO to train a model to identify barren land in the analysed images as well as in real time video distinguishable objects. The findings need to evaluate the use of appropriate quantitative barren land detection and type metrics.

The precision of a seed-liberating operation is still a work in progress. Nevertheless, there may be usually space for enhancements. Make it more surroundings pleasant by putting in transportable and smaller solar panels or rechargeable pads, capable of providing sufficient power assets. Provide thermal sensors that can stumble on feasible forest fires and scan for uncomplicated areas in need of reforestation. On occasion and autonomously patrolling and sending notifications to the floor station in case of excessive temperature trade and consequently possible woodland fires.

Figure 5. Working of SDS Program

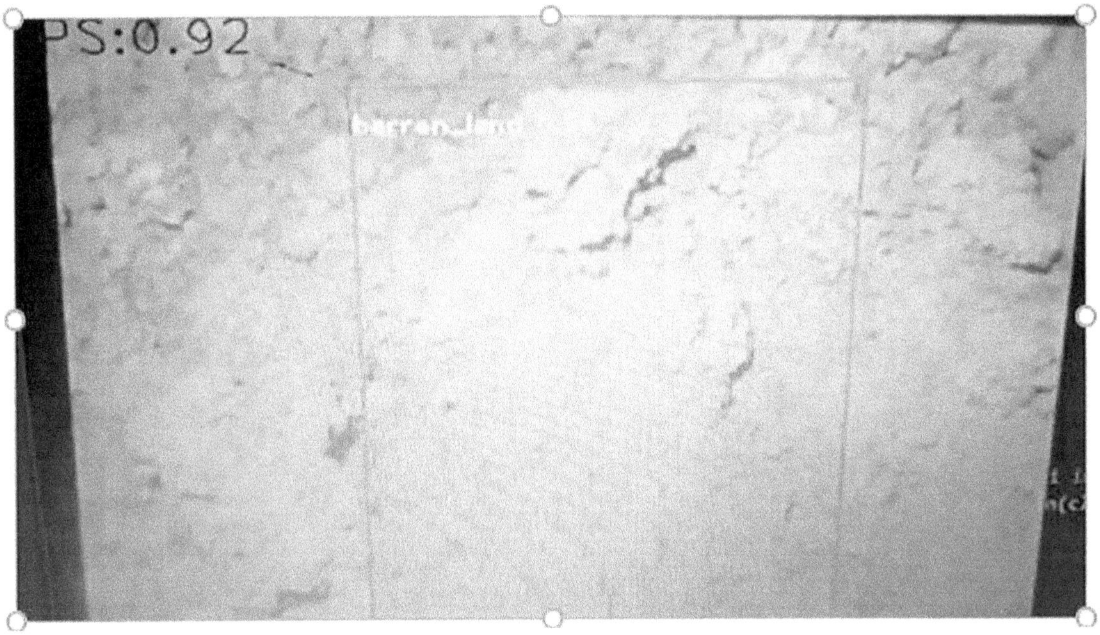

Figure 6. Drone Frame with its landing leg

Figure 7. Drone Setup

REFERENCES

Abdullahi, H. S., Sheriff, R., & Mahieddine, F. (2017). Convolution neural network in precision agriculture for plant image recognition and classification. In *2017 Seventh International Conference on Innovative Computing Technology (INTECH)* (Vol. 10, pp. 256-272). Ieee

Andrio, A. (2019, August). Development of UAV technology in seed dropping for aerial revegetation practices in Indonesia. *IOP Conference Series. Earth and Environmental Science, 308*(1), 012051. doi:10.1088/1755-1315/308/1/012051

Cheng, H., Peng, H., & Liu, S. (2013, March). An improved K-means clustering algorithm in agricultural image segmentation. In PIAGENG: Image Processing and Photonics for Agricultural Engineering (Vol. 8761, pp. 86-90). SPIE.

Dickson, M. A., Ni, B., & Han, S. (2004). *U.S. Patent No. 6,686,951*. Washington, DC: U.S. Patent and Trademark Office.

Diwate, S. K., Nitnaware, V. N., & Argulwar, K. (2018). Design and development of application specific drone machine for seed sowing. *International Research Journal of Engineering and Technology, 5*(05), 4003–4007.

Fortes, E. P. (2017). Seed plant drone for reforestation. *The Graduate Review, 2*(1), 13–26.

Jawahar, R., & Umarani, R. (2019). Development of Seed Cube Technology with Enhanced Seeds of Albizia lebbeck for Rapid Propagation in Fallow Lands. *International Journal of Current Microbiology and Applied Sciences, 8*(6), 1603–1613.

Latha, M., Poojith, A., Reddy, B. A., & Kumar, G. V. (2014). Image processing in agriculture. *International journal of innovative research in electrical, electronics, instrumentation and control engineering, 2*(6).

Radhakrishnan, V. (n.d.). Seed Ball Technique. *Bangalore conference.*

Saha, A. K., Saha, J., Ray, R., Sircar, S., Dutta, S., Chattopadhyay, S. P., & Saha, H. N. (2018, January). IOT-based drone for improvement of crop quality in agricultural field. In *IEEE 8th Annual Computing and Communication Workshop and Conference (CCWC)* (pp. 612-615). IEEE. 10.1109/CCWC.2018.8301662

Singh, H. (2019). *Practical Machine Learning and Image Processing: For Facial Recognition, Object Detection, and Pattern Recognition Using Python*. Apress. doi:10.1007/978-1-4842-4149-3

Vibhute, A., & Bodhe, S. K. (2012). Applications of image processing in agriculture: A survey. *International Journal of Computers and Applications, 52*(2).

Chapter 7
Drone Harvester:
Detect and Collect Ripen Fruit and Vegetables

Ritu Gupta
https://orcid.org/0000-0003-4616-3262
Galgotias University, India

Deepika Kataria
Galgotias University, India

B. S. Tripathi
Indian Institute of Technology, Delhi, India

ABSTRACT

Drone technology brings a revolutionary change in every sector of production, supply, and service. It came as a new revolution in the field of agriculture also. Being a light weight, easy-to-handle, and smart device, drone devices along, with IOT and sensors, help make the smart agriculture to rise; it can also be considered as the agriculture 4.0 revolution. This chapter demonstrates the potential application of drone technology that are currently in use at the countries like US for smart cultivation. Usage of drone technology not only helps to improve the production; it can also help to avoid the wastage of the crops and fruits grown in places that are harder to be accessible by the human being due the geographical conditions. Considering the same problem, the author proposes a smart drone-based harvester that can be used to pluck the fruits and vegetable at high altitude or difficult to reach places. This harvester is also able to measure the ripening condition of the fruits in order to avoid any type of wastage.

INTRODUCTION

The New Era in farming sector is at a landmark, with highly demand worldwide supply and ware costs at an unmatched low due to rising demands for food production and consumption. There is a more noteworthy need than ever before for and agronomists across the globe to further develop asset the

DOI: 10.4018/978-1-6684-6413-7.ch007

Copyright © 2023, IGI Global. Copying or distributing in print or electronic forms without written permission of IGI Global is prohibited.

board because of fixing financial plans, while the "farm to fork" development has seen rising strain for upgraded item recognizability, as buyers become keener on the beginning of the merchandise they buy and how they were developed.

The advancement of new innovations has carried alleviation to ranchers. Throughout recent years, the utilization of robots has expanded, and they are utilized in pretty much every area of the economy. In any case, the utilization of robot innovation in horticulture is blasting. As per a few reports, "the rural robot market is supposed to develop from a $1.2 billion (USD) industry in 2018 to $4.8 billion of every 2024."

Agriculture is going up against various money related hardships concerning effectiveness and cost-ampleness, and the rising work inadequacy deficiently due to the end of nation locales, too. Among these broader faults, it is necessary to acknowledge the population growth in general, urbanism, an obviously polluted environment, the rise in the consumption of animal proteins as a result of human and economic development, and, undoubtedly, ecological change (Puri, Nayyar, and Raja, 2017; Mogili and Deepak, 2018). In addition, strong acknowledgment, exact distinctive confirmation and proper assessment of organisms and various factors impacting both plant and animal prosperity, are essential to be checked to reduce monetary utilizations, trade aggravations and, shockingly, human prosperity possibilities. Hence, a further evolved cultivation ought to be set, portrayed by the gathering of improvised creation cycles, developments and instruments got from sensible advances, inventive work works out.

Recently, drones—often referred to as automated aeronautical vehicles—have been used for agricultural reasons, which is said to be the most recent development in this area. The huge occupation of robots in agriculture is to help confirmation-based orchestrating and assembling remarkable data. Normally meanders were made to be used in military, humane assistance, catastrophe the chiefs, etc. The usage of these robotized vehicles is extending incredibly speedy into the yield creation region. The significant regions where it is being used are: Agriculture, forestry, fisheries, disaster risk reduction, and wildlife conservation.

The drone has a compelling way to deal with manageable horticultural life. This permits agronomists, agrarian architects, and ranchers to smooth out their tasks and gain compelling bits of knowledge into crops. It has previously become piece of enormous scope cultivating.

Drones can enormously affect the plant business, as they can screen a couple of parts of developing that individual can't accomplish in isolation. Robots can perform airborne preparation, giving an obvious image of the hard and fast size of a yield field, as well as showing possibly underutilized areas of land. They can similarly, whenever equipped with the legitimate camera gear, screen the strength of plants to the extent that temperature, chlorophyll levels, new poisons and even leaf thickness. This information can allow a farmer to change the significant limits of their agrarian communication to resolve the issues before they become more far reaching.

The usage of robots in the agribusiness business is reliably creating as a component of a convincing method for managing common sense plant organization that grants agronomists, rustic planners, and farmers to help with streamlining their undertakings, using lively data assessment to secure strong encounters into their harvests. Crop checking, for example, is simplified by using drone data to plan and make constant improvements, similar to the usage of channels and propelling excrement applications definitively. Things can be unequivocally followed from residence to fork including GPS regions for each point in the journey, rather than more standard time and work serious data variety.

Figure 1. Drone based harvesting system

Potential of Drones in Agriculture

A UAS is mostly used to assist farmers in gaining a higher perspective of their yields so they can spot obvious changes that are difficult to spot while inspecting crops up close. These devices' sensors attached to them allow them to construct multispectral images. The photos are combined to create an exceptional reflectance band that can be used to calculate data for the Normalized Difference Vegetation Index (NDVI), Leaf Area Index (LAI), and Photochemical Reflectance Index (PRI). The NDVI looks at various soil biomass levels, water pressure, nutrient requirements, insect invasion, crop disorders, and other factors. The NDVI's information is derived from field research and meticulous calculations. These projections serve as a companion while committing fertiliser to the yields. The fairly identified information facilitates accelerating the most well-known method for coordinating harvest inventories and yield checks. Additionally, the development of UAS can be utilised to report unauthorised exploitation of resources and land.

Improving Agricultural Productivity

Robot development and significant level image data analysis can be especially useful in resolving any complications between ongoing creation and the premium for the future. The robot's imaging data is utilised to assess the quality of the harvests. With the final goal of properly inspecting the yields, the NDVI assessment aids in comprehending the strength of sun-combined radiation. This development aids in the early identification of aggravations and diseases by checking clear-cut plant parts. Advanced geospatial NDVI items are used to identify the occurrence of calamitous events or the loss of yields. The two most important cycles that run throughout the yield generating cycle with the use of sensors are arranging and imaging.

The part is additionally coordinated as follows: Section 2 exhibits the connected work and explores led in this field. Section 3 expounds the agrarian robot innovation and its connected perspectives and sorts of robots. Section 4 makes sense of about the impending innovation "drone-based gatherer for recognizing and gathering ready leafy foods". Followed by Section 5, which makes sense of the outcome and perceptions connected with this innovation. Segment 6 comprises of result conversation and ends. Followed by Section 7, making sense of the Future extent of this innovation.

RELATED WORK

Drone technology introduced in order to make the farming easier and more capable. With the growing involvement of technology in our day-to-day life, people start working in a smart way with help of smart technology and devices. In agriculture field we use traditional bulking machines for farming purposes at a large scale. Introduction of UAV technology in agriculture domain brings a huge revolution in the farming and related processes. Many researchers and technologists worked in this direction; some related studies are discussed in this section.

Puri, Nayyar, and Raja (2017) investigated the significance of drones in Agriculture and has featured the different drones accessible for assorted horticulture applications alongside specialized particulars. It also specifies the application area of a particular drone based on its technical specifications.

Mogili and Deepak (2018) investigated a drone-based crop monitoring system that uses a multispectral camera mounted on a UAV to determine the areas that need pesticide application. In one flight the camera takes pictures and investigates by the geographic marker. In view of the outcomes, it very well may be not difficult to track down the area where to splash the pesticides. The UAV spraying framework automatically navigated using GPS coordinates to spray pesticides on contaminated areas where there was no vegetation indicated by the NDVI.

Budiharto et al. (2019) states that for future cultivating, drones are a fundamental device in accuracy horticulture, as they permit ranchers to screen harvest and domesticated animals' conditions via air.

Joshi et al. (2020) presents a study about the agricultural revolution of Indian cultivation technology through drone-based harvesting. it also states how the use of agri-drone technology can fill the gap between the current and future needs of agricultural production. Supporting environments would work with the development of numerous inventive new businesses giving farming insight involving drones and other arising innovations as a support of country communities.

Rachmawati et al. (2021) proposes a study about the role of image processing in drone-based harvesting, i.e. how images taken by agri-drones can be used for the purpose of monitoring and mapping of data in the farms. With the help of drone technology, agri-drones can be used to assess the health of planted agricultural plants to determine whether they are ready for harvest or still need time to mature.

Shivaji et al. (2017) have provided insight into the use of an agricultural drone for a pre-programmed showering component. The researchers mention the problem of pesticide poisoning and fatalities as their prime concern, for this study. In this essay, the authors propose the measures and precautions to be taken while using the ranchers for the showring components. They also explain the other safety precautions a rancher should follow in order to prevent the harmful effects of pesticides and their treatments; the study also explains how practical innovation may be used to manage farming robots with components like PIC microcontrollers.

Meivel et al. (2017) proposes another study related to the pesticide spraying system using drone technology. The authors in this study proposes the use of a quadcopter UAV for the farming purposes. The proposed model consists of a UAV system with sprayer modules along with multispectral cameras. The cameras are used here to find the green vegetation field and its edges for making the effective pesticide spraying with human intervention. These proposed model can used effectively in hard-to-reach regions.

Korlahalli et al. (n.d.) proposes automated aerial pesticide sprayer, the proposed model takes the measure like load balancer, GPS, speed, remote control etc. into account in order to make a perfect automated device as pesticide sprayer. For this purpose the authors uses different modules mlike FCB (flight controlled load), that works as a load balancer and avoids the conditions of overloading; Brushless DC engine, more effective and efficient to be used in UAVs; ESC(electronic speed controller), to control the running speed of the device; remote handset, casing, propellers, battery, and other factors associated with remote robots.

Nayyar, Nguyen, and Nguyen (2020) states how the robot innovation can be joined with web of things to furnish improved results with higher precision likewise the utilization of IOT with drone innovation assists with further developing adequacy and proficiency of UAV in agrarian applications.

The Green Water Drone (GWD) project (Ren et al., 2020) proposed is a pilot execution of an inventive and independent framework (installed micrometeorological station) that distinguishes a rancher's continuous water system needs through the CWSI to gauge each harvest's water status. Immediate assessment of CWSI relies on direct spatial information away from the field, primarily on crop shelter infrared temperature.

With numerous drones, it is feasible to a great extent lessen the mission time, which gives incredible advantages for drone administration suppliers. In Shivaji et al., (2017) the method of drone fleet deployment for tremendous extension agribusiness and officer administration concentrating on issue was basically penniless down. Likewise, a few critical viewpoints in drone armada sending were figured out in nuances, including fleet of robots, consideration disintegration, pleasant way expecting ordinary and flighty locale, weather patterns effect, correspondence and data taking care of.

AGRICULTURAL DRONE TECHNOLOGY

Crops and agrarian organizations can increment crop creation by saving time and pursuing better essential choices by consolidating drone development. Rural robots had furnished with different kinds and state of the art developments. In this way, infrared sensors and GPS are instances of private, modern, and military UAVs. Distant correspondence frameworks are utilized to screen drones, transfer new orders and explicit rules. Additionally, it had every now and again used to watch out for the different streaming substance from UAV cameras.

The robot and the route framework are the two parts of the robot organization. The identifiers and movement regulators are all in the Aircraft gag. There is no requirement for capacity to help people as the main part of the climate is loaded up with the robot data framework.

Figure 2. Illustrates the architectural diagram of a drone. Essential components of a drone system can be described as follows:

1. **Propellers:** Propellers are key pieces of the robot. They push the air downwards to stay aware of the level during the flight. Propellers similarly assist with moving all over.

Figure 2. Drone -Essential Components Architecture Diagram

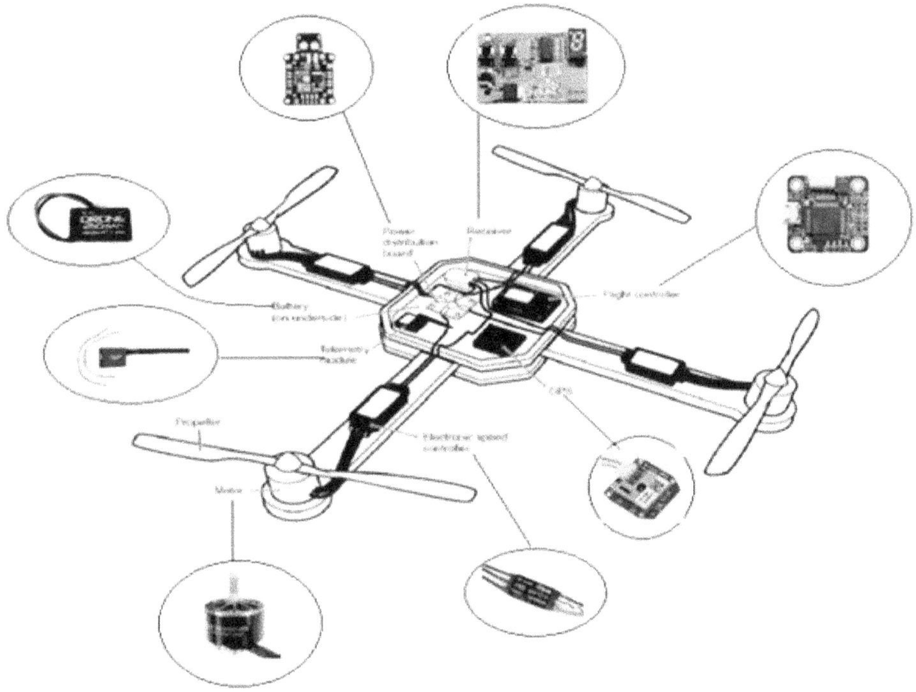

2. **Drone Motor:** Motors are a significant piece of a robot and are basic for the working of the robot, as they keep it in the skies. In current scenario, brushless motors are used.
3. **Electronic Speed Controller:** The electronic speed controller is the circuit to investigate the motor's speed as a matter of fact. It moreover controls the course of the robot and fills in as a strong brake. ESC moreover switches the prompt current over totally to a substitute current to run the brushless motor of the robot.
4. **GPS Module:** GPS Module is utilized to give longitude, scope, and compass routes from a consistent gadget. GPS Module is likewise expected for the waypoint, return to home, and numerous other high level independent flight modes.
5. **Receiver:** The receiver is one of the primary parts, it receives the signs from the distant regulator during the flight, and signals give the guidance to the robot. Generally, a standard RC radio recipient is utilized in the robots.
6. **Battery:** Lithium Polymer (LiPo) batteries are regularly utilized in the robot. It offers extraordinary power, thickness, and solidness in the robot's life.
7. **Telemetry Module:** The telemetry module in drone allows an easy two-way telemetry connection between drone and ground station.
8. **Power Distribution Board:** A drone power distribution board (PDB) is just a printed circuit board (PCB) used to supply power from the source to the different pieces of a robot.
9. **Flight Controller:** Flight controller is a circuit board with a scope of sensors that recognize development of the robot, as well as client orders. Utilizing this information, it then, at that point, controls the speed of the engines to make the specialty move as instructed.

Uses of Drone Technology in Agriculture

Crop Surveillance

Drones are essentially used during the time spent crop perception. It is the oversight of yield progress from the time seeds are planted to the hour of accumulate. The robots are outfitted with remarkable imaging gear called the Normalized Difference Vegetation Index (NDVI). The NDVI shows the plant's prosperity with distinct assortment information. This helps farmers with checking crops as they grow, so any issues can be overseen quickly with the eventual result of saving the plant.

Drone Field Monitoring

Checking field conditions is significant for a solid manor. Yet, it becomes hard for ranchers to screen such a huge area of land. The farming robot screens the soundness of the dirt and field conditions. The robot gives exact field data, remembering rise data that assists with tracking down anomalies for the field.

Planting and Seeding

Ten drones are equipped for plating 4,000,000 trees per day. Computerized drone seeders are for the most part utilized in the ranger service industry at the present time, however they have the potential for more far-reaching use not too far off. With the assistance of robots, one can arrive at the areas without imperilling laborers.

Spray Treatment

Agro-showering in Southeast Asia and South Korea utilizing drones is far and wide. Drone sprayers save laborers from being required to explore the field with knapsack sprayers, which is dangerous to their wellbeing. They convey exceptionally fine shower applications that can be designated to explicit regions to augment effectiveness and save compound expenses. In certain nations, similar to Canada, they are not presently legitimate as need might arise to be finished to comprehend the effect of splash float.

Security

Observing the region, which used to require long stretches of strolling, can be finished shortly with the assistance of a robot. Aside from horticulture, the robot business is additionally developing quickly. It is thought of as incredibly valuable for ranch the board. The robots minute the furthest reaches of a homestead without arriving. These recoveries significant time and takes into account more continuous observing of hard-to-arrive at regions.

The drone camera gives an outline of homestead activities over the course of the day to guarantee that the tasks are chugging along as expected. They are additionally used to find gear, finding absent or harmed creatures in far-away looking regions.

Types of Drones Used in Agriculture

There are modern drones that had specially manufactured for the production and effectiveness of crops.

Rotary Drones

The complete of axles or propellers on a revolving gadget had likewise used to distinguish it. For crop creation, a turning drone is an exceptional spotting machine. Since a robot can set down a plane sideways, field streets and stopping regions can use as landing cushions. Turning drones are easy to zoom around a scene and can drift over pain points.

Fixed Wing Drone

Fixed-wing airplane use lift power to stay upward, and like airplane, most fixed-wing drones just had one rotor. Fixed-wing robots can likewise travel quicker than rotational robots. fixed-wing robots would require an arrival region near an air terminal landing area. They had furnished to drop by steering across the field.

The Benefits of Agricultural-Drones

- As the drones are worked via prepared pilots, there is no way of their abuse.
- They can expand ROI (Return on Investment).
- They work at twofold the speed of human work.
- Drones use ULV (Ultra-low volume) splashing innovation, consequently it saves water and abuse of synthetic substances.
- They are low in cost and require negligible upkeep.
- They help to build ranchers' efficiency.

DRONE BASED HARVESTER FOR DETECTING AND COLLECTING RIPE FRUITS AND VEGETABLES

As per research data by United Nations Food and Agriculture Organization 40% quantity of perishable meals is wasted at source. Currently in most of the agricultural field heavy machine harvesters are used. In most of the cases, these bulky machine harm the agricultural field while harvesting it. Furthermore, there is no automated process to detect ripe fruits and vegetables before plucking it from high altitude trees in large Agricultural/ Terrestrial fields. Also, the traditional systems are bulky mechanism contributing to the air and noise pollution.

A Drone Based Harvester for Detecting and Collecting ripe Fruits and Vegetables, is proposed here that works to pluck the consumable products from high altitude Trees, in areas where human intervention is difficult like mountain-hills, rain forests etc. The system proposed here is a Machine Learning (ML) Based Drone Harvester as shown in fig.1, comprised of Fruit and Vegetable Sensor, in order to detect the availability of food. Also, this system consists of a camera mounted over it, that helps to take real-time pictures of available fruit and vegetables. The fruit and vegetable sensor and drone are trained using

Figure 3. Machine learning based drone harvester

machine learning, on the real time datasets to detect and identify fruits and vegetable that are pluckable and ready to be consumed in next 5 days. This system enables the sensing mechanism to detect various stages of fruits and vegetables. The fruit and vegetable sensor emits light in different colours like red, orange and yellow to identify ripe stage.

The framework utilizes an Artificial Intelligence (AI) based Computer Vision and Perception calculation is utilized to find and identify the consumable foods grown from the ground. This framework will empower the detecting instrument to recognize the different phases of the products of the soil. This framework likewise utilizes a laser procedure i.e., CO2 laser method. This method is protected and satisfies the FDA guideline. Utilizing this method, we can cull the leafy foods from high height trees at the right stage.

In Figure 4, we made sense of the Drone structure. There is a Drone with its expanding arms. A capacity assortment box is mounted on the Drone that is utilized to gather the products of the soil culled from the high height trees. Drone based compass is utilized for the adjustment reason i.e., to adjust and fix distance from the high height trees. The whole course of recognizing and gathering the collect will naturally push the information on the cloud. The information is reflected in an application dashboard continuously. Robot's stretching out arms are utilized to get the products of the soil from the trees.

The sensor having indicators emits light in either red, orange, or yellow light to identify the ripening stage. The Co2 based laser is safe and meets the FDA standards to pluck the fruits and vegetables at the right stage. The working of this drone-based harvester can be explained with the help of data flow diagram mentioned in Fig.5.

CO2 Laser Cutting Technique

Till now the utilization of laser cutting and laser radiates is restricted to the material enterprises and the laser cutting methodology are wide utilized for fine and sharp modern cutting perspectives. In any case, lately barely any laser radiates that are of low power began to be utilized in the food business as well. Among numerous laser radiates accessible to cut, CO2 Laser is viewed as the most appropriate for the

Figure 4. Detecting, Plucking and Collecting ripe fruits and vegetables

food items and consumable things. As a result of its frequency, the CO2 laser permits you to handle materials of natural beginning effectively.

The laser cutting process comprises in utilizing an engaged laser shaft to eliminate the skin of the produce. This is potential because of the raised energy thickness that the laser figures out how to focus on a tiny region. The cycle causes the prompt vanishing of a layer of material.

Regardless of whether the energy and strain would be extremely high, they focus on a minuscule piece of the natural material. Thus, the zone impacted by the laser is scant and extremely engaged, which is vital for the handling of food items. The outcome is that the organoleptic properties of food, like flavour, newness, surface and variety, are not the slightest bit adjusted by the laser's activity.

Hardware and Software Specifications Required

1. **Fruit and Vegetable Sensor:** A newness sensor has been created by Kemiklioglu and Ozen (2018), which works with the rule of estimation of particle fixation change happening in foods grown from the ground.
2. **Compass Calibration:** Drone based compass calibration to balance and fix distance from the high-altitude trees.
3. **Cloud Platform:** The whole course of distinguishing and gathering the reap will naturally push the information on the cloud. The information is reflected in an application dashboard progressively. For this reason, cloud stage like Azure and AWS can be utilized.
4. **Camera:** Agricultural drones with multispectral imaging camera sensors help farmers manage crops, soil, fertilising, and irrigation more efficiently. Green, Red, Red-Edge, and Near Infrared wavebands of multispectral camera remote sensing imaging technologies are used to record both visible and invisible images of crops and plants.
5. **Extending Arms:** Extending arms of the UAV will help to collect the plucked fruit or vegetable and put that in collection box.
6. **Collection Box:** The robot weighs 18.5kg and can hold up to 16L of fluids and since it can shower up to 6.5 meters in width it's quite possibly of the most effective horticultural robot available, having the option to cover 10 hectares each hour.

Figure 5. Data Flow Diagram representing the working of Drone based Harvester

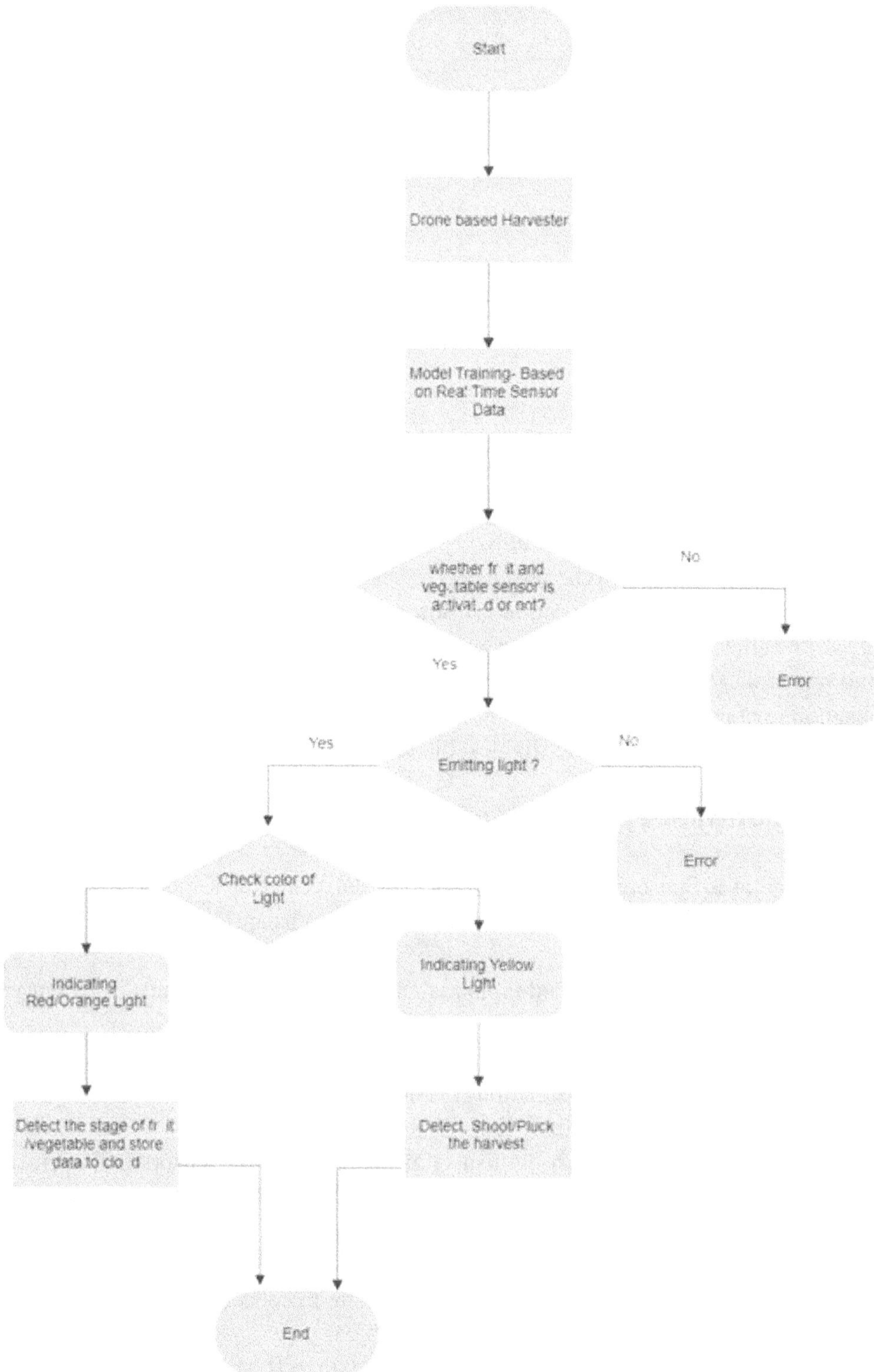

RESULTS AND OBSERVATIONS

Following Observations can be considered as the resultant outcomes while using the proposed ML based drone harvester for finding and collecting ripen food products:

- Reduces Heavy Food wastage at source
- Reduces the need of extensive manpower and other non-renewable resources in the agriculture fields
- Easily operated with minimum efforts and understanding.
- The system notifies the time needed for a fruit to be ripened.
- Reduces efforts, time and cost.
- Provides alternative to manual identification, plucking, collecting and processing of the harvest in large agricultural and terrestrial farming
- Reduces emission of carbon dioxide in the air

CONCLUSION

In recent years, there has been major areas of strength for an in the purported accuracy farming, especially the observing perspective, not exclusively to further develop efficiency, yet in addition to full-fill the need of a developing populace. At an enormous scope, exact observing of developed fields is a very difficult errand. Thus, this paper means to propose a review on methods, applied to accuracy farming checking, using drones furnished with multispectral, warm and noticeable cameras. For every application, the fundamental limits are featured and the boundaries to be considered before to play out a flight are accounted for.

Right now, the pragmatic programs for drones are developing from experts to groups and specific areas like pictures and so on. It is ordinary that the Drones marketplace can touch two hundred billion dollars with the aid of using 2020. Out of all these specific auspicious areas, Agriculture is the most important one, wherein numerous assortments with highlighted crammed places of work are required beating some problems of ranchers for higher harvest production (Kirkpatrick, 2019; Budiharto et al., 2019; Joshi et al., 2020).

Coming up next are the different Applications cum Advantages of involving drones in smart farming being sent for everyday farming errands:

1. **Horticulture Farm Research:** Drones are excellent quality trustworthy instruments soaring over and can be used by farmers to explore the residence condition around the beginning of any yield year. Drones produce three layered guides for soil assessment which is important for farmers to take care during seed wrinkling. Soil and field examination through meanders moreover gives data significant to the water framework and directing the nitrogen level of fields for improved yield advancement.
2. **Effective efficiencies:** Ranchers with bunches of hectares of land view it as attempting to expose up at every area of expertise and nook of the sector for survey occasionally. Drones does this undertaking with almost not anything surprising as farmers can do standard air seeing of the sector to realize what goes on with their yields at standard time spans.

Figure 6. Application of Drone in Advance Farming

3. **Higher farming Yield:** The exactness use of pesticides, water and use of fertilizers definitively noticed with the aid of using robots will on this manner amplify the yield and typical first-rate may be idea about.
4. **GIS Mapping Integration:** GIS Mapping has as of past due proven its really well worth throughout the growing enterprise to manipulate assets, yield increment, enter fee association, higher enterprise the board and this is most effective a sprinkle of some thing bigger. With GIS organizing solidified with Drones, the ranchers can draw area borders for careful flight plan.
5. **Imaging of Crop Health Status:** With drones, crop prosperity imaging have to be doable the usage of Infrared, NVDI and multispectral sensors similarly growing the farmers tune the energy of yield, occurring quotes and daylight consistency standards, etc.

Future Scope

Future examples in this assessment field go toward the use of humble business downsized or small-scale robots. The use of robotic technologies like drones in agriculture will bring the era of automation technologies in the field of cultivation. Use of drones may result as a muti-beneficiary way to monitor and regulate your fields and crops position and requirements. Regardless, in doing in that capacity, the assessment accuracy subtleties are attempting to be tended to and a couple of issues arise. For instance, for that drones the breeze impact, the low GPS precision and the solid float of the INS play destructing impacts in flight security and picture procurement.

REFERENCES

Budiharto, W., Chowanda, A., Gunawan, A. A. S., Irwansyah, E., & Suroso, J. S. (2019, December). A review and progress of research on autonomous drone in agriculture, delivering items and geographical information systems (GIS). In *2nd world symposium on communication engineering (WSCE)* (pp. 205-209). IEEE.

Joshi, E., Sasode, D. S., Singh, N., & Chouhan, N. (2020). Revolution of Indian Agriculture through Drone Technology. *Biotica Research Today, 2*(5 Spl.), 174-176.

Kemiklioglu, E., & Ozen, O. (2018). Design of a sensor to detect fruit freshness. *International Journal of Scientific and Technological Research*, *4*(01), 1–6.

Kirkpatrick, K. (2019). Technologizing agriculture. *Communications of the ACM*, *62*(2), 14–16. doi:10.1145/3297805

Korlahalli, K. B., Hangal, M.A., Jituri, N., Rego, P.F., Raykar, S.M. (n.d.). An Automatically Controlled Drone Based Aerial Pesticide Sprayer. *Project Reference No.39S_BE_0564*

Meivel, S., Maguteeswaran, D. R., Gandhiraj, N., & Srinivasan, G. (2016). Quadcopter UAV Based Fertilizer and Pesticide Spraying System *International Academic Research Journal of Engineering Sciences.*

Mogili, U. R., & Deepak, B. B. V. L. (2018). Review on application of drone systems in precision agriculture. *Procedia Computer Science*, *133*, 502–509. doi:10.1016/j.procs.2018.07.063

Nayyar, A., Nguyen, B. L., & Nguyen, N. G. (2020). The internet of drone things (IoDT): future envision of smart drones. In *First international conference on sustainable technologies for computational intelligence* (pp. 563-580). Springer.

Puri, V., Nayyar, A., & Raja, L. (2017). Agriculture drones: A modern breakthrough in precision agriculture. *Journal of Statistics and Management Systems*, *20*(4), 507–518. doi:10.1080/09720510.2017.1395171

Rachmawati, S., Putra, A. S., Priyatama, A., Parulian, D., Katarina, D., Habibie, M. T., ... Valentino, V. H. (2021, August). Application of Drone Technology for Mapping and Monitoring of Corn Agricultural Land. In *International Conference on ICT for Smart Society (ICISS)* (pp. 1-5). IEEE.

Ren, Q., Zhang, R., Cai, W., Sun, X., & Cao, L. (2020, February). Application and development of new drones in agriculture. In *IOP Conference Series: Earth and Environmental Science* (Vol. 440, No. 5, p. 052041). IOP Publishing.

Shivaji, C. P., Tanaji, J. K., Satish, N. A., & Mone, P. P. (2017). Agriculture drone for spraying fertilizer and pesticides. *Int. J. Res. Trends Innov*, *2*, 34–36.

ADDITIONAL READING

Alexandris, S., Psomiadis, E., Proutsos, N., Philippopoulos, P., Charalampopoulos, I., Kakaletris, G., Papoutsi, E.-M., Vassilakis, S., & Paraskevopoulos, A. (2021). Integrating Drone Technology into an Innovative Agrometeorological Methodology for the Precise and Real-Time Estimation of Crop Water Requirements. *Hydrology*, *8*(3), 131. doi:10.3390/hydrology8030131

Ayamga, M., Tekinerdogan, B., & Kassahun, A. (2021). Exploring the challenges posed by regulations for the use of drones in agriculture in the African context. *Land (Basel)*, *10*(2), 164. doi:10.3390/land10020164

De Clercq, M., Vats, A., & Biel, A. (2018). Agriculture 4.0: The future of farming technology. *Proceedings of the World Government Summit, Dubai, UAE*, 11-13.

Dutta, G., & Goswami, P. (2020). Application of drone in agriculture: A review. *IJCS*, *8*(5), 181–187. doi:10.22271/chemi.2020.v8.i5d.10529

Hafeez, A., Husain, M. A., Singh, S. P., Chauhan, A., Khan, M. T., Kumar, N., Chauhan, A., & Soni, S. K. (2022). Implementation of drone technology for farm monitoring & pesticide spraying: A review. *Information Processing in Agriculture*. doi:10.1016/j.inpa.2022.02.002

Johnson, C. J. (2021). *Drone Technology in Agriculture Appraisal.* Utah State University.

Kitonsa, H., & Kruglikov, S. V. (2018). Significance of drone technology for achievement of the United Nations sustainable development goals. *R-economy*, *4*(3), 115–120. doi:10.15826/recon.2018.4.3.016

Kumar, A., Rani, M., & Kumar, P. (2022). Drone Technology in Sustainable Agriculture: The Future of Farming Is Precision Agriculture and Mapping. In *Agriculture, Livestock Production and Aquaculture* (pp. 3–12). Springer. doi:10.1007/978-3-030-93262-6_1

Kumari, R., & Saini, K. (2021, March). Advanced Automobile Manufacturing: An Industry 4.0. In *2021 8th International Conference on Computing for Sustainable Global Development (INDIACom)* (pp. 899-904). IEEE.

Liang, M., & Delahaye, D. (2019, October). Drone fleet deployment strategy for large scale agriculture and forestry surveying. In Intelligent Transportation Systems Conference (ITSC) (pp. 4495-4500). IEEE. doi:10.1109/ITSC.2019.8917235

Marzuki, O. F., Teo, E. Y., & Rafie, A. S. M. (2021). The mechanism of drone seeding technology: A review. *Malays Forester*, *84*(02), 349–358.

Mogili, U. R., & Deepak, B. B. V. L. (2020). An intelligent drone for agriculture applications with the aid of the mavlink protocol. In *Innovative product design and intelligent manufacturing systems* (pp. 195–205). Springer. doi:10.1007/978-981-15-2696-1_19

Park, S., Yun, S., Kim, H., Kwon, R., & Ganser, J. (2018, June). Forestry monitoring system using lora and drone. In *Proceedings of the 8th International Conference on Web Intelligence, Mining and Semantics* (pp. 1-8). 10.1145/3227609.3227677

Patel, P. (2016). Agriculture drones are finally cleared for takeoff [News]. *IEEE Spectrum*, *53*(11), 13–14. doi:10.1109/MSPEC.2016.7607013

Pobkrut, T., Eamsa-Ard, T., & Kerdcharoen, T. (2016, June). Sensor drone for aerial odor mapping for agriculture and security services. In *13th International Conference on Electrical Engineering/Electronics, Computer, Telecommunications and Information Technology (ECTI-CON)* (pp. 1-5). IEEE. 10.1109/ECTICon.2016.7561340

Raj, P., Saini, K., & Surianarayanan, C. (2022). *Edge/Fog Computing Paradigm: The Concept, Platforms and Applications*. Academic Press.

Reinecke, M., & Prinsloo, T. (2017, July). The influence of drone monitoring on crop health and harvest size. In *1st International conference on next generation computing applications (NextComp)* (pp. 5-10). IEEE. 10.1109/NEXTCOMP.2017.8016168

Saini, K. (2019). Evolution of blockchain technology in business applications. [JETIR]. *Journal of Emerging Technologies and Innovative Research*, *6*(9), 240–244.

Saini, K., Mishra, I., & Srivastava, S. (2021, December). Farmer's E-mart: An E-Commerce Store For Crops. In *2021 3rd International Conference on Advances in Computing, Communication Control and Networking (ICAC3N)* (pp. 346-350). IEEE.

Saini, K., & Raj, P. (Eds.). (2021). *Advancing Smarter and More Secure Industrial Applications Using AI, IoT, and Blockchain Technology*. IGI Global.

Shafi, U., Mumtaz, R., Iqbal, N., Zaidi, S. M. H., Zaidi, S. A. R., Hussain, I., & Mahmood, Z. (2020). A multi-modal approach for crop health mapping using low altitude remote sensing, internet of things (IoT) and machine learning. *IEEE Access: Practical Innovations, Open Solutions*, *8*, 112708–112724. doi:10.1109/ACCESS.2020.3002948

Wynn, C. R. (2019). *Drone Technology: Is It Worth the Investment in Agriculture* [Doctoral dissertation, The University of Nebraska-Lincoln, USA].

Chapter 8
5G and IoT for Smart Farming

Prerna Ajmani
https://orcid.org/0000-0003-2268-1188
Vivekananda Institute of Professional Studies, Technical Campus, India

Pooja Saigal
Vivekananda Institute of Professional Studies, Technical Campus, India

ABSTRACT

Since the advent of technology, it is growing by leaps and bounds. The thirst to be connected to the world has led to the emergence of 5G in 2019. Mobile technologies are evolving rapidly to address various concerns related to speed, efficiency, connectivity, power consumption, bandwidth, and latency. As compared to its predecessors, 5G provides a greater speed and higher bandwidth to exchange data, in lesser time. Thus, it can cater to many advanced communication applications and fully supports the current modern technology and solves its intended purpose. 5G seems promising in connecting any device, object, or human to any other node, at any location and at all times. It is expected that by 2030, the network traffic and dependency on wireless communication will experience a massive increase and 5G is ready to support such a huge volume of traffic. This chapter will present a systematic and detailed study of 5G, a comparative analysis of 5G with its predecessor technologies, its potential applications, and current use cases of 5G, with a special focus on the agricultural sector.

INTRODUCTION

Around the globe, many people's subsistence is farming and major countries' financial systems rely primarily on agriculture. Agriculture is a vast term that includes the cultivation, crop yield, and breeding of animals to cater to the growing need for food, clothes, and medicines across the planet. It not only gives livelihood but also contributes significantly to the economy of underdeveloped nations. Many forms of agriculture practices are adopted in different areas around the globe with the primary goal to provide nourishing food to nourish society. A revolution in agriculture was experienced in the 18th century where significant changes in methods of farming were introduced to boost the yield. Subsequently, many changes were introduced which have revolutionized farming further. But despite so many efforts,

DOI: 10.4018/978-1-6684-6413-7.ch008

Copyright © 2023, IGI Global. Copying or distributing in print or electronic forms without written permission of IGI Global is prohibited.

Figure 1. World Population Forecast

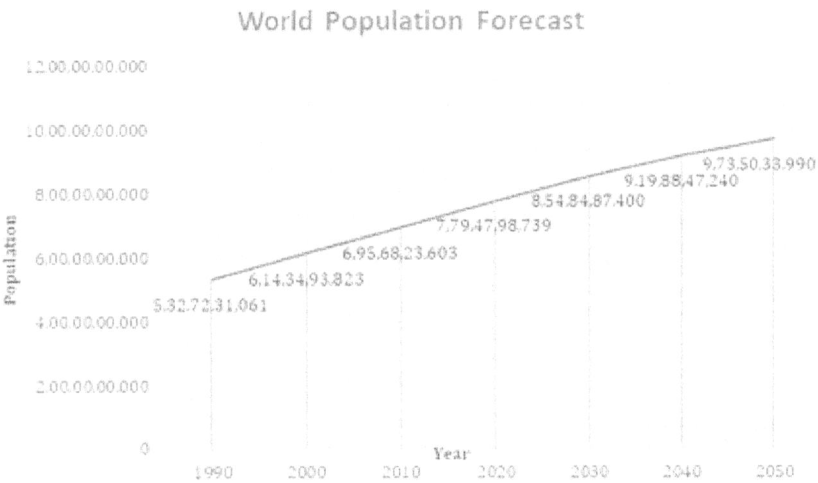

traditional farming methods could not control abrupt yield, excessive use of resources, and high wastage. Figure 1 of the estimate by the World Population Clock (2020) clearly illustrates that there is a continual growth in population, necessitating a massive increase in food production without wasting resources in order to fulfil the ongoing demand.

However, traditional farming methods result in unpredictable yield production, inefficient resource utilization, and more wastage. To overcome these challenges farmers, need to use modern technologies which provide good assistance in increasing production with less labor and minimal wastage involved. At this point, automation plays a vital role.

Smart farming (Grogan, 2012) based on advanced technology is the key to elevating production as well as enhancing the quality of products with negligible wastage. This technology is smart and less labor intensive. ICT and Precision farming-based agriculture leads to the fourth revolution in the agriculture sector. Many researchers (Neufeldt et al., 2013; Steenwerth et al., 2014; Bekele and Drake, 2003; Lipper et al., 2014; Gilbert, 2011; Wallace, 2000; Beuchelt and Badstue, 2013; Johnston et al., 2009; Molden, 2013) explained that with good communication network smart farming can address numerous problems of unpredictable weather, environment change, adulterated soil, water deficiency, high demand of food supplies and many more. Many IoT sensors get connected to the internet to perform various automatic functions for smart farming and also sense data which will help further in making decisions regarding the improvement of crop yield (Atzori et al., 2010; Ojha et al., 2015; Lee et al., 2013; Ray, 2017). Furthermore, smart farming uses machine learning algorithms to analyze the condition of the soil, to identify the crops for plantation based on numerous factors such as climate, water availability, and soil condition, to discover diseases in vegetation at a premature level so that it can be treated (Sadowski and Spachos, 2020; Magomadov, 2019; Infosys, 2018; Khan et al., 2020; Kamilaris and Prenafeta-Boldú, 2018). Plant issues can be identified and solved easily and quickly by analyzing the real-time data which is sensed by the drones or sensors installed on farms. Smart farming using these cutting-edge technologies can increase production, improves quality, and involves less labor.

To sense real-time farm data, sensors need to effectively communicate with each other or external devices. It was observed that 4G could connect various sensors and devices for data exchange without the hassle of traffic and ensures less delay or no loss which in turn helps in improving production by reducing time, labor, and cost (Payaswini and Manjaiah, 2014; Tong et al., 2019; Zhao et al., 2016). In 2020 the agricultural IoT devices have reached nearly 75 million and it is rapidly increasing at the rate of approximately 20% yearly.

Although 4G offers great connection and even high speed, researchers have found that using this network interconnection between smart devices in remote areas with low-cost installation is not possible. However, 5G dissolves this problem and makes it much simpler for devices to interconnect and communicate. The use of 5G in agriculture to increase productivity and its effects will also be covered in this chapter.

SMART FARMING AND IoT

Smart farming is highly dependent on IoT sensors that can automatically do several tasks and revolutionized the agriculture industry. They are installed over the entire field and sense information about climate, plant condition, soil, farm breeds, etc. thus giving an insight into the farm to the farmers. Entire information about the farm like the type of soil, mapping appropriate crops, pesticides required, acidic levels, manure condition, nutrient value, appropriate fertilizers to be used, appropriate seeds, detection of a plant infection, and harvest prediction is automatically sensed by the IoT sensors or devices (Adamchuk et al., 2004). Drones or sensors can be deployed to obtain geographical maps of the entire agricultural field and also to retrieve the details of different resources such as water, seeds, soil, etc. which are accessible in the particular farm area. This entire comprehensive study of the farm will help the farmer to plan and do plantation accordingly which helps in increasing quality production and not only will save the time, and laborious task but also a huge amount of money (Sureephong et al., 2017; Zhang et al., 2017; Sanchez-Iborra et al., 2018; Payero et al., 2017; Placidi et al., 2020). Furthermore, this sensed information can be used along with already present historical data to determine the quality of soil, suitable crops, fertilizers, pesticides, and water level. Smart farming creates digital soil maps and is very useful for determining a variety of crops suitable for plantation in a particular soil. These maps use artificial intelligence and recommend to farmers about the soil texture, water retention properties, acidic nature of the soil, depth to be considered while plantation, etc. which helps to prevent soil erosion, degradation, pollution, acidification, and crop death, etc.

Owing to the shortage of water supply, IoT in irrigation is pure bliss because this system allows precise water flow in the fields and prevents wastage of water. Therefore, it is more meaningful and suitable than conventional irrigation methods (Harun et al., 2015; Feng, 2011; Gupta et al., 2020). The crop water stress index (CWSI), developed by (Alderfasi and Nielsen, 2001), is intended to evaluate crop tension levels and regulate other factors that have an impact on plants, such as temperature and water availability. To collectively determine the CWSI of a certain irrigation area, many sensors are installed to collect temperature data, crop photos, infrared thermometry, etc. (Khanal et al., 2017; Lu et al., 1997; Labb´e et al., 2012). Also, appropriate sensors can optimize the irrigation time period according to the exact need of the crop (Hedley and Yule, 2009; Evans et al., 2013; Dukes and Perry, 2006).

Further, these IoT-based sensors can evaluate the fertilizers chemical proportion such as potassium, phosphorus, and nitrogen contents needed to secure the fertility of the plant. These IoT-based devices help determine the accurate fertilizer amount needed for a good yield as extra fertilizers may affect the crop adversely. According to the Food and Agriculture Organization (FAO) (FAO,2020), approximately 20-40% of crops die annually because of excessive pesticides, fertilizers, and diseases. Robots, drones, and sensors are installed on farms to identify the pest and diseases, allowing appropriate use of pesticides. These IoT- devices help immensely to curtail cost and wastage. Many software apps, such as Agronote, FARMDOK, Tania, etc. integrate with sensors and can give relevant and important information about the yield such as quality of produce, yield estimation as well as live harvest monitoring. The data collected can be easily stored on servers for recurrent usage.

Challenges of 3G/ 4G leading to 5G

In the agriculture domain, 3G/4G technology seems promising and aims at providing good connection among IoT-based devices primarily for sharing the data, accurate evaluation and precise estimation, etc. (Hassebo et al., 2018; Dahlman et al., 2011; Akyildiz et al., 2010; L.Huang et al., 2012). Despite this current technology has shown significant changes in connectivity but it suffers from constraints that affect or limits its practical and significant usage in the agriculture field. The biggest limitation is that current wireless technology provides no connection in remote places or even in cities with large skyscrapers. Regarding 4G networks quality of service (QoS) is also quite challenging (Payaswini and Manjaiah, 2014) because of variable data rates, channel condition, hand off among heterogeneous networks, resource assignment, and utilization. Many IoT devices such as sensors and drones used in smart farming run on battery and often the battery drains out quickly thus, they cannot be operational for sensing data of the crops for a longer period. Though active research is in progress to improve the speed, battery, intelligence, and secure communication capabilities of sensors used in smart farming but presently to get good performance from IoT devices installed in farms high connectivity is needed. And, 4G (LTE) is IP- based so it cannot support high speed (Lu Zhaogan et al., 2007; L.Martin et al., 2011). The above-mentioned problems of the existing generations of cellular networks lead to the transition to a further generation: 5G.

Growth of 5G

With the urge to develop high speed network 5G came into picture in 2017 and 3GPP defined its standard. This wireless network seems promising in terms of providing high speed and less latency as it uses high-band spectrum for communication (Andrews et al., 2014; GSMA, 2018; 3GPP Release 16, 2020).In addition to ultra-fast speed of nearly 10 Gbps and huge data capacity,5G has higher bandwidth thus ensures much larger spectrum to connect billions of devices together. The upload and download speed are ultrafast in 5G when compared to 4G or 4G LTE. A two-hour movie would take six minutes to download using 4G as an example. The download and playback of the same movie would be claimed by 5G to take around four film seconds. In 2020, technical standards are laid down by the International Telecommunication Union (ITU) (IMT-2020). 5G is so powerful that it can connect approximately one-million IoT devices per square kilometer, and can extend support to high-speed moving devices as well. Another important advantage of 5G is low latency of approximately 1 ms which will overcome the issue of jitter and lags that can be felt in present network services (Yu et al., 2017; Sutton, 2018; Gupta and Jha, 2015; Huawei, 2018; Next Generation Mobile Networks Alliance 5G Initiative, 2015).

Figure 2. 5G Small Cell

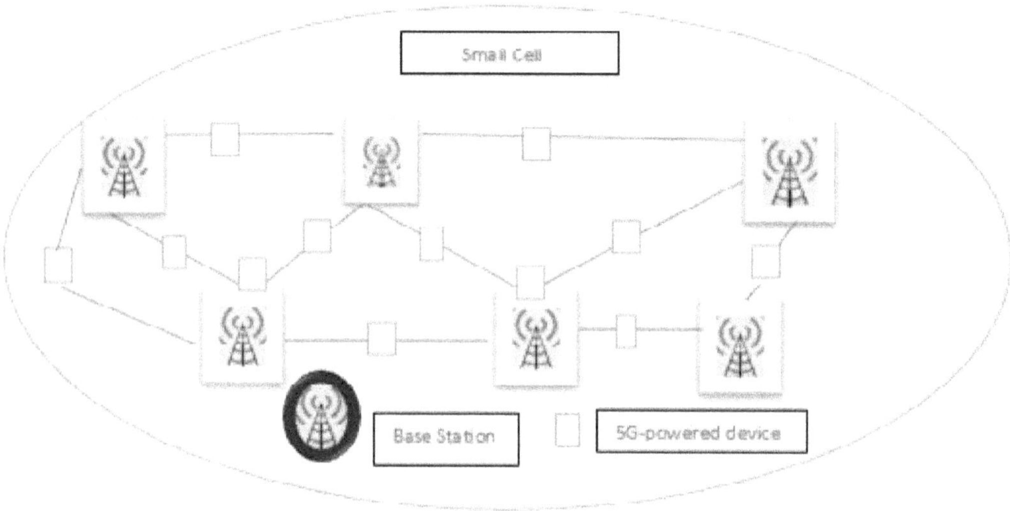

In 5G network millimeter waves (MMWs) are used as wireless carriers which cannot travel to larger distance. Therefore, to provide uninterruptable connection, smaller cells are used in which base stations or cell towers are in close proximity of approximately 250 meter which can be extended to give the greater coverage. To provide connectivity to larger areas these small cell towers can be installed at anyplace (from lamppost to vehicle roofs etc.). Figure 2 illustrates the deployment of 5G to cover any range for smooth communication (DotEcon Ltd & Axon Partners, 2018).

Similar to a 4G network, a 5G network sends and receives multiple data signals concurrently using an improved multiple-input/multiple-output (MIMO) technology (MIMO) technique (Jungnickel et al., 2014; Elijah et al., 2016). For this type of transmission and reception MIMO uses multiple transmitters and receiver antennas. The massive MIMO suffer from a critical problem that often signals can easily interfere with each other. However, beamforming technique is used to mitigate this issue and to enhance the efficiency of transmission. As the name implies, beamforming enables antenna to emit focused radio wave beams in a defined direction for each particular device. 5G has good spectrum range because it operates at very high frequencies that permits superfast and more data exchange. Unlike 4G network, 5G uses cognitive radio technique to distinguish each device and allocate dynamic channels to them.

According to (Tang et al., 2021) key advantages of 5G are:

- Ultra-high speed
- Greater capacity
- Low latency
- More device connections
- Network slicing
- Dynamic channel allocation
- Good throughput
- Large Spectrum
- Smooth/Uninterrupted connectivity

- Large area coverage
- Energy efficient

In comparison to the previous generation networks, 4G permitted HD video streaming and video calls while on the move. However, 4G faced worse network congestion and has now used all available spectrum. The ITU categorizes the most recent 5G technology into three groups: massive machine type communications (mMTC), improved mobile broadband (eMBB), and international and regional levels low latency communications, which all appear to be promising (URLLC) (Anand et al., 2020; Popovski et al., 2018). The main aim of mMTC is to provide a high-speed network to the high-density region at a lower cost with minimum power consumption. Thus, this is used in many applications such as logistics, fleet management, making hotspots, precision/smart farming, etc. On the other hand, eMBB focuses on providing larger bandwidth to cater to AR/VR applications. However, URLLC covers mission-critical applications.

Though 5G shows a remarkable improvement in spectrum bandwidth and connectivity speed, its less coverage requires additional installation of stations. Unlike the current generation, the 5G base station suffers from the problem that it can send data in a shorter-range area thus it requires the installation of additional base stations. Despite the fact that these stations can be put anywhere, there are significant infrastructural costs. The good news is that although this cost first appears exorbitant, it will actually be much less expensive when it is carried out on a large scale. Cost of base stations, throughput density, and recurrent rate of interest all play a significant role in determining 5G network costs (Bouras et al., 2017). By lowering these costs, 5G technology should be more affordable for small businesses and the agricultural industry. Given that 5G and 4G LTE share the same frequency band, it can reduce data prices even more than the present network costs. Also, 5G is compatible with the existing 4G LTE supported devices. Therefore, with the power of 5G the current generation chipset can enjoy faster speed and less delay.

Smart Farming: IoT Use Case

The idea of smart farming is to make the process of farming more productive and to use more efficient ways for farm-to-market transfers, with minimum wastage. Smart farming, which is also known as "Smart Agriculture", "Agricultural technology" or "Ag-tech" in short, makes use of IoT enabled sensors to integrate various aspects of farming. It is expected that 5G, due to its high bandwidth, is going to have a huge positive impact on smart farming. Ag-Tech aims to improve the farm yield by entrusting the farmers with timely data, which enables them to take good decisions.

Ag-tech solutions help farmers reduce the consumption of resources like water, seeds, fertilizers and other chemicals. They can be used to reduce soil erosion and would add efficiency in product distribution. A hydroponic farm in Bengaluru, India, under the name Clover, monitors the growth of greenhouse plants through wireless sensors. They aggregate the data from different locations and use it to produce healthier plants, which further yield a generous harvest. Ag-tech solutions not only manage farming activities, but also provide IoT solutions to monitor and manage livestock on farms. It provides current information on temperature, soil conditions, water quality and availability, the presence of pollutants, and many other topics. Information Technology applications designed for fixed and mobile devices would require cellular IoT connectivity. This would enable them to collect remote data and manage activities in the field. These days, various smart farming solutions have been developed that use wireless technologies of a shorter range, such as Wi-Fi and Bluetooth. Whereas cellular technology with a long range is needed to cover larger areas and for remote monitoring. 5G would enable us to develop applications that could augment or replace short-range applications. 5G mobile networks supports farming due to their wide area coverage, high-bandwidth, low energy consumption and low-cost sensors and devices (Tang, et al., 2021).

Currently, 5G is in its early stages of development and is focusing on enabling high bandwidth connectivity. The 5G Rural First project started in 2017, which was able to successfully sow seeds, irrigate and harvest a crop using autonomous tractors. It employed drones for crop monitoring and application of pesticides and fertilizers. No human laborer's were involved at any stage in this process. Centralized farms could be the most practical use case of 5G, wherein a big corporate farming set up can have a virtually private 5G network for enabling high bandwidth use cases like crop management with the help of drones and collecting data through various IoT sensors. With the growth of cellular 5G technology, it is expected that precision farming can benefit a lot. With the expansion of 5G coverage, agricultural producers would be able to manage their farm operations and livestock remotely. Due to the high-speed data, large bandwidth and low latency, 5G would enable the use of sensors at a dense level and would provide access to groundbreaking innovation in farming. (Adamchuk, Hummel, Morgan, & Upadhyaya, 2004)

In the following section, some of the potential current and future use cases of 5G in smart farming are listed.

Data Collection and Real-time monitoring: Wireless sensors monitor the environmental conditions repeatedly at the farm, like humidity, soil quality, temperature, and can communicate the data collected to distant processors through the internet (Prathibha, Hongal, & Jyothi, 2017). Based on the input provided by the sensors, the situation is assessed and a corrective action is initiated for the healthy growth of the plant. Since data provided by various sensors has different formats, appropriate interfaces are required for interoperability. Different application program interfaces, notably Open Ag Data Alliance (OADA, 2020), are used to manage data transactions on storage platforms in order to overcome this incompatibility. Software specifically designed for weather monitoring and data analytics, such as SmartElement and Pycno, are utilised to monitor the farming parameters. Numerous open-source software platforms are available for managing farm equipment, collecting data, and other tasks. ZigBee or Wi-Fi are used for short-range data transmission, whereas mobile communication networks use LoRaWAN, Bluetooth, and LPWAN protocols for long-range transmission. Low power wide area (LPWA) technology can be utilised in 5G mobile networks to provide wide-area coverage, high energy efficiency, and high bandwidth. (Sinha, Wei, & Hwang, 2017). In a 5G cellular network, a single tower can communicate with around a million sensors or devices per square kilometer. Farmers could do this to install additional Internet of Things sensors to carry out the duties effectively. With a large number of sensors, it is likely that decisions will be made more quickly and effectively by anticipating and predicting uncommon changes in the farms. If the sensors must have long-range connectivity and each sensor must be power-efficient using solar or battery power, a 5G network can be energy-efficient. The low power transmission and dependability of 5G technologies are advantageous for sensors. As a result, if data is transferred at a lower energy level and retransmission is avoided, the sensor's lifetime can be dramatically extended.

5G technology has great prospects for managing operations at a large farm by collecting and aggregating the data at a centralized location. These corporate farms could build a virtual private 5G network that can accumulate data from a number of systems that do micro-monitoring for crop management.

This would provide more efficient and real-time monitoring of the activities like controlled irrigation and managing crop support systems.

Predictive Analytics: Due to huge data aggregation with the use of 5G technology, big commercial farms can exploit this data for predictive analytics. With the help of historical and current data about various environmental conditions, soil type, water availability, usage of pesticides, etc., the analytics software can make predictions, which can help farmers in decision-making. Since 5G provides real-time dense data, the predictions given by the analytical model are more precise, and it maximizes the production and efficiency of the farm. With the use of feedback from numerous 5G-enabled sensors and predictive analyses, the farmer is given timely information about any impending problems and any malfunctioning parts. This allows him to carry out corrective actions like repairs or replacements without halting any operation. This reduces unplanned downtime that would otherwise occur due to faulty equipment or machine breakdown.

Data collected through various sources located on the farm, like IoT enabled sensors, robots and drones, is sent to a data repository on the cloud. 5G enables real-time transfer of data to the cloud, and the data is analyzed to automate and streamline farming operations. Data analytics applications include disease detection, weather forecasting to plan activities, soil analysis, fertilizer and pesticide recommendations, automated irrigation of fields, crop yield estimation, and so on. The huge amount of data thus collected must be sent to the cloud, and the results of analysis must be sent back to the farmers and experts for appropriate actions. For instance, as soon as the robot captures a photograph of a crop infested by a pest, it quickly sends it to the cloud for verification using image processing methods. The farmer and the automated robot, which is programmed to execute the control action, are then supplied the results.

To lessen the computational complexity, edge computing in the cloud is combined with robots. Here, the robots' navigation and data processing services are stored in the cloud, which also acts as a data centre. Edge computing, which integrates the GPU in the cloud edge server in the robot, eliminates the need to use the graphics processing unit (GPU). Only 5G can deliver the processing bandwidth since it is so high. The robots' size, complexity, cost, and energy usage are also decreased as a result. The entire control infrastructure for navigation, facility management, and processing of data is housed in the cloud and is operated by data centers or specialized hosts. Therefore, 5G can significantly enhance the data transfer experience. At speeds of up to 30 gigabits per second (Gbps), a significant volume of data can be effectively exchanged between numerous devices with little data loss. This reduces the amount of connection downtime and prevents the needless retransmission of data. All of this data may be accessed securely and instantly because of its cutting-edge decreased latency. Cloud computing fully utilizes 5G technology to give the best possible productivity of a smart farm thanks to faster data processing in the cloud and minimum round-trip latency across numerous 5G linked devices.

Virtual consultation: Farmers can benefit from having a virtual consultation with domain experts or agriculture consultation services regarding various aspects of farm management. They can use high-definition video streaming for the meeting. Additionally, 5G gives professionals and advisors the ability to access real-time status or live feed directly from the monitoring sensors or automated drones. This would provide a thorough understanding of the current circumstance and prompt assistance to the farmers in order to maximize agricultural produce. There are various agricultural operations that can be improved through online consultation, which includes soil sampling, environmental planning, livestock health updates, precision agriculture, pest management and many others. Due to the low latency and fast speed of 5G technology, the condition of farm machinery and equipment can be checked in real time.

This enables timely repairs, refueling, and fault management and prevents delays or halting of planned activities.

Drone enabled operations: Drones, also known as unmanned aerial vehicles, are used by farms for land management and effective decision making. They are used for both arial and on-ground tasks. The most frequent farm operations where drones can be employed are monitoring of land and crops, analysis of soil, planting of seeds, spraying of fertilizers and pesticides, managing livestock, health assessment and delivery of products. In order to scan a large area for 3D mapping, drones use a variety of sensors, including multispectral imaging, thermal infrared imaging, hyperspectral, and lidar (Ampatzidis, Partel, & Costa, 2020). They can also count plants, calculate vegetation indices like the Normalized Difference Vegetation Index and the Normalized Difference Red Edge, and measure soil temperature and water levels. (Gaˇsparoviˊc, Zrinjski, Barkoviˊc, & Radoˇcaj, 2020). Drones are less expensive in comparison to the traditional methods of monitoring through tractors and provide targeted information about crop damage or any other issue. Due to the use of high bandwidth 5G technology, drones are able to collect high-quality video data in less time.

For covering larger land areas, multiple drones can be used, but this would lead to an increase in data traffic. Drones use a variety of sensors, including multispectral imaging, thermal infrared imaging, hyperspectral, and lidar, to scan large areas for 3D mapping, count plants, calculate vegetation indices like the Normalized Difference Vegetation Index and the Normalized Difference Red Edge, and record plant health data like soil temperature and water levels (Ivancic, Murawski, Matheou, & Downey, 2019). As an alternative, the distance problem is reduced by using a 4G network, which enables a drone to fly many kilometers distant from its controller. However, as the use of drones has grown, it has become more difficult and time-consuming to control and identify them (Lin, et al., 2018). The integration of drone traffic management and the facilitation of low-altitude drone communication are better suited for the 5G mobile network. This boosts the security and safety of drone operations while also providing coverage both above and below the ground.

Drones can be utilized across a great distance with the help of 5G technology, either manually or through pre-programmed checkpoints. They are manageable from any location in the world (Faraci, Raciti, Rizzo, & Schembra, 2018). In comparison to previous-generation mobile networks, the 5G cellular network enables seamless real-time transfer of high-definition video feeds, sensory data, and telemetry acquired by drones. With 5G technology, the drones can transfer all the data to the cloud for processing more quickly. Drones have the ability to communicate with one another, enabling synchronized autonomous flying over a wide area. With less energy usage and data transmission loss, they can complete a variety of activities. This makes it possible to operate cost-effectively beyond the line of sight for extended periods of time while airborne. Consequently, a substantial amount of obtained data can be archived and examined for purposes of making decisions (Li, Fei, & Zhang, 2019), (Razaak, 2019). Drones can also be used to transport agricultural goods by flying through the air. Drone traffic management, flights outside of visual range, and sensor data transmission are essential criteria for connecting drones utilizing 5G cellular communications. Drone traffic management systems can be integrated into the 5G cellular network, increasing drone operations' safety and lowering the cost of beyond visual line of sight drone operations by providing connection.

Augmented reality (AR) and virtual reality (VR): Consumer-grade augmented reality and virtual reality kits help farmers by giving them information such as crop status, cattle updates, drone live feed and telemetry, weather conditions, soil moisture, etc. Smartphones and wearable technology are used to identify pests, identify plant and animal diseases, and more. The farmer can use augmented reality

(AR) glasses to virtually inspect the crops and gather important information without being present on the field. They can do activities on the farm more efficiently, which saves time, money, and crop losses while ensuring higher-quality yields. The deployment of 5G can address the high bandwidth requirements of VR and AR for video streaming. Real-time multimedia rendering in the cloud is made possible via cloud VR, which also lessens the computing burden on the VR kit. The prospect for highly interactive VR services for teaching and practice in agriculture education is also made possible by 5G cloud VR.

FUTURE PROSPECTS

Apparently, on paper, 5G is the fastest network among all the existing mobile networks. To achieve the optimum performance, several telecom companies are conducting experiments to push policies to their limits. For the Internet of Things to facilitate connections between numerous devices, a dense network must be constructed Telecom companies have identified several areas where 5G technology will be beneficial, among which health care, agriculture, and entertainment are on the top list. Particularly in the agricultural sector 5G technology has a considerable scope and can change the outlook of farming as "connected farms" which is needed in reality. The productivity of smart farming can be enhanced to manifolds by fetching the three categories of services from telecom companies such as better connectivity, partnership, and vertical integration. Not only connectivity telecom companies can go beyond by offering edge-to-edge solutions in farming, which in turn results in increased growth. At different levels, various services of IoT devices can be endorsed, such as device installation and management, security, location-based services, data analysis, etc. The live feed and important data related to crops can be streamed into farmers mobile phones with the help of 5G and cloud computing. So, farm management becomes easy and food safety can be ensured. With the help of 5G, the live statistics of a farm can help the farmers remotely manage the farm, and even without being present on the farms, experts can do crop inspections accurately. In addition to the above, the consumer can be informed about the quantity of produce available.

Some prominent use cases of the current happening of 5G technology in agriculture that leads to the transformation:

1. To evaluate the use of 5G, AI, and cloud computing in farming, Huawei Technologies chose to launch a partnership with XAG, the producer of agricultural drones (XAG, 2019).
2. Zhejiang province in East China constructed a smart greenhouse that uses 5G to grow tomatoes. Numerous sensors are placed to perform a variety of duties, like regulating humidity and temperature, feeding plants, and even playing gentle music. The pest management technology used in this greenhouse is also made possible by 5G. (Soumik_Roy, 2019).
3. In 2019, a 5G-based collar was created by the British Agricultural Technology Research Center that was fitted to 180 cows to keep an eye on their physical movements and food consumption patterns. Dairy calves were given smart sensor-based collars in Somerset, an English cattle farm, to enable communication with something like a mechanized milking system. (Reuters, 2019).
4. Self-driven hydrogen-powered smart 5G technology-based tractor (model: ET504- H) was launched by CHIAIC in Luoyang, Henan. The real-time condition of the tractor and its surroundings may be obtained from this smart tractor, which helps the farming operation.

CONCLUSION

In this era, technology has become an essential component and plays a significant role in almost all the domains of life, namely vehicular industry, construction, telecom, aerospace, military, healthcare, and many more. Nowadays even the traditional fields like agriculture also showed higher demand for technology to increase production with less labor in a lesser time. Therefore, the concept of smart farming or precision farming evolved. Though, sensor-based smart farming expects higher investments, more coverage, connectivity, and higher frequency bandwidth to manage the huge data of farms gathered by IoT devices deployed remotely on the farm yet it is worth it because it helps the farmers in many ways and provides practical solutions to increase the food production The 4G network can offer the necessary capacity and coverage, but it is unable to exchange significant amounts of real-time data among a large number of devices. This crucial need is being met by the development of 5G, which is also required for smart farming to increase crop output with fewer labor. Based on studies, various initiatives have already been taken in this context, and in the future, almost everywhere 5G-based smart farming practices will be adopted. With the telecom operators aiming at providing 5G network everywhere, it is expected that the initial high investment costs for smart farming will considerably reduce further, which is a boon for agriculture. Smart farming ability to identify, and predict crop diseases helps farmers remotely manage the farms using their smartphones. Telecom companies will bring the revolution in farming by providing 5G facilities to farms so that sensors can be connected and exchange data. Additionally, deep learning systems can assess this sensed data and assist farmers in making the right choices to quickly boost productivity and food quality. Despite the fact that 5G has a variety of uses in agriculture, it will alter the occupations available. On clever farms, the inferior position will be offered. The entire impact of 5G won't be felt right now; it will take some time to find all of its potential uses.

REFERENCES

Adamchuk, V. I., Hummel, J. W., Morgan, M. T., & Upadhyaya, S. K. (2004). On-the-go soil sensors for precision agriculture. *Computers and Electronics in Agriculture*, *44*(1), 71–91. doi:10.1016/j.compag.2004.03.002

Akyildiz, I.F., Gutierrez-Estevez, D.M., Reyes, E.C. (2010). The evolution to 4G cellular systems: LTE-Advanced. *Phys. Commun.* *3*(4), 217–244. doi:.phycom.2010.08.001. doi:10.1016/j

Alderfasi, A. A., & Nielsen, D. C. (2001). Use of crop water stress index for monitoring water status and scheduling irrigation in wheat. *Agricultural Water Management*, *47*(1), 69–75. doi:10.1016/S0378-3774(00)00096-2

Ampatzidis, Y., Partel, V., & Costa, L. (2020). Agroview: Cloud-based application to process, analyze and visualize UAV-collected data for precision agriculture applications utilizing artificial intelligence. *Computers and Electronics in Agriculture*, *174*, 105457. doi:10.1016/j.compag.2020.105457

Anand, A., de Veciana, G., & Shakkottai, S. (2020). Joint Scheduling of URLLC and eMBB Traffic in 5G Wireless Networks. *IEEE/ACM Transactions on Networking*, *28*(2), 477–490. doi:10.1109/TNET.2020.2968373

Andrews, J. G., Buzzi, S., Choi, W., Hanly, S. V., Lozano, A., Soong, A. C. K., & Zhang, J. C. (2014). What Will 5G Be? *IEEE Journal on Selected Areas in Communications*, *32*(6), 1065–1082. doi:10.1109/JSAC.2014.2328098

Atzori, L., Iera, A., & Morabito, G. (2010). The Internet of Things: A survey. *Computer Networks*, *54*(15), 2787–2805. doi:10.1016/j.comnet.2010.05.010

Bekele, W., & Drake, L. (2003). Soil and water conservation decision behavior of subsistence farmers in the Eastern Highlands of Ethiopia: A case study of the Hunde-Lafto area. *Ecological Economics*, *46*(3), 437–451. doi:10.1016/S0921-8009(03)00166-6

Beuchelt, T. D., & Badstue, L. (2013). Gender, nutrition- and climate-smart food production: Opportunities and trade-offs. *Food Security*, *5*(5), 709–721. doi:10.100712571-013-0290-8

Bouras, C., Kollia, A., & Papazois, A. (2017). Dense Deployments and DAS in 5G: A TechnoEconomic Comparison. *Wireless Personal Communications*, *94*(3), 1777–1797. doi:10.100711277-016-3711-0

Dahlman, E., Parkvall, S., & Skold, J. (2013). *4G: LTE/LTE-advanced for mobile broadband*. Academic press.

Dell'Uomo, L., & Scarrone, E. (2002, October). An all-IP solution for QoS mobility management and AAA in the 4G mobile networks. In The 5th International Symposium on Wireless Personal Multimedia Communications (Vol. 2, pp. 591-595). IEEE.

DotEcon Ltd & Axon Partners. (2018). Study on Implications of 5G Deployment on Future Business Models. Axon Partners Group, p. 116.

Dukes, M. D., & Perry, C. (2006). Uniformity testing of variable-rate center pivot irrigation control systems. *Precision Agriculture*, *7*(3), 205–218. doi:10.100711119-006-9020-y

Elijah, O., Leow, C. Y., Rahman, T. A., Nunoo, S., & Iliya, S. Z. (2016). A Comprehensive Survey of Pilot Contamination in Massive MIMO—5G System. *IEEE Communications Surveys and Tutorials*, *18*(2), 905–923. doi:10.1109/COMST.2015.2504379

Evans, R. G., LaRue, J., Stone, K. C., & King, B. A. (2013). Adoption of site-specific variable rate sprinkler irrigation systems. *Irrigation Science*, *31*(4), 871–887. doi:10.100700271-012-0365-x

FAO. (2019, April 3). New standards to curb the global spread of plant pests and diseases. FAO. https://www.fao.org/news/story/en/item/1187738/icode/

Faraci, G., Raciti, A., Rizzo, S., & Schembra, G. (2018). A 5G platform for Unmanned Aerial Monitoring in Rural Areas: Design and Performance Issues. *Conf. Netw. Softwarization Work. NetSoft 2018* (pp. 356–361). IEEE.

Gašparović, M., Zrinjski, M., Barković, Đ., & Radočaj, D.Gaˇsparovi'c. (2020). An automatic method for weed mapping in oat fields based on UAV imagery. *Computers and Electronics in Agriculture*, *173*, 105385. doi:10.1016/j.compag.2020.105385

Gilbert, N. (2011, March). Climate-smart agriculture is needed. *Nature*. doi:10.1038/news.2011.131

Grogan, A. (2012, July). Smart farming. *Eng. Technol.*, *7*(6), 38–40. doi:10.1049/et.2012.0601

GSMA. (2018). Road to 5G: Introduction and Migration. *Gsma*, p. 54.: https://www.gsma.com/futurenetworks/wp-content/uploads/ 2018 /04/Road-to-5G-Introduction-and-Migration_FINAL.pdf

Gupta, A., & Jha, R. K. (2015). A Survey of 5G Network: Architecture and Emerging Technologies. *IEEE Access: Practical Innovations, Open Solutions*, *3*, 1206–1232. doi:10.1109/ACCESS.2015.2461602

Harun, A. N., Kassim, M. R. M., Mat, I., & Ramli, S. S. (2015, May). Precision irrigation using wireless sensor network. In *International Conference on Smart Sensors and Application (ICSSA)* (pp. 71-75). IEEE. 10.1109/ICSSA.2015.7322513

Hassebo, A., Obaidat, M., & Ali, M. A. (2018). Commercial 4G LTE cellular networks for supporting emerging IoT applications. In *Advances in Science and Engineering Technology International Conferences (ASET)*, pp. 1–6. 10.1109/ICASET.2018.8376832

Hedley, C. B., & Yule, I. J. (2009). Soil water status mapping and two variable-rate irrigation scenarios. *Precision Agriculture*, *10*(4), 342–355. doi:10.100711119-009-9119-z

Huang, J., Qian, F., Gerber, A., Mao, Z. M., Sen, S., & Spatscheck, O. (2012, June). A close examination of performance and power characteristics of 4G LTE networks. In *Proceedings of the 10th international conference on Mobile systems, applications, and services* (pp. 225-238). 10.1145/2307636.2307658

Iera, A., Molinaro, A., Polito, S., & Ruggeri, G. (2005). End-to-end QoS provisioning in 4G with mobile hotspots. *IEEE Network*, *19*(5), 26–34. doi:10.1109/MNET.2005.1509949

Infosys. (2018). AI for Healthcare: Balancing Efficiency and Ethics. *Infosys*. https://www.infosys.com/smart-automation/docpdf/ai-healthcare.pdf

Ivancic, W., Murawski, R., Matheou, K., & Downey, A. (2019). Flying Drones beyond Visual Line of Sight Using 4G LTE: Issues and Concerns. *Integr. Commun. Navig. Surveill. Conf.* (pp. 1-13). 10.1109/ICNSURV.2019.8735246

Johnston, A. E., Poulton, P. R., & Coleman, K. (2009). *Soil Organic Matter*.

Jungnickel, V., Manolakis, K., Zirwas, W., Panzner, B., Braun, V., Lossow, M., Sternad, M., Apelfrojd, R., & Svensson, T. (2014). The role of small cells, coordinated multipoint, and massive MIMO in 5G. *IEEE Communications Magazine*, *52*(5), 44–51. doi:10.1109/MCOM.2014.6815892

Kamilaris, A., Prenafeta-Boldú, F.X. (2018). Deep learning in agriculture: A survey. *Comput. Electron. Agric. 147*, 70–90. doi:.compag.2018.02.016. doi:10.1016/j

Khan, A., Sohail, A., Zahoora, U., & Qureshi, A. S. (2020). A survey of the recent architectures of deep convolutional neural networks. *Artificial Intelligence Review*, *53*(8), 5455–5516. doi:10.100710462-020-09825-6

Khanal, S., Fulton, J., & Shearer, S. (2017). An overview of current and potential applications of thermal remote sensing in precision agriculture. *Computers and Electronics in Agriculture*, *139*, 22–32. doi:10.1016/j.compag.2017.05.001

Labbé, S., Lebourgeois, V., Jolivot, A., & Marti, R. (2012). Thermal infra-red remote sensing for water stress estimation in agriculture. *Options Méditerranéennes. Série B: Études et Recherches*, *67*, 175–184.

Lee, M., Hwang, J., & Yoe, H. (2013). Agricultural Production System Based on IoT. In *16th International Conference on Computational Science and Engineering*, pp. 833–837. doi:10.1109/CSE.2013.126

Li, B., Fei, Z., & Zhang, Y. (2019). UAV Communications for 5G and Beyond: Recent Advances and Future Trends. *IEEE Internet Things J.*, *6*(2), 2241–2263.

Lin, X., Yajnanarayana, V., Muruganathan, S., Gao, S., Asplund, H., Maattanen, H.-L., & Wang, Y.-P. (2018). The Sky Is Not the Limit: LTE for Unmanned Aerial Vehicles. *IEEE Communications Magazine*, *56*(4), 204–210.

Lipper, L., Thornton, P., Campbell, B. M., Baedeker, T., Braimoh, A., Bwalya, M., & Torquebiau, E. F. (2014). Climate-smart agriculture for food security. *Nature Climate Change*, *4*(12), 1068–1072.

Lu, Y.-C., Daughtry, C., Hart, G., & Watkins, B. (1997). The current state of precision farming. *Food Rev. Int. 13*(2), 141–162. 87559129709541104. doi:10.1080/

Magomadov, V. S. (2019). Deep learning and its role in smart agriculture. *Journal of Physics: Conference Series*, *1399*, 044109. https://doi.org/10.1088/1742-6596/1399/4/044109

Martin, J., Amin, R., Eltawil, A., & Hussien, A. (2011, June). Limitations of 4G wireless systems. In *Proceedings of Virginia Tech Wireless Symposium* (Blacksburg, VA).

Molden, D. (2013). *Water for food water for life: A comprehensive assessment of water management in agriculture*. Routledge.

Neufeldt, H., Jahn, M., Campbell, B. M., Beddington, J. R., DeClerck, F., De Pinto, A., & Zougmoré, R. (2013). Beyond climate-smart agriculture: Toward safe operating spaces for global food systems. *Agriculture & Food Security*, *2*(1), 1–6.

Next Generation Mobile Networks Alliance. (2015). 5G Initiative. *5G white paper.* Next generation mobile networks.

OADA. (2020, Jul 28). OADA/oada-docs: OADA API Related Documentation and References. *Github.* https://github.com/oada/oada-docs#readme

Ojha, T., Misra, S., & Raghuwanshi, N. S. (2015). Wireless sensor networks for agriculture: The state-of-the-art in practice and future challenges. *Computers and Electronics in Agriculture*, *118*, 66–84. https://doi.org/10.1016/j.compag.2015.08.011

Payaswini, P., & Manjaiah, D. H. (2014). Challenges and issues. In *4G Networks Mobility Management.* https://arxiv.org/abs/1402.3985

Payero, J. O., Mirzakhani-Nafchi, A., Khalilian, A., Qiao, X., & Davis, R. (2017). Development of a Low-Cost Internet-of-Things (IoT) System for Monitoring Soil Water Potential Using Watermark 200SS Sensors. *AIT 07*(03), 71–86. doi:10.4236/ait.2017.73005

Placidi, P., Gasperini, L., Grassi, A., Cecconi, M., & Scorzoni, A. (2020, June). Characterization of Low-Cost Capacitive Soil Moisture Sensors for IoT Networks. *Sensors (Basel)*, *20*(12), 3585. https://doi.org/10.3390/s20123585

Popovski, P., Trillingsgaard, K. F., Simeone, O., & Durisi, G. (2018). 5G Wireless Network Slicing for eMBB, URLLC, and mMTC: A Communication-Theoretic View. *IEEE Access: Practical Innovations, Open Solutions*, *6*, 55765–55779. https://doi.org/10.1109/ACCESS.2018.2872781

Prathibha, S. R., Hongal, A., & Jyothi, M. P. (2017). IOT Based Monitoring System in Smart Agriculture. *International Conference on Recent Advances in Electronics and Communication Technology, ICRAECT.*

Ray, P. P. (2017). Internet of things for smart agriculture: Technologies, practices and future direction. AIS 9 (4), 395–420. doi:10.3233/AIS-170440

Razaak, M. (2019). An integrated precision farming application based on 5G, UAV and deep learning technologies. *Communications in Computer and Information Science*, *1089*, 109–119.

Reuters. (2019, April 11). 5G-connected cows test milking parlor of the future. *Reuters*. https://www.reuters.com/article/us-telecoms-5g-cows/5g-connected-cows-test-milking-parlor-of-the-future-idUSKCN1RN1IY (accessed Jun. 30, 2020).

Roy, S. (2019). A peek into China's smart IoT and 5G-enabled tomato farms in Wuzhen. *Techwire Asia*. https://techwireasia.com/2019/10/a-peek-int o-chinas-smart-iot-and-5g-enabled-tomato-farms-in-wuzhen/

Sadowski, S., & Spachos, P. (2020). Wireless technologies for smart agricultural monitoring using internet of things devices with energy harvesting capabilities. *Computers and Electronics in Agriculture*, *172*, 105338. https://doi.org/10.1016/j.compag.2020.105338

Sanchez-Iborra, R., Sanchez-Gomez, J., Ballesta-Viñas, J., Cano, M. D., & Skarmeta, A. F. (2018). Performance evaluation of LoRa considering scenario conditions. *Sensors (Basel)*, *18*(3), 772.

Sinha, R., Wei, Y., & Hwang, S.-H. (2017). A survey on LPWA technology: LoRa and NBIoT. *ICT Express*, *3*(1), 14–21.

Steenwerth, K. L., Hodson, A. K., Bloom, A. J., Carter, M. R., Cattaneo, A., Chartres, C. J., & Jackson, L. E. (2014). Climate-smart agriculture global research agenda: Scientific basis for action. *Agriculture & Food Security*, *3*(1), 1–39.

Sureephong, P., Wiangnak, P., & Wicha, S. (2017, March). The comparison of soil sensors for integrated creation of IOT-based Wetting front detector (WFD) with an efficient irrigation system to support precision farming. In *International Conference on Digital Arts, Media and Technology (ICDAMT)* (pp. 132-135). IEEE.

Sutton, A. (2018). 5G network architecture. *J. Inst. Telecommun. Prof.*, *12*, 8–15. https://doi.org/10.1016/b978-0-08-102267-2.00001-4

Tang, Y., Dananjayan, S., Hou, C., Guo, Q., Luo, S., & He, Y. (2021). A survey on the 5G network and its impact on agriculture: Challenges and opportunities. *Computers and Electronics in Agriculture*, *180*, 105895. https://doi.org/10.1016/j.compag.2020.105895

Tong, W., Feng, X., & Chen, X. J. (2019). Research on Remote Control and Management Based on '4G Network' In Modern and High Efficiency Agriculture, pp. 233–241

Wallace, J. S. (2000). Increasing agricultural water use efficiency to meet future food production. *Agric. Ecosyst. Environ. 82*(1-3), 105–119. S0167-8809(00)00220-6. doi:10.1016/

World Population Clock. (2020) 7.8 billion People. *Worldometer*. https://www. worldometers.info/world-population/#table-forecast .

XAG. (2019). Huawei cooperate on 5G-powered smart agriculture. *Geospatial News Portal.* https://www.thegeospatial.in/xag-huawei-cooperate-on-5g-powered-smart-agriculture .

Yu, H., Lee, H., & Jeon, H. Oct. 2017. What is 5G? Emerging 5G Mobile Services and Network Requirements. *Sustainability 9*(10), 1848. su9101848. doi:10.3390/

Zhang, X., Zhang, J., Li, L., Zhang, Y., & Yang, G. (2017, February). Monitoring Citrus Soil Moisture and Nutrients Using an IoT Based System. *Sensors (Basel)*, *17*(3), 447. https://doi. org/10.3390/s17030447

Zhao, G., Lianjun, Y., Yu, H., Wang, G., Sui, Y., & Zhang, L. (2016). Design and Implementation of Greenhouse Remote Monitoring System Based on 4G and. Virtual Network. 455–462.

Zhaogan, L., Taiyi, Z., Xiaodong, S., & Xiaohe, L. (2007). Limitations of Current 4G Systems and Its Substitute Schemes with TDD/TDMA. *WSEAS Transactions on Communications*, *6*(11), 850–860.

Chapter 9
Smart System and Services Using Artificial Intelligence and Machine Learning Algorithms:
Sky of AI

K. Sai Bhuvaneshwari
Malla Reddy College of Engineering and Technology, India

ABSTRACT

The concept is devoted to the study of the legal regulation of the use of artificial intelligence in deep learning, machine learning and services. A new service system classification that requires minimum human interaction but works especially based on things interaction. Efficient results can be acquired in the fields of information technology, IoT security, healthcare, business, agriculture, etc. by using AI in smart systems and services. We can use AI algorithms and techniques to solve the issues surrounding regular smart based services, as the future is a clear sky of AI. The goal is to investigate the issues and drawbacks of existing methods and to improve the service's efficiency in order to achieve the best results.

INTRODUCTION

We are witnessing the rapid evaluation of cities using smart systems and services across the world. Before going into the topic, let's have a glance at the exact meaning of smart systems, services, and artificial intelligence. A smart system can be defined as the use of technology to manage consumption and indirectly manage to reduce operational costs and improve efficiency. Artificial intelligence (AI) is concerned with building smart machines capable of performing tasks that typically require human intelligence. The use of AI in smart systems will improve the performance of the device and make tasks easier.

Smart Cities, Smart Manufacturing, and Smart Homes have gained attention not only because they are being addressed by industry practitioners, but also because they are being widely discussed in the scientific literature. Generally, these concepts are associated with advances within specific fields through the development, implementation, and application of intelligent devices and artificial intelligence (AI).

DOI: 10.4018/978-1-6684-6413-7.ch009

Copyright © 2023, IGI Global. Copying or distributing in print or electronic forms without written permission of IGI Global is prohibited.

In upcoming sections we will learn about the history of AI, algorithms and their applications etc.

Objective

Computer science disciplines like artificial intelligence and system simulation are utilized to represent complicated systems. Programming techniques for artificial intelligence provide more accurate and reliable simulation models and support user development, execution, and interpretation of simulation experiments. Artificial intelligence (AI) has the potential to speed up, lower costs, and eliminate mistakes when creating a highly complex, interconnected smart system.

More and more people are thinking about how AI might help with the creation and growth of smart systems. In the simulation of smart systems, AI approaches can be used for a number of purposes, such as knowledge representation in models, decision-making within simulations, rapid model prototyping, data analysis of simulator-generated outputs, and model augmentation and maintenance.

Background of Smart system and Services

Already, smart systems have a significant impact on how we utilize technology in our daily lives. Examples include telephones, computers, home appliances, surveillance equipment, and many other things. Researchers studying smart systems are at the forefront of a bigger computational revolution that is currently taking place. This technical revolution has seen the introduction of new types of tools, automation, services, and access to information and communication. Efficiency gains made possible by computing advancements have already surpassed human imagination from just two decades ago. It is obvious that throughout the coming decades, researchers studying smart systems will drive a number of advancements.

The general consensus among researchers was one of intense enthusiasm and optimism over the potential global impact of (A.I.) smart systems research and development. Many facets of human existence, including healthcare, transportation, education, commerce, information retrieval, and scientific study and discovery, are predicted to benefit significantly from smart systems research.

The History of Artificial Intelligence

The concept of artificial creatures, mechanical men, and other automatons existing or having the potential to exist was first raised by thinkers in antiquity, which is when artificial intelligence first emerged.

Throughout the 1700s and beyond, early thinkers helped artificial intelligence become more real. Philosophers wondered whether it was possible to artificially automate and control non-human machine intelligence. Classical philosophers, mathematicians, and logicians first investigated the mechanical manipulation of symbols, which eventually sparked interest in AI and led to the development of the Atanasoff Berry Computer (ABC), the first programmable digital computer, in the 1940s. The development of an "electronic brain" or artificially intelligent entity was spurred forward by this particular invention.

Before AI icons contributed to our current knowledge of the topic, about ten years had passed. Alan Turing, a mathematician among other things, created a test that determined how well a machine could mimic human behavior. The phrase "artificial intelligence" was first used by computer and cognitive scientist John McCarthy in the middle of that decade during a summer conference held at Dartmouth College.

Many researchers, programmers, logicians, and theorists contributed to the development of the current understanding of artificial intelligence as a whole starting in the 1950s. Every decade brought new discoveries and inventions that altered people's fundamental understanding of artificial intelligence and how historical developments have propelled AI from an impractical fantasy to a viable possibility for both the present and the future.

BASIC ALGORITHMS

Convolution Neural Networks

Convolutional neural networks (CNN or ConvNet) are a subclass of deep neural networks that are mostly employed for object identification, object recognition, and image categorization.

With time, one particular algorithm—a convolutional neural network—has been developed and polished, principally leading to breakthroughs in computer vision with deep learning.

Convolutional neural networks outperform other neural networks when given inputs such as images, voice, or audio, for example. There are three basic categories of layers in them:

1. Convolutional layer
2. Pooling layer
3. Fully connected (FC) layer

A convolutional network's first layer is the convolutional layer. The fully connected layer is the last layer, even though convolutional layers, further convolutional layers, or pooling layers, can come after it. The CNN becomes more complicated with each layer, detecting larger areas of the image. Early layers emphasize basic elements like colors and borders. The larger features or shapes of the object are first recognized when the visual data moves through the CNN layers, and eventually the intended object is recognized.

Convolutional Layer

The convolutional layer, which makes up the core of a CNN, is also where the majority of processing occurs. Among other things, it requires input data, a filter, and a feature map. Assume that the input will be a 3D pixel matrix made up of a color picture. The input will therefore have three dimensions—height, width, and depth—which are comparable to RGB in a picture. We also have a feature detector, sometimes known as a kernel or filter, that will traverse the image's receptive fields and decide if the feature is there or not. This technique is described as convolution.

As seen in the graphic above, it is not necessary to tie every output value in the feature map to every pixel value in the input image. It just has to be linked to the receptive field or the region where the filter is being applied. Due to the fact that not all input values must be perfectly mapped to output values, convolutional (and pooling) layers are commonly referred to as "partially connected" layers. This characteristic is also known as local connection.

1. The depth of the output is influenced by the number of filters. Three distinct filters, for instance, would result in three different feature maps, giving a depth of three.
2. The kernel's stride is the number of pixels it travels over the input matrix. Despite the rarity of stride values of two or higher, a longer stride results in a lesser output.

A CNN performs a Rectified Linear Unit (ReLU) transformation on the feature map following each convolution operation, adding nonlinearity to the model.

The convolutional layer ultimately transforms the image into numerical data, enabling the neural network to understand and extract pertinent patterns.

Pooling Layer

Down sampling, sometimes referred to as pooling layers, carries out dimensionality reduction and lowers the number of parameters in the input. The pooling operation sweeps a filter across the entire input similarly to the convolutional layer, with the exception that this filter lacks weights. Instead, the kernel populates the output array by applying an aggregation function to the values in the receptive field. There are principally two forms of pooling:

- Maximum pooling the filter chooses the input pixel with the highest value to send to the output array as it advances across the input. As a side note, this method is applied more frequently than average pooling.

    1. Average pooling: As the filter moves across the input, it calculates the average value within the receptive field to send to the output array.

Fully Connected Layer

The fully connected layer is exactly what its name implies. As was already noted, partially connected layers do not have a direct connection between the input image's pixel values and the output layer. In contrast, every node in the output layer of the fully connected layer is directly connected to a node in the layer above it.

NAIVE BAYES ALGORITHM

It is a classification technique founded on predictor independence and the Bayes Theorem. Simply expressed, a Naive Bayes classifier thinks that the existence of one feature in a class has no bearing on the presence of any more features.

The Naive Bayes model is simple to construct and is very beneficial for really big data sets. Along with being straightforward, Naive Bayes is known to perform better than even the most complex classification techniques. From P(c), P(x), and P(x|c), the posterior probability P(c|x) can be calculated using the Bayes theorem. Look at the following equation:

Bayes Theorem-

Likelihood — Class Prior Probability

$$P(c \mid x) = \frac{P(x \mid c)P(c)}{P(x)}$$

Posterior Probability — Predictor Prior Probability

$$P(c \mid X) = P(x_1 \mid c) \times P(x_2 \mid c) \times \cdots \times P(x_n \mid c) \times P(c)$$

How Does it Work?

I've included a training set of weather data and a goal variable called "Play" below (suggesting possibilities of playing). We must now categorize whether participants will participate in games based on the weather. Let's carry it out by following the steps:

- Step 1: Create a frequency table from the data set as the first step.
- Step 2: Compile the probabilities into a likelihood table, such as the probability of rain = 0.29 and the probability of playing = 0.64.

Weather	Play
Sunny	No
Overcast	Yes
Rainy	Yes
Sunny	Yes
Sunny	Yes
Overcast	Yes
Rainy	No
Rainy	No
Sunny	Yes
Rainy	Yes
Sunny	No
Overcast	Yes
Overcast	Yes
Rainy	No

Frequency Table		
Weather	No	Yes
Overcast		4
Rainy	3	2
Sunny	2	3
Grand Total	5	9

Likelihood table				
Weather	No	Yes		
Overcast		4	=4/14	0.29
Rainy	3	2	=5/14	0.36
Sunny	2	3	=5/14	0.36
All	5	9		
	=5/14	=9/14		
	0.36	0.64		

- Step 3: Next, get the posterior probability for each class using the Naive Bayesian equation. The result of the prediction is the class with the highest posterior probability.

Even though it's sunny outside, people will still play. Is this a true statement?
The posterior probability method, which was previously explained, can be used to resolve it.

P (Yes I Sunny) is equivalent to P (Yes I Sunny) * P (Yes) / P. (Sunny)

Here, P (Sunny | Yes) = 3/9 = 0, P (Sunny) = 5/14 = 0, and P (Yes) = 9/14 = 0.33, 0.36, and 0.64 respectively.

The chance is now P (Yes | Sunny) = 0.33 * 0.64 / 0.36 = 0.60, which is higher.

A similar strategy is used by Naive Bayes to forecast the probability of various classes based on various attributes. This algorithm is mostly used in text classification and with problems having multiple classes.

K MEANS ALGORITHM

The goal of the Iterative K Means technique is to partition the dataset into K distinct, non-overlapping clusters, each of which contains one data point. It seeks to make the intra-cluster data points as comparable as possible while preserving the clusters as distinct (far) apart as practical. In order to minimize the sum of the squared distances between the data points and the cluster centroid, which represents the average value of all the data points in the cluster, it assigns data points to clusters in a particular pattern. As the degree of variance inside the cluster reduces, the homogeneity (similarity) of the data points within the cluster rises.

The k means algorithm operates in the manner described as follows:

- Indicate how many clusters there are, K.
- Initialize centroids by randomly choosing K data points for the centroids after shuffling the dataset and making no replacements.
- Continue iterating until the centroids remain unchanged. In other words, the clustering of data points remains the same.

$$J = \sum_{i=1}^{m} \sum_{k=1}^{K} w_{ik} \left\| x^i - \mu_k \right\|^2 \tag{1}$$

where wik = 1 for data point xi if it belongs to cluster k; otherwise, wik = 0. Also, k is the centroid of Xi's cluster.

DECISION TREE

One of the most popular and useful methods for supervised learning is the decision tree. Both regression and classification problems may be solved with it, albeit the latter is more frequently used in real-world settings.

There are three different sorts of nodes in this tree-structured classifier. The initial node, known as the Root Node, represents the complete sample and may be divided into other nodes. Interior nodes stand in for a data set's characteristics, while branches stand in for the decision-making processes. Finally, the result is shown by the leaf nodes. For dealing with issues involving decisions, this method is highly helpful.

Figure 1. Decision tree

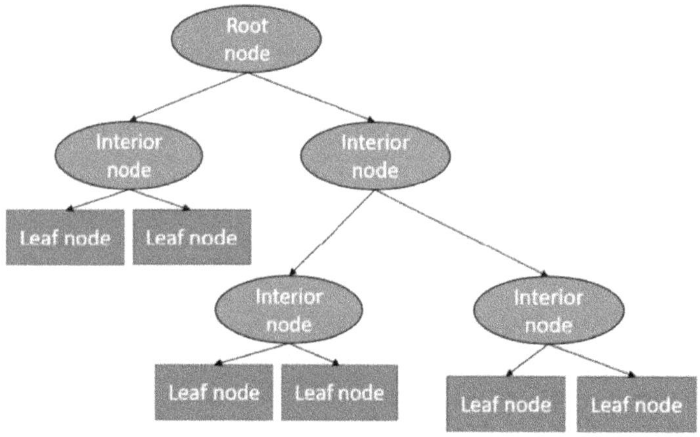

Decision trees offer the advantages of being simple to grasp, requiring minimal data cleansing, performing well even when non-linearity is present, and having essentially no hyper-parameters to modify.

LOGISTIC REGRESSION

A machine learning technique called logistic regression is used to predict the likelihood of a binary outcome. A binary result can only have one of two outcomes. Examples include true/false, 0/1, and yes/no.

Figure 2. Logistic regression

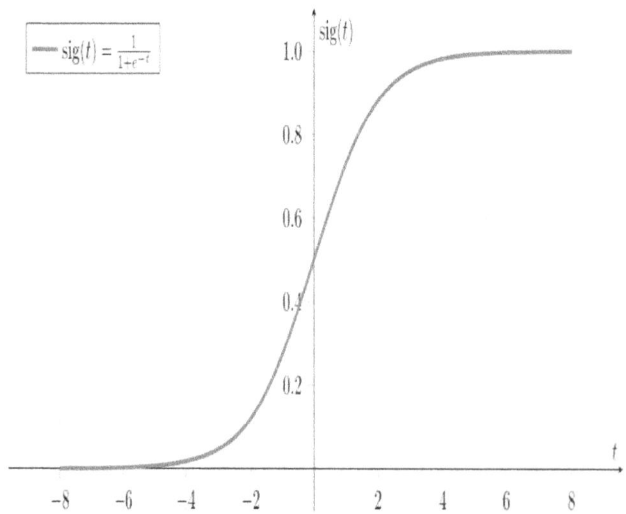

In predictive modeling, logical regression is used to examine big datasets where one or more independent factors might affect the outcome. A dichotomous variable with one of two outcomes is used to express the result.

Logistic regression essentially calculates the likelihood that a given instance belongs to a certain class or not.

The Sigmoid function is used in logistic regression to map expected predictions and their probability. On a graph, the model will predict that the instance belongs to that class if the calculated probability exceeds a predetermined acceptance threshold. The model will predict that the instance does not belong to the class if the estimated probability is less than the predetermined threshold on the graph. There are three fundamental forms of logistic regression in statistics: Predicting the connection between a binary dependent variable (Y) and an independent variable is possible using binary logistic regression (X). When there are two or more discrete outcomes for the dependent variable, and it is irrelevant which outcomes happen first, multinomial logistic regression can be used to make predictions. When the dependent variable comprises more than two discrete outcomes and the order of the outcomes has some relevance, ordinal logistic regression can be used to make predictions.

SUPPORT VECTOR MACHINE

The SVM algorithm, sometimes referred to as the support vector machine algorithm, is, in theory, linear. The SVM algorithm is unique from other algorithms in that it uses an SVM classifier and an SVM regressor to solve classification and regression problems. In contrast, the SVM classifier is the cornerstone of the support vector machine idea and, generally speaking, the best solution for dealing with classification issues.

If the SVM classifier's operation is to be mathematically understood, then it may be comprehended in the following ways:

- **Step 1**: The classifications are predicted by the SVM algorithm. The courses are labeled with a 1 for one and a -1 for the other.
- **Step 2:** Since each machine learning technique first converts a business issue into a mathematical equation with unknowable variables, these unknowable's are then identified by converting the issue into an optimization issue. Since optimization problems often aim at maximizing or lowering something while studying and accounting for the unknowns, the hinge loss function is used in the case of the SVM classifier and tweaked to find the widest margin
- **Step 3:** This loss function can alternatively be thought of as a cost function to make things easier, with a cost of 0 when no class is incorrectly predicted. A loss or mistake is then identified if this is not the case. The problem with the existing state of affairs is that there is a tension between maximizing margin and the significant loss that follows from doing so. These concepts become a theory when a regularization parameter is included.
- **Step 4:** As with the majority of optimization problems, weights are optimized by computing gradients using partial derivatives, an intermediate calculus mathematical concept
- **Step 5:** When there is no classification error, the gradients are updated just using the regularization parameter; when misclassification occurs, the loss function is also utilized
- **Step 6:** The gradients are only updated using the regularization parameter while there is no classification error, and when misclassification occurs, the loss function is also employed.

Support Vector Machines (SVM), Support Vector Regression (SVR), and other classification techniques have long been employed in machine learning and data science applications. SVM/SVR is simply one of several algorithms that may be employed, whether in the areas of image, text, NLP, or anything else. Therefore, based on the complexity and use cases, choose the best (and simplest) algorithm is important to eventually save time and money.

Applications of Smart Systems and Services using AI Algorithms

Medical Image Computing

The sector of healthcare is where all the most advanced technologies are tested.

Try implementing a given technology in a healthcare setting to evaluate its applicability. The same applies to image recognition. The most intriguing use of picture recognition is in the field of medical imaging. A significant amount of additional data analysis is included in the medical image as a result of first image recognition. Better than the human eye, the CNN medical image classification can identify anomalies in X-ray and MRI pictures.

Figure 3.

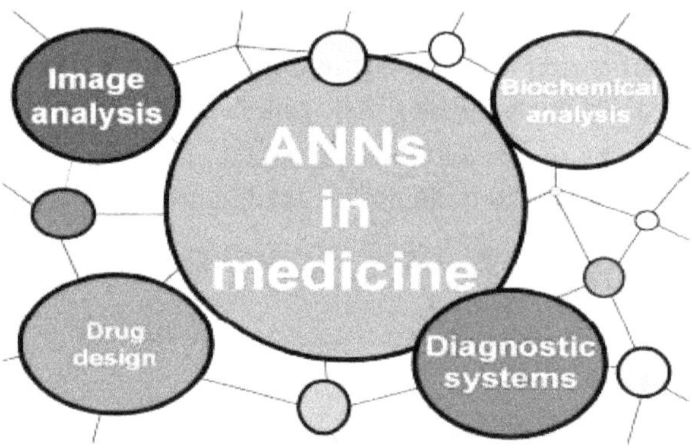

These devices can show both the differences between a series of images and the photos themselves. Future predictive analytics will be enabled by this capability.

The classification of medical images is dependent on large databases, such as public health records. It serves as a training ground for the algorithms, as well as for private patient information and test results. They collaborate to develop an analytical platform that tracks the patient's current condition and projects outcomes. Data is processed by CNN using a grid topology technique, which is a collection of spatial correlations between data points. For photos, the grid is only two dimensional. For time series textual data, the grid is one dimension. Next, various characteristics of the input are identified using the convolution process. Consider the variances in the input. Find sparse interactions between the variables. For all of a model's functionalities, use the same configurations

Neural networks show how much can be done with their help even in the simplest applications. A lot can be learned about the design and implementation of the graphics from how CNN detects images. Contrarily, convolutional neural networks discover innovative drugs, which is only one of many incredible examples of how artificial neural networks are improving the world.

Test Classification

An essential component of natural language processing is text classification. In the current digital era, text is everywhere—on our social media accounts, in advertisements, on websites, in e-books, etc. Since most of this text data is unstructured, categorizing it can be quite helpful.

There are several uses for text classification. Popular applications include

Email spam detection, Online content tagging, language detection, such as in Google Translate, sentiment analysis of online reviews and etc.

It might be a good idea to use Naive Bayes text classifiers, especially if there is a lack of training data and computational resources. If features are well engineered, results are typically quite competitive in terms of performance.

Figure 4. Text classification

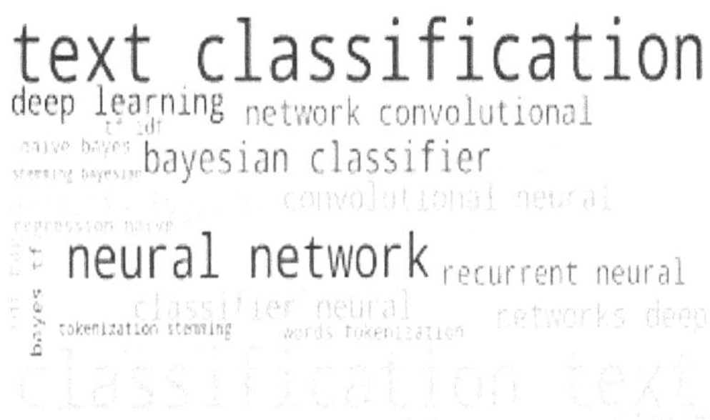

Customer Segmentation

Customer segmentation is the dividing of a market into distinct groups of customers who have common traits. Client segmentation can be an effective tool for locating and meeting unmet customer demands.

Figure 5. Customer segmentation

Using the aforementioned information, businesses can subsequently surpass their rivals by creating distinctively appealing goods and services.

Demographic data, which varies depending on the company's scope, including gender, age, familial and marital status, income, education, and occupation. If a business is localized, this information may apply to particular towns or counties. For bigger businesses, it could refer to a customer's home city, state, or even nation.

Organizing data points into discrete, non-overlapping subgroups is the aim of K means. One of the main uses of K means clustering is the segmentation of consumers to gain a deeper understanding of them, which might then be used to boost the company's income.

Sentiment Analysis

Sentiment analysis is primarily concerned with a text's polarity (positive, negative, or neutral), but it also extends beyond polarity to identify certain moods and emotions (angry, joyful, sad, etc.), urgency (urgent, not urgent), and even intents (interested v. not interested).

Figure 6. Sentiment analysis

Beyond polarity, emotion detection sentiment analysis enables you to find emotions like happiness, frustration, wrath, and sadness.

Brands may discover what makes consumers happy or irritated by automatically evaluating customer feedback, such as comments in survey replies and social media conversations, in order to modify products and services to suit their requirements.

Sentiment analysis focuses on the polarity of a text (positive, negative, neutral), but it also goes beyond polarity to identify particular sentiments and emotions (angry, pleased, sad, etc), urgency (urgent, not urgent) and even intents (interested v. not interested) (interested versus not interested).

Beyond polarity, emotion detection sentiment analysis enables you to find emotions like happiness, frustration, wrath, and sadness. Brands may discover what makes consumers happy or irritated by automatically evaluating customer feedback, such as comments in survey replies and social media conversations, in order to modify products and services to suit their requirements.

Numerous corporate functions, such as brand monitoring, product analytics, customer service, and market research, might benefit from sentiment analysis. Leading companies (not to mention entire cities) may operate more quickly, more accurately, and toward more beneficial goals by integrating it into their current systems and analytics.

Facial Recognition

Using a combination of algorithms, facial recognition technology can recognize individuals in a static image or video. Although this technology has been around for a while, it has recently grown significantly in popularity and innovation.

The incorporation of artificial intelligence (AI) into facial recognition systems is one such advancement. Intelligent software powered by AI can instantly search face databases and match them to one or more faces found in a scenario. You may receive very accurate findings immediately; generally, systems provide 99.5% accuracy rates on open standard data sets.

Figure 7. Facial recognition

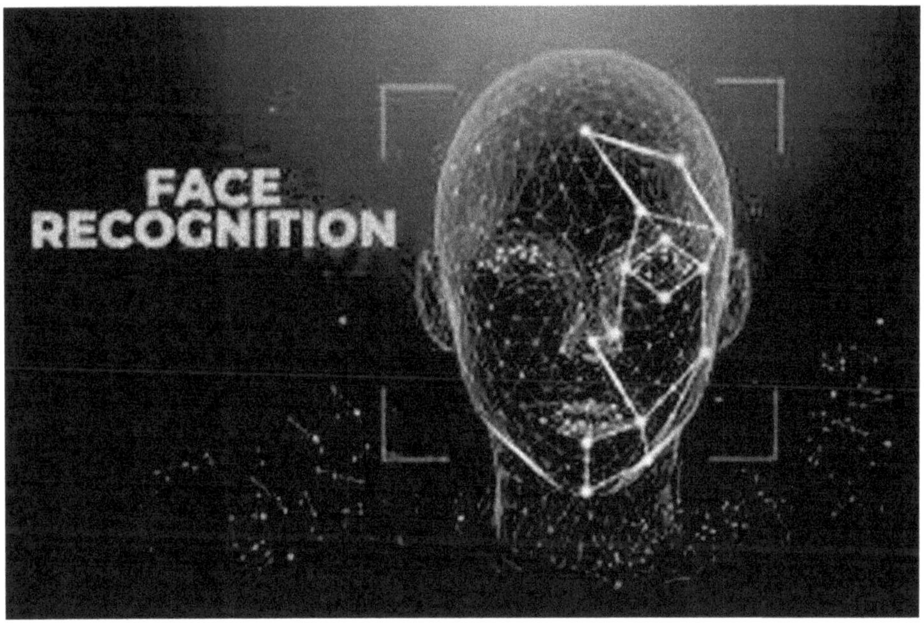

However, even well-trained AI facial recognition systems don't have real-world context and can be fooled. If you see a colleague who is wearing a face mask, sunglasses, and a baseball cap, you may still recognize them. An AI system, however, might not. It depends on the level of training the neural network. Even though AI facial recognition systems are more superficially accurate, it is also easier for them to blunder under less-than-ideal conditions. For security and surveillance purposes, a model can compare those calculations to other facial calculations located within a database. But any AI facial recognition system has to train with a lot of face picture data, independent of the use case. AI models need to be trained on facial photos that differ in terms of age, ethnicity, lighting, angles, and other elements. Although AI facial recognition is effective, there are several ethical concerns. Facial recognition is used in Augmented reality, cashless payments, security gates, patient identification, attendance monitoring, running cardless ATM transactions and etc.,.

- **Water:** Enhanced service delivery, better resource management, and deeper understanding of infrastructure health. AI can forecast soil moisture and nutrient levels on farms in conjunction with satellites to increase agricultural yields while using less water.
- **Power:** By using AI-controlled smart power grids, renewable energy sources will be used more effectively and introduced more quickly;
- **Social Services and Governance:** Better insights into urgent concerns may be provided by predictive analytics, which will increase the effectiveness with which they are handled. enhanced assistance to those in need, improved resource coordination, and a general simplification of bureaucracy
- **Housing:** AI-powered tools can gather and evaluate building data from several data points, including electricity, humidity, carbon dioxide, temperature, light, and many more. The study of short- and long-term trends, the need for upgrades and repairs, the reduction of waste, and the use of energy may all be improved with the help of this data.
- **Police/Fire/Ambulance Service:** By using mathematical models that address spatiotemporal relationships between various places, neural networks are able to anticipate traffic in all locations. This may then be utilized to better track crime patterns, possible dangers, and reaction information.

Smart technologies are also anticipated to significantly increase citizens' access to services, mostly through easing interactions between the public and administrative bodies. We've all faced drawn-out lineups and challenging procedures at places like the Department of Motor Vehicles, for example. Citizens should be able to do a variety of basic tasks via an automated contact center or online portal thanks to advances in intelligent technology like natural language processing and neural networking, thereby lowering wait times for those in need of more complicated services.

Issues with Smart things

The majority of smart gadgets lack built-in security or privacy measures to safeguard the transmission of sensitive data. The relatively few that do have controls generally don't have them set to be protected by default, which leads users to fail to install security controls because they wrongly think that security is enabled by default because they were touted as having security built in. The users unknowingly expose themselves to unwanted access as a result.

we have met with and conducted assessments for hundreds of IoT app and device developers over the past few years, but I have yet to come across a single maker of smart devices that had all of the following security and privacy protections pre-installed and turned on by default as Strong encryption for data in storage and in transit, Multi-factor authentication, activity logging, device management user interface etc.

Online connections provide for access to devices. Those widely used IoT devices, including many that are bought to increase physical security, lack authentication and encryption and are readily discoverable using tools like Shodan. This enables potential attackers to connect directly to these devices, circumventing any firewall constraints. Numerous devices also have flaws that enable unwanted cyberstalkers to peer inside.

Just like a coin with two sides, smart devices and services also have two sides, therefore we must accept this fact and consider the benefits of technology.

CONCLUSION

This research should demonstrate how AI advances social intelligence. We found that AI technology supports device management, energy management, healthcare, intelligent interaction, security, entertainment systems, and personal robots by utilizing activity recognition and data processing, decision making, image recognition, prediction, and voice recognition. Numerous studies in the literature and many products have been reviewed to achieve this goal. Literature and goods are behind one another, with the latter emphasizing relatively straightforward techniques like speech and picture recognition. The machine learning literature focuses on trickier methods like activity detection and prediction-making.

This chapter provides a comprehensive overview of AI-based intelligent systems that are an important part of society. This article describes research motivations and how artificial intelligence has evolved in many application areas. Subsequently, important technologies in this area will be considered in many aspects. This in-depth analysis explores 10 categories of popular AI technologies such as machine learning, deep learning, natural language processing, knowledge discovery, and expert system modeling. It can be applied to different applications depending on your current needs. From a machine intelligence perspective, complex learning algorithms need to be trained with data and knowledge from targeted applications before the system can help intelligent decisions used in intelligent systems and services.

ACKNOWLEDGMENT

There is a race of existence in the current competitive environment where those who have the desire to advance succeed. Between theoretical and actual activity, a project acts as a bridge. I voluntarily became a member of this specific chapter. First and foremost, I want to express my gratitude to the highest power for my sister, who is undoubtedly the one who has always led the way in helping me find the proper route in life. This chapter would not have been possible without her grace. Next to him are my parents, to whom I owe a great deal for loving and supporting me as I grew up to this point. I felt compelled to thank Dr. Gouri Ganesh in person for his assistance.

Additionally, I feel obligated to take this time to express my gratitude to all of the teachers at Mallareddy College for their kind dispositions and demeanors. Last but not least, I want to express my gratitude to all of my friends for supporting and encouraging me throughout the year. Although I am speechless in gratitude, my heart is still overflowing with gratitude for all the kindnesses I have gotten from everyone.

ADDITIONAL READING

Appleton, J. (2021, April 6). The Use Of Ai For Smart Urban Services In Smart Cities. *Bee Smart City*. https://www.beesmart.city/en/solutions/the-use-of-ai-for-smart-urban-services-in-smart-cities

Kumari, R., & Saini, K. (2021, March). Advanced Automobile Manufacturing: An Industry 4.0. In *2021 8th International Conference on Computing for Sustainable Global Development (INDIACom)* (pp. 899-904). IEEE.

Lee, D., & Park, J. H. (2019). Future trends of AI-based smart systems and services: Challenges, opportunities, and solutions. *Journal of Information Processing Systems*, *15*(4), 717–723.

Raj, P., Saini, K., & Surianarayanan, C. (2022). *Edge/Fog Computing Paradigm: The Concept, Platforms and Applications*. Academic Press.

Ray, S. (2015). Six Easy Steps to Learn Naive Bayes Algorithm (with code in Python). *Analytics Vidhya*. https://www. analyticsvidhya. com/blog/2015/09/naive-bayes

Saini, K. (2019). Evolution of blockchain technology in business applications. [JETIR]. *Journal of Emerging Technologies and Innovative Research*, *6*(9), 240–244.

Saini, K., Mishra, I., & Srivastava, S. (2021). Farmer's E-mart: An e-commerce store for crops. *3rd International Conference on Advances in Computing, Communication Control. And Networking* (ICAC3N), (pp. 346–350). doi:10.1109/ICAC3N53548.2021.9725783

Saini, K., & Raj, P. (2021, December). *Advancing smarter and more secure industrial applications using AI, IoT, and blockchain technology*. IGI Global. doi:10.4018/978-1-7998-8367-8

KEY TERMS AND DEFINITIONS

AI Algorithm: An AI algorithm is essentially an expanded subset of machine learning that instructs the computer how to learn to function independently. In turn, the machine keeps learning to enhance procedures and carry out duties more effectively.

Artificial Intelligence: The replication of human intelligence functions by machines, particularly computer systems, is known as artificial intelligence. Expert systems, natural language processing, speech recognition, and machine vision are some examples of specific AI applications.

Clustering: We frequently group instances in machine learning as a first step to comprehend a subject (data set) in a machine learning system. Clustering is the process of collecting unlabeled samples.

Deep Learning: Artificial intelligence (AI) and machine learning techniques called deep learning model how people acquire specific types of information. Data science, which also encompasses statistics and predictive modelling, contains deep learning as a key component.

Machine Learning: Without being specifically designed to do so, machine learning (ML), a form of artificial intelligence (AI), enables software systems to become better at making predictions about the future.

Regression: The statistical link between a dependent variable and one or more independent variables can be ascertained using regression. The independent variable's change is connected to the independent variables' change.

Smart Services: A smart service is a digital service that responds to data that has been gathered and evaluated using platforms and systems that are networked and intelligent.

Smart System: Smart systems combine sensing, actuation, and control capabilities to describe and analyze a condition, make decisions that are predictive or adaptive, and take smart actions as a result. Most of the time, the autonomous operation based on closed loop control, energy efficiency, and networking capabilities can be used to explain the system's "smartness."

Chapter 10
Machine and Deep Learning (ML/DL) Algorithms, Frameworks, and Libraries

Jigna Bhupendra Prajapati
Ganpat University, India

Savan Patel
Achaya Motibhai Patel Intitute of Computer Studies, India

Richesh Gaurav
Achaya Motibhai Patel Intitute of Computer Studies, India

Dhvanil Nileshkumar Prajapati
LDRP Institute of Technology and Research, India

Kavita Saini
https://orcid.org/0000-0003-4058-8622
Glgotias University, India

ABSTRACT

Each primary, secondary, tertiary, quaternary, and the quinary sector has huge or very huge incremental data from large-scale, small-scale industries, medium industries, or cottage industries. The data associated with each of them are very crucial from every point of view. The complex problems are increasing day by day in real-time execution which can be addressed using current trends of technology like machine learning and deep learning. Machine learning is a subset of artificial intelligence. ML is functioning for image & speech recognition, mail filtering, Facebook tagging mechanism, and many others. Deep learning is an advanced technology that is a subset of machine learning with the capacity to learn more intelligently on a large set of data. Deep learning works with multiple hidden layers to produce the predicted outcomes. Deep learning algorithms include convolutional neural networks, recurrent neural networks, long short-term memory networks, stacked auto-encoders, deep boltzmann machines &, etc.

DOI: 10.4018/978-1-6684-6413-7.ch010

Copyright © 2023, IGI Global. Copying or distributing in print or electronic forms without written permission of IGI Global is prohibited.

INTRODUCTION

Machine learning is technology which assist to solve very complex. It empowers system to learn itself with using Previously collected data. ML works with Multiple algorithms using mathematical base models & historical data. As human leans from his/her own experience, as human make decision for certain known-unknown matters, similarly machine learns from historical data and make decision accordingly in machine leaning mechanism. Machine learning is subset of artificial intelligence. ML is functioning for image & speech recognition, mail filtering, Facebook tagging mechanisms, and many others. The training dataset is very important to predict accurate results with the concern appropriate mathematical base model. The performance of decision making will be increase with correct combination of chosen algorithm and collected & train dataset. Machine learning combines the statistical & computer science approaches for generating various models. There are many popular machine learning algorithms as below

- Regression-Algorithms
- Instance-based Algorithms
- Regularization Algorithms
- Decision Tree Algorithms
- Bayesian Algorithms
- Clustering Algorithms
- Association Rule Learning Algorithms,
- Artificial Neural Network Algorithms

The regression algorithms includes various algorithms as ordinary least squares regression, linear regression, logistic regression, stepwise regression, multivariate adaptive regression splines & etc. The instance-based algorithms includes k-nearest neighbour, learning vector quantization, self-organizing map, support vector machines & etc. The regularization algorithms includes ridge regression, least absolute shrinkage and selection operator, elastic net, least-angle regression & etc. The decision tree algorithms includes classification and regression tree, decision stump, conditional decision trees & etc. The bayesian algorithms includes naive bayes, gaussian naive bayes, multinomial naive bayes, averaged one-dependence estimators, bayesian belief network, bayesian network & etc. The clustering algorithms includes k-means, k-medians, expectation maximisation (em), hierarchical clustering. the association rule learning algorithms indues apriori algorithm, eclat algorithm & etc

The artificial neural network algorithms includes perceptron, multilayer perceptron's, back-propagation, stochastic gradient descent, Hopfield network & etc. Some other algorisms are available to perform on complex problem.

MACHINE LEARNING

A developing technique called machine learning makes it possible for computers to learn autonomously from historical data. Machine learning employs a variety of techniques to create mathematical models and make predictions based on previous knowledge or data (Ayodele, 2010; Kotsiantis, 2007). Currently, it is utilised for many different things, including recommender systems, email filtering, Facebook auto-tagging, picture identification, and speech recognition (Wang, 2003).

Machine also learn from experiences or past data like a human does. In the actual world, individuals who are capable of learning from their experiences are all around us, and computers or other things that follow our commands similarly Machine does.

Machine learning functioning various approaches including supervised, unsupervised, and reinforcement learning. The sequential models, hidden Markov models, clustering techniques, and regression and classification models are important factors for same.

Machine learning system gets new data and on the base of it, ML forecasts the outcome using the prediction models. The quantity of data used determines how well the output is anticipated, as a larger data set makes it easier to create a model that predicts the outcome more precisely (Dove et al., 2017).

Sometimes very difficult problem that requires certain predictions. Instead of creating code for it, we can just input the data to generic algorithms, and the machine would develop the logic according to the data and forecast the result. The perspective on the issue has altered as a result of machine learning.

Supervised Learning

Machines are taught using training data that has been carefully "labelled" and then anticipate the outcome using supervised learning techniques. The term "labelled data" refers to information that has already been associated with the desired outcome (Ayodele, 2010; Moore, 2019).

The training data that is given to the computers acts as the supervisor in supervised learning, teaching them how to anticipate the output properly. The same idea that a pupil learns under a teacher's guidance is applied.

Incorrect input and output data are provided to the machine learning model during supervised learning. A supervised learning algorithm's goal is to identify a mapping function that will connect the input variable (x) and the output variable (y). Supervised learning may be applied in the real world for tasks including spam filtering, fraud detection, risk assessment, and image categorization (Tracey, Duraisamy, and Alonso, 2013).

Regression trains on and predicts a continuous-valued response, for example predicting real estate prices. Classification attempts to find the appropriate class label, such as analyzing positive/negative sentiment, male and female persons, benign and malignant tumors, secure and unsecure loans etc (Nasteski, 2017; Sen, 2020).

Unsupervised Learning

Unsupervised learning is used to categorise clients who exhibit similar behaviours for a sales campaign or to identify abnormalities and outliers, such as fraud or damaged equipment. In contrast to supervised learning, it. This data is not labelled (Barlow, 1989; Ayodele, 2010).

It is up to the coder or the algorithm to figure out the structure of the underlying data, to detect hidden patterns, or to decide how to characterise the data when learning data simply provides certain signals without any descriptions or labels. Unlabelled data is the name for this type of learning (Ghahramani, 2003).

Semi-Supervised Learning

Semi-supervised learning is when some learning samples have labels while others do not. For training, it uses a lot of unlabeled data, and for testing, it uses a lot of labelled data. When labelling a small subset is more practical than labelling the entire dataset since it is more expensive, semi-supervised learning is used. For instance, labelling some remote sensing photos frequently takes specialised expertise, and finding oil at a certain site generally involves several field operations, yet obtaining unlabeled data is quite simple (Ayodele, 2010).

Reinforcement Learning

Reinforcement Learning provides feedback so that the system may adapt to changing circumstances in order to fulfil a certain goal (Kaelbling, 1996). Based on the feedback answers, the system assesses its performance and responds appropriately. The most well-known examples include self-driving vehicles and the AlphaGo chess programme (Dayan, 2008).

ML Application

Figure 1 represent the popular machine Learning Application used in various industries as automobile, healthcare, language processing, finance, education, marketing and many more.

SUPERVISED LEARNING

Regression

Regression is a technique is broadly popular for two concepts. First, regression analyses are frequently used for forecasting and prediction, two areas where machine learning and their application have a lot in common. Second, in some circumstances, causal relationships between the independent and dependent variables can be ascertained using regression analysis. Regressions alone, it is important to note, only reveal relationships between a dependent variable and a defined dataset of other variables (Maulud, 2020).

A key idea in the study of machine learning is regression analysis. The system is learned using both input characteristics and output labels, which is supervised learning. By calculating how one variable affects the other, it aids in establishing a link between the variables. If there is a correlation between the input and output variables, regression procedures are applied (Cunningham, 2008).

Linear Regression

One of the simplest and most used machine learning methods is linear regression. It is a method of performing predictive analysis that is mathematical. Projections of continuous, real, or mathematical variables are possible using linear regression. In 1894, Sir Francis Galton was the first to propose the idea of linear (Weisberg, 2005).

Figure 1. Application of ML

It is grounded on the mathematical idea of a linear equation, which establishes the relationship between the dependent variable "Y" and the independent variable "X," and the general formula for a linear equation is (i.e., in case of only one independent variable)

$$\Upsilon = MX + C$$

Typically, linear regression uses a best-fit straight line to determine the linear connection between the dependent variable and one or more independent variables. Typically, a linear model simply calculates the weighted sum of the input characteristics plus a constant known as the bias term to create a prediction (also called the intercept term). This method uses a continuous dependent variable, one or more continuous or discrete independent variables, and a linear regression line (Montgomery, 2021). The prediction using linear regression is stated as follows mathematically:

$$\Upsilon = \Theta_0 + \Theta_1 \chi_1 + \Theta_2 \chi_2 + \cdots + \Theta_n \chi_n$$

Here, Υ is the predicted value,

n is the total number of input features,
χ_i is the input feature for ith value,
Θ_i is the model parameter (Θ_0 is the bias and the coefficients are Θ_1, Θ_2, Θ_n).

The slope of the line in the regression equation, which is M in the instance of Y=MX+C, is referred to as a regression coefficient. It changes depending on the matching input property, changing the resultant value. In the context of linear regression analysis, bias is the inaccuracy introduced by equating a poten-

Figure 2.

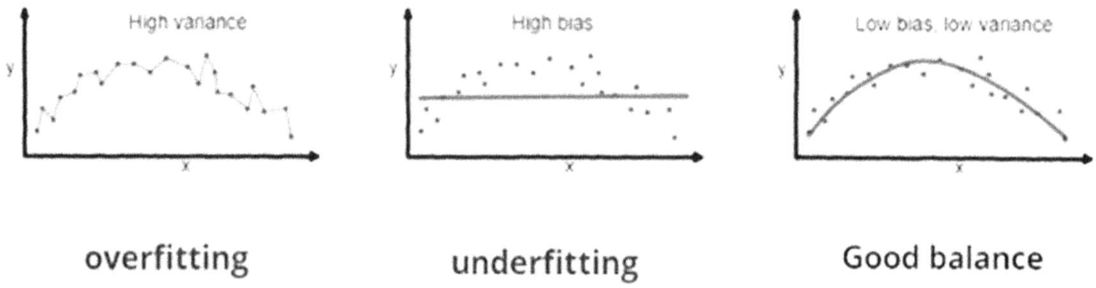

tially complex real-world situation with a relatively simpler model. For the predictions we make, bias is a divergence from the line equation "y = mx". Bias is the propensity of an algorithm to always learn the incorrect thing by ignoring all the information in the input. A model must have low bias in order to be accurate. It gauges how much a dataset's expected or projected value deviates from the actual value.

Variance

A measure of variability is the variance. Variance is the difference in target function estimation when different training sets are employed. The relationship between the input and output variables is established by the target function f. The goal function must maintain stability and low variance when a new dataset is utilised since the model should be versatile enough to handle any type of data. In essence, variance estimates the average amount of variation around the predicted value (between datasets).

THE BIAS-VARIANCE TRADE-OFF - The trade-off between bias and variance never ends. When the variance is low, the bias is large and vice versa when the variance is high. If the trade-off is handled appropriately, it will eventually bubble up to the target curve, which is what we need in terms of variance and bias to be as modest as feasible.

A model should have low variance, low bias, and low error in order to be considered optimal. We must divide the dataset into train and test datasets in order to do this (i.e., usually 40:60 ratio for training and testing datasets respectively). Following pattern learning using the training dataset, the model's performance is assessed using the test dataset. We develop an error function to lower the error as the model learns. The accuracy on the test dataset is poor since the curve created from the training model would then pass across all the data points. High variance leads to overfitting, which is what this is. On the other hand, underfitting occurs when a model exhibits good performance on test data but poor accuracy on training data (Su, 2012).

- **Multiple Linear Regression:** There are several independent variables, but this is identical to simple linear regression. There is a value of the dependent variable y for each value of the independent variable x. The best-fit line is a plane since it is a multi-dimensional representation. It may be mathematically stated as: $Y = \beta_0 + \beta_1 X_{1i} + \beta_2 X_{2i} + \cdots + \beta_k X_{ki}$

The parameters for multiple regression may be determined in the same manner as for basic linear regression by applying gradient descent to minimise the cost function (Eberly, 2007; Tranmer, 2008).

- **Multivariate Linear Regression:** Multivariate linear regression, as the name suggests, deals with several output variables. When there are multiple independent variables, multivariate regression is used because simple linear regression is ineffective. Multiple variables or characteristics are present in real-world data, and when they are, multivariate regression is necessary for a more thorough examination (Uyanık, 2013).

Logistic Regression

A classification algorithm is logistic regression. On the basis of a number of independent factors, it is used to forecast a binary result. There are only two outcomes that may occur in a binary situation: either the event occurs (1) or it doesn't (0). Independent variables are those variables or elements that might have an impact on the result (LaValley, 2008).When dealing with binary data, the proper form of analysis to utilise is logistic regression. When the output or dependent variable is dichotomous or categorical in nature, or if it falls into one of two categories (such as "yes" or "no," "pass" or "fail," and so on), you know you are working with binary data (Menard, 2002).

Binary Logistic Regression

When the dependent variable is binary in nature, it is the statistical method used to forecast the connection between the dependent variable (Y) and the independent variable (X). The result might be Success/ Failure, 0/1, True/False, or Yes/No, for instance (King, 2008).

Multinomial Logistic Regression

When a categorical dependent variable has two or more unordered levels, it is utilised (i.e., two or more discrete outcomes). Except that there are more than two possible outcomes here, it is extremely similar to logistic regression. Consider the scenario where you wish to foretell the most popular mode of transportation in 2030. The dependent variable will be the mode of transportation, with the train, bus, tram, and bicycle as potential outputs (Kwak, 2002; Böhning, 1992).

Ordinal Logistic Regression

When the dependent variable (Y) is ordered, it is employed (i.e., ordinal). The dependent variable has more than two categories or levels and a meaningful order. Such variables may include things like t-shirt sizes (XS/S/M/L/XL), responses to surveys (Agree/Disagree/Neutral), or test results (Poor/Average/ Good)(Harrell, 2015 #361).

Instance-Based Algorithms

It is a class of learning algorithms that compares new issue cases with examples encountered during training that have been memorised rather than doing explicit generalisation. sometimes Memory-based learning is the name for it (Aha, 1991).

Machine learning systems that fall under the category of instance-based learning are those that memorise training examples before generalising to fresh cases using some sort of similarity metric. Because it constructs the hypotheses from the training cases, it is known as instance-based. It is sometimes referred to as sluggish learning or memory-based learning. As an illustration, if we were to develop a spam filter using an instance-based learning algorithm, in addition to identifying emails that have already been classified as spam, our spam filter would also be trained to detect emails that are strikingly similar to those emails. A degree of similarity between two emails is necessary for this. The same sender, frequent usage of the same terms, or another characteristic may be used to compare two emails (Brighton, 2002).

K Nearest Neighbour (KNN)

The Supervised Learning subcategory includes the K Nearest Neighbor algorithm, which is used for regression and classification most frequently. It is a flexible approach that may also be used to resample datasets and impute missing values. The algorithm uses K Nearest Neighbors (Data points) to forecast the class or continuous value for a new Datapoint, as the name of the algorithm indicates [18]. When you are undertaking a pattern recognition task for classifying objects based on various attributes, the KNN method is helpful (Shouman, 2012).

Classification

The process of predicting a class or category from observed values or provided data points is known as classification. The output can be classed in the following ways: "Black" or "White"; "spam" or "no spam". Calculating an approximate mapping function (f) from input variables (X) to output variables is the problem of classification in mathematics (Y). It essentially falls under the category of supervised machine learning, where objectives are supplied in addition to the input data set (Kotsiantis, 2007).

Random Forest

Popular machine learning algorithm Random Forest is a part of the supervised learning methodology. It may be applied to ML issues involving both classification and regression. It is built on the idea of ensemble learning, which is a method of integrating many classifiers to address difficult issues and enhance model performance (Biau, 2016).

Random Forest, as the name implies, is a classifier that uses a number of decision trees on different subsets of the provided dataset and averages them to increase the dataset's prediction accuracy. Instead than depending on a single decision tree, the random forest uses forecasts from each tree and predicts the result based on the votes of the majority of predictions (Rigatti, 2017). The greater number of trees in the forest leads to higher accuracy and prevents the problem of overfitting (Oshiro, Perez, and Baranauskas, 2012).

Decision Trees

One of the most widely used machine learning algorithms nowadays is the decision tree algorithm; it is a supervised learning method used to categorise situations. Both categorical and continuous dependent variables may be classified well using this method. Based on the most important characteristics/independent variables, this method divides the population into two or more homogenous groupings (Kingsford, 2008). It allows us to carefully consider all of the potential effects of a choice. It gives us a framework to evaluate the worth of outcomes and the likelihood that they will materialise. It enables us to base our judgments on the most accurate predictions and available facts (Quinlan, 1986).

Support Vector Machines

Using the SVM technique, you may classify data by plotting the raw data as dots in an n-dimensional space (where n is the number of features you have). It is therefore simple to categorise the data because each feature's value is connected to a specific coordinate (Steinwart, 2008; Meyer, 2015).

- **Linear SVM:** Linear SVM is used for data that can be divided into two classes using a single straight line. This type of data is called linearly separable data, and the classifier employed is known as a Linear SVM classifier. Non-linear SVM: Non-Linear SVM is used for non-linearly separated data. If a dataset cannot be classified using a straight line, it is considered non-linear data, and the classifier employed is referred to as a Non-linear SVM classifier (Chen, 2005; Kecman, 2005).

UNSUPERVISED LEARNING

Unsupervised learning is a subcategory of machine learning in which models are trained using unlabeled datasets and are free to operate on the data without being checked by a human observer. Because unlike supervised learning, we have the input data but no corresponding output data, unsupervised learning cannot be used to solve a regression or classification issue directly. Finding the underlying structure of a dataset, classifying the data into groups based on similarities, and representing the dataset in a compressed manner are the objectives of unsupervised learning (Ghahramani, 2003; Hastie, 2009).

Clustering

Using the clustering technique, items are grouped into clusters so that those who have the most similarities stay in one group and share little to none with those in another. The data items are classified based on the existence or lack of similarities discovered by cluster analysis (Al-Omary, 2006).

Density-Based

In density-based clustering, data is categorised into regions with high data point densities that are encircled by regions with low densities. In essence, the algorithm identifies areas with a high density of data points and refers to those areas as clusters (Campello, 2020).

Distribution-Based

Using a distribution-based clustering method, each data point is regarded as a component of a cluster according to the likelihood that it is one. A centre point is present, and as a data point gets farther from the centre, its likelihood of becoming a part of a cluster reduces (Kameshwaran, 2014).

K-Means Clustering

The most used clustering algorithm is K-means clustering. It is the most straightforward unsupervised learning approach and is centroid based. This technique seeks to reduce data point variation inside a cluster. Additionally, it's how the majority of people first encounter unsupervised machine learning. Because K-means iterates through all of the data points, it is best applied to smaller data sets. As a result, if the data set contains a lot of data points, classifying them will take longer (Likas, 2003; Kanungo, 2002).

DBSCAN Clustering

Density-based spatial grouping of applications with noise is referred to as DBSCAN. Unlike k-means, it uses a density-based clustering technique. This approach works well for locating outliers in a data collection. Based on the density of data points in various places, it discovers clusters of any shape. In order to identify outliers between the high-density clusters, it divides regions into low-density zones (Meng'Ao et al., 2015). When dealing with data that is irregularly shaped, this method performs better than k-means. DBSCAN employs the minPts (the minimal number of data points that must be clustered together for an area to be considered high-density) and eps parameters to decide how clusters are created (the distance used to determine if a data point is in the same area as other data points). The success of this method depends on selecting the appropriate initial parameters.

Gaussian Mixture Model

The fact that the data must be organised in a circular pattern is one of the issues with k-means. Non-circular data is incorrectly grouped because the method k-means calculates the distance between data points involves a circular journey. Gaussian mixture models can solve this problem. The data need not be circular for it to function properly (Yang, 2012).The Gaussian mixture model fits data with arbitrary shapes by combining several Gaussian distributions.

In this hybrid model, there are many single Gaussian models that function as hidden layers. In order to determine which cluster a data point will go under; the model determines the likelihood that the data point belongs to a certain Gaussian distribution.

Association

Association rule is used to uncover the connections among the variables in a sizable database. It establishes the group of items that co-occur in the collection. Marketing strategy is more effective when the association rule is used. People who buy X (let's say, bread) also frequently buy Y (let's say, butter or jam). Market Basket Analysis is a classic illustration of an association rule (Mannila, 1996).

Apri Algorithm

To produce association rules, this method makes advantage of frequent datasets. It is intended to operate on transaction-containing databases. The itemset is effectively calculated by this approach using a breadth-first search and a hash tree.

It aids in understanding the items that may be purchased in combination and is mostly used for market basket analysis. To detect medication responses for patients, it may also be applied in the healthcare industry.

Eclat Algorithm

Eclat algorithm stands for Equivalence Class Transformation. This algorithm uses a depth-first search technique to find frequent item sets in a transaction database. It performs faster execution than Apriori Algorithm (Yu, 2014).

F-P Growth Algorithm

The F-P growth algorithm stands for Frequent Pattern, and it is the improved version of the Apriori Algorithm. It represents the database in the form of a tree structure that is known as a frequent pattern or tree. The purpose of this frequent tree is to extract the most frequent patterns (Borgelt, 2005).

DEEP LEARNING

Machine learning, a subfield of artificial intelligence, is the basis of deep learning. Deep learning will function as neural networks mimic the human thought processes. Nothing is explicitly coded in deep learning. In essence, it is a kind of machine learning that does feature extraction and modification using a huge proportion of nonlinear processing units. Each subsequent layer takes the output from the layer below as its input (Learning, 2020).

Deep learning models are quite effective in resolving the dimensionality issues to focus on the accurate features by themselves with additional programming assistance. Particularly when there are many inputs and outputs, deep learning methods are employed (Lauzon, 2012).

Artificial Neural Network

Artificial neural networks are a subfield of AI influenced by biology and fashioned after the brain. A computational network based on biological neural nets, which create the aspects of the human brain, is usually referred to as an artificial neural network (Zhang, 2000). Artificial neural networks also feature neurons that are linked to each other in different levels of the networks, just as neurons in a real brain. Nodes are the name for these neurons. Artificial neural networks have a numerical value that allows them to carry out several tasks at once. Traditional programming does not employ a database; instead, it stores data on the entire network. The ANN continues to function even if some data disappears from one location temporarily (Wang, 2003). Determining the instances and motivating the network in accordance with the intended output by showing it these examples is crucial for ANN to be able to adapt. If the event can't be seen by the network in all of its aspects, it might provide erroneous output since the network's succession is directly proportional to the selected occurrences.

Figure 3. Deep Learning Applications

Recurrent Neural Network

One important type of artificial neural network is the recurrent network, which is primarily designed to find patterns in data sequences like text, genomes, handwriting, spoken language, numerical times series data from sensors, stock markets, and governmental organisations (Zaremba, 2014). A backpropagation technique is used to train recurrent neural networks, however because backpropagation occurs at every timestamp, it is sometimes referred to as backpropagation across time. The parameters are shared by RNNs at various time steps. This practise is referred to as parameter sharing. As a result, there are fewer parameters to learn, which lowers the computational cost. The disappearing and exploding gradient problem, which affects all varieties of neural networks, also affects deep RNNs (RNNs with a lot of time steps) (Arras, 2019).

Convolutional Neural Network

Classifications used during neural networks for image identification is popular at convolutional neural network. Convolutional neural networks are often applied in a wide variety of applications, including scene labelling, object identification, face recognition, and others (Albawi et al., 2017). CNN accepts a parts of pictures as input and categorises and processes it using terms like "x," "y," "z," etc. The image resolution dictates how well the computer understands it as an array of pixels. It will perceive as h * w * d, where h = height, w = width, and d = dimension, depending on the image resolution. Each input picture in CNN is processed by a series of convolutional layers, pooling, fully connected layers, and filters/ kernels. The Soft-max function will then be used to categorise an item using probabilistic values between 0 and 1 (O'Shea, 2015).

MACHINE LEARNING FRAMEWORK

TensorFlow

TensorFlow is an end-to-end open-source platform for machine learning. It has a comprehensive, flexible ecosystem of tools, libraries, and community resources that lets researchers push the state-of-the-art in ML and developers easily build and deploy ML-powered applications. Google designed and distributed TensorFlow, a Python library for fast numerical computing. Tensor Flow was built to scale since Google has more than just data; they also have the world's most powerful computer. TensorFlow is a machine learning and deep neural network research library created by the Google Brain Team.

TensorFlow takes inputs as a multi-dimensional array called Tensor and allows you to design dataflow graphs and structures to specify how data goes through a graph. It enables you to create a flowchart of operations that may be performed on these inputs, which goes on one end and returns as the output on the other.

Tenser and Graph are the foundations of TensorFlow. A tensor is an n-dimensional vector or matrix that can represent any type of data. A tensor's values all have the same data type with a known (or partially known) shape. The dimensionality of the matrix or array determines the shape of the data. And a graph framework is used by TensorFlow.

To give a specific example, AI can help Google users have a faster and more sophisticated search experience. When a user types a keyword into Google's search field, the search engine suggests the next word (A).

PyTorch

Deep learning frameworks have traditionally prioritized either usability or speed, rather than both. PyTorch is a machine learning library that demonstrates that these two goals can coexist: it was built from the ground up to support an imperative and Pythonic programming style that supports code as a model, simplifies debugging, and is consistent with other popular scientific computing libraries, all while remaining efficient and supporting hardware accelerators like GPUs (Paszke, 2019).

Tensor (torch. Tensor) is a PyTorch class for storing and manipulating homogenous multidimensional rectangular arrays of numbers. PyTorch Tensors are comparable to NumPy Arrays, however, they may be used on an NVIDIA GPU with CUDA support. Support for additional GPU systems, such as AMD's ROCm and Apple's Metal Framework, has also been added to PyTorch (B).

scikit-learn

Scipy is a Python module that integrates a wide range of cutting-edge machine learning techniques for supervised and unsupervised issues on a medium scale. This package focuses on using a general-purpose high-level language to bring machine learning to non-specialists. Ease of use, performance, documentation, and API consistency are all prioritised. It has few dependencies and is distributed under the BSD licence, making it suitable for usage in both academic and commercial environments. Scikit-learn takes advantage of this rich environment to deliver state-of-the-art implementations of many well-known machine learning techniques while preserving a simple, Python-based intface (Pedregosa, 2011).

Scikit-learn is distinct from other Python machine learning toolboxes for several reasons: i) It's released under a BSD licence. ii) It employs compiled code for efficiency, unlike MDP (Zito, 2009) and pybrain (Schaul, 2010); iii) It just uses numpy and scipy for easy distribution, unlike pymvpa (Hanke, 2009), which includes optional dependencies like R and shogun; and iv) It concentrates on imperative programming, unlike pybrain, which uses a data-flow paradigm. LibSVM (Chang, 2011) and LibLinear (Fan, 2008) are C++ libraries that provide reference implementations of SVMs and generalised linear models with comparable licences.

MACHINE LEARNING LIBRARIES

Numpy

The NumPy array is a data structure for storing and accessing multidimensional arrays (Van Der Walt, 2011 #9) that allows for a wide range of scientific computations. It consists of a memory pointer and information, such as 'data type,''shape,' and'strides,' which are used to interpret the data stored there. NumPy comes with two basic objects: an N-dimensional array object (ndarray) and a universal function object (ufunc). An N-dimensional array is a collection of "things" that are all indexed with N integers. An N-dimensional array is defined by two basic pieces of information: 1) the array's shape, and 2) the type of item that makes up the array (Oliphant, 2006).

NumPy arrays are access by users using 'indexing,' 'operators,' and 'array-aware functions,' which together give a readily readable, expressive, high-level API for array programming while NumPy handles the underlying mechanics of making operations quick (Harris, 2020). NumPy additionally has array-aware functions for reshaping, concatenating, and padding arrays, as well as searching, sorting, and counting data and reading and writing files. It has significant support for generating pseudorandom numbers, includes a variety of probability distributions, and performs accelerated linear algebra using one of several backends suited for the CPUs at hand, such as OpenBLAS (Wang et al., 2009) or Intel MKL.

SciPy

SciPy is a Python-based library of numerical procedures that provides fundamental building blocks for modelling and solving scientific problems. SciPy has algorithms for optimization, integration, interpolation, eigenvalue issues, algebraic equations, differential equations, and a variety of other problems, as well as specific data structures like sparse matrices and k-dimensional trees.

SciPy is based on NumPy (Oliphant, 2006; Van Der Walt, 2011), which provides array data structures and related fast numerical routines, while SciPy serves as the foundation for higher-level scientific libraries such as scikit-learn (Pedregosa, 2011) and scikit-image (Van der Walt, 2014). SciPy is used by scientists, engineers, and others all over the world. Several subpackages of SciPy are used in published scripts (Nitz, 2018) for gravitational wave (Abbott, 2016, 2017) research, and the M87 black hole imaging project uses SciPy (Akiyama, 2019).

Fundamental algorithms for scientific computing are provided by SciPy. The guide to available mathematical software (GAMS) classification system was used to determine the scope of the project (Boisvert, 1991). SciPy strives to give comprehensive coverage in fields that move slowly, such as linear algebra. In other areas, it seeks to give basic building blocks that work well with other packages that specialise in that area (Virtanen, 2020).

Theano

Theano is a Python library that allows to define, optimize, and evaluate mathematical expressions involving multi-dimensional arrays efficiently. It has been one of the most widely used CPU and GPU mathematical compilers, particularly in the machine learning community, since its inception, and has demonstrated constant performance gains. Since 2008, Theano has been actively and consistently developed, with numerous frameworks built on top of it and it being used to create many cutting-edge machine learning models.

The following is a breakdown of the structure of this article. The first section introduces the Theano programme and community. Section II discusses Theano's main features and how to utilise them, as well as how they compare to other similar projects. Section III focuses on newly announced features and enhancements. On numerous machine learning models, Section IV compares Theano's performance against that of Torch7 and TensorFlow. Section V examines Theano's present limitations as well as proposed improvements (Al-Rfou, 2016).

Keras

Keras is a high-level Python library for deep learning that runs on top of TensorFlow and is small and easy to use (or Theano or CNTK). It allows developers to concentrate on the core principles of deep learning, such as layer creation for neural networks, while the nitty-gritty details of tensors, their forms, and their mathematical subtleties are taken care of. Keras' back end must be TensorFlow (or Theano or CNTK). Keras can be used for deep learning applications without having to deal with the more complicated TensorFlow (or Theano or CNTK). The sequential API and the functional API are the two main types of framework. The sequential API is built on the concept of a layer sequence; this is the most typical application of Keras and the most straightforward element of it.A linear stack of layers can be compared to the sequential model (Manaswi, 2018).

CONCLUSION

Both the TensorFlow and Theano libraries are wrapped with Keras, which abstracts their capabilities and hides their complexity. Keras is built for simplicity and modularity, allowing you to create deep learning models quickly. Keras deep learning models are built by defining, constructing, and fitting models, which are subsequently assessed or utilised to make predictions (Brownlee, 2016).

REFERENCES

Aha, D. W., Kibler, D., & Albert, M. K. (1991). Instance-based learning algorithms. *Machine Learning*, *6*(1), 37–66. doi:10.1007/BF00153759

Al-Omary, A. Y., & Jamil, M. S. (2006). A new approach of clustering based machine-learning algorithm. *Knowledge-Based Systems*, *19*(4), 248–258. doi:10.1016/j.knosys.2005.10.011

Albawi, S., Mohammed, T. A., & Al-Zawi, S. (2017). *Understanding of a convolutional neural network*.

Arras, L., Osman, A., Müller, K.-R., & Samek, W. (2019). Evaluating recurrent neural network explanations. doi:10.18653/v1/W19-4813

Ayodele, T. O. (2010). Types of machine learning algorithms. *New advances in machine learning, 3*, 19-48.

Barlow, H. B. (1989). Unsupervised learning. *Neural Computation, 1*(3), 295–311. doi:10.1162/neco.1989.1.3.295

Biau, G., & Scornet, E. (2016). A random forest guided tour. *Test, 25*(2), 197–227. doi:10.100711749-016-0481-7

Böhning, D. (1992). Multinomial logistic regression algorithm. *Annals of the Institute of Statistical Mathematics, 44*(1), 197–200. doi:10.1007/BF00048682

Borgelt, C. (2005). *An Implementation of the FP-growth Algorithm.*

Brighton, H., & Mellish, C. (2002). Advances in instance selection for instance-based learning algorithms. *Data Mining and Knowledge Discovery, 6*(2), 153–172. doi:10.1023/A:1014043630878

Campello, R. J. G. B., Kröger, P., Sander, J., & Zimek, A. (2020). Density-based clustering. *Wiley Interdisciplinary Reviews. Data Mining and Knowledge Discovery, 10*(2), e1343. doi:10.1002/widm.1343

Chen, P. H., Lin, C. J., & Schölkopf, B. (2005). A tutorial on ν-support vector machines. *Applied Stochastic Models in Business and Industry, 21*(2), 111–136. doi:10.1002/asmb.537

Cunningham, P., Cord, M., & Delany, S. J. (2008). Supervised learning. In *Machine learning techniques for multimedia* (pp. 21–49). Springer. doi:10.1007/978-3-540-75171-7_2

Dayan, P., & Niv, Y. (2008). Reinforcement learning: The good, the bad and the ugly. *Current Opinion in Neurobiology, 18*(2), 185–196. doi:10.1016/j.conb.2008.08.003 PMID:18708140

Dove, G., Halskov, K., Forlizzi, J., & Zimmerman, J. (2017). UX design innovation: Challenges for working with machine learning as a design material. *Conference on Human Factors in Computer Systems.*

Eberly, L. E. (2007). Multiple linear regression. *Topics in Biostatistics*, 165-187.

Ghahramani, Z. (2003). Unsupervised learning. *Advanced Lectures on Machine Learning.*

Harrell, F. E. (2015). Ordinal logistic regression. In *Regression modeling strategies* (pp. 311–325). Springer. doi:10.1007/978-3-319-19425-7_13

Hastie, T., Tibshirani, R., & Friedman, J. (2009). Unsupervised learning. In *The elements of statistical learning* (pp. 485–585). Springer. doi:10.1007/978-0-387-84858-7_14

Kaelbling, L. P., Littman, M. L., & Moore, A. W. (1996). Reinforcement learning: A survey. *Journal of Artificial Intelligence Research, 4*, 237–285. doi:10.1613/jair.301

Kameshwaran, K., & Malarvizhi, K. (2014). Survey on clustering techniques in data mining. *International Journal of Computer Science and Information Technologies, 5*(2), 2272–2276.

Kanungo, T., Mount, D. M., Netanyahu, N. S., Piatko, C. D., Silverman, R., & Wu, A. Y. (2002). An efficient k-means clustering algorithm: Analysis and implementation. *IEEE Transactions on Pattern Analysis and Machine Intelligence, 24*(7), 881–892. doi:10.1109/TPAMI.2002.1017616

Kecman, V. (2005). Support vector machines–an introduction. In *Support vector machines: theory and applications* (pp. 1–47). Springer. doi:10.1007/10984697_1

King, J. E. (2008). Binary logistic regression. *Best practices in quantitative methods*, 358-384.

Kingsford, C., & Salzberg, S. L. (2008). What are decision trees? *Nature Biotechnology*, *26*(9), 1011–1013. doi:10.1038/nbt0908-1011 PMID:18779814

Kotsiantis, S. B., Zaharakis, I., & Pintelas, P. (2007). Supervised machine learning: A review of classification techniques. *Emerging artificial intelligence applications in computer engineering, 160*(1), 3-24.

Kwak, C., & Clayton-Matthews, A. (2002). Multinomial logistic regression. *Nursing Research*, *51*(6), 404–410. doi:10.1097/00006199-200211000-00009 PMID:12464761

Lauzon, F. Q. (2012). An introduction to deep learning. *International Conference on Information Science, Signal Processing and their Applications.*

LaValley, M. P. (2008). Logistic regression. *Circulation*, *117*(18), 2395–2399. doi:10.1161/CIRCULATIONAHA.106.682658 PMID:18458181

Learning, D. (2020). *Deep learning*. High-Dimensional Fuzzy Clustering.

Likas, A., Vlassis, N., & Verbeek, J. J. (2003). The global k-means clustering algorithm. *Pattern Recognition*, *36*(2), 451–461. doi:10.1016/S0031-3203(02)00060-2

Mannila, H. (1996). Data mining: machine learning, statistics, and databases. *Proceedings of 8th International Conference on Scientific and Statistical Data Base Management.*

Maulud, D. H., & Abdulazeez, A. M. (2020). *A Review on Linear Regression Comprehensive in Machine Learning*.

Menard, S. (2002). Applied logistic regression analysis. *Sage (Atlanta, Ga.).*

Meng'Ao, L., Dongxue, M., Songyuan, G., & Shufen, L. (2015). *Research and improvement of DBSCAN cluster algorithm*.

Meyer, D., & Wien, F. T. (2015). Support vector machines. *The Interface to libsvm in package e1071, 28*, 20.

Montgomery, D. C., Peck, E. A., & Vining, G. G. (2021). *Introduction to linear regression analysis*. John Wiley & Sons.

Moore, M. M., Slonimsky, E., Long, A. D., Sze, R. W., & Iyer, R. S. (2019). Machine learning concepts, concerns and opportunities for a pediatric radiologist. *Pediatric Radiology*, *49*(4), 509–516. doi:10.100700247-018-4277-7 PMID:30923883

Nasteski, V. (2017). An overview of the supervised machine learning methods. *Horizons. b, 4*, 51-62.

O'Shea, K., & Nash, R. (2015). An introduction to convolutional neural networks.

Oshiro, T. M., Perez, P. S., & Baranauskas, J. A. (2012). *How many trees in a random forest?* Quinlan, J. R. (1986). Induction of decision trees. *Machine Learning*, *1*(1), 81–106.

Rigatti, S. J. (2017). Random forest. *Journal of Insurance Medicine (New York, N.Y.)*, *47*(1), 31–39. doi:10.17849/insm-47-01-31-39.1 PMID:28836909

Sen, P. C., Hajra, M., & Ghosh, M. (2020). Supervised classification algorithms in machine learning: A survey and review. In *Emerging technology in modelling and graphics* (pp. 99–111). Springer. doi:10.1007/978-981-13-7403-6_11

Shouman, M., Turner, T., & Stocker, R. (2012). Applying k-nearest neighbour in diagnosing heart disease patients. *International Journal of Information and Education Technology (IJIET)*, *2*(3), 220–223. doi:10.7763/IJIET.2012.V2.114

Steinwart, I., & Christmann, A. (2008). *Support vector machines*. Springer Science & Business Media.

Su, X., Yan, X., & Tsai, C. L. (2012). Linear regression. *Wiley Interdisciplinary Reviews: Computational Statistics*, *4*(3), 275–294. doi:10.1002/wics.1198

Tracey, B., Duraisamy, K., & Alonso, J. (2013). *Application of supervised learning to quantify uncertainties in turbulence and combustion modeling*.

Tranmer, M., & Elliot, M. (2008). Multiple linear regression. [CCSR]. *The Cathie Marsh Centre for Census and Survey Research*, *5*(5), 1–5.

Uyanık, G. K., & Güler, N. (2013). A study on multiple linear regression analysis. *Procedia: Social and Behavioral Sciences*, *106*, 234–240. doi:10.1016/j.sbspro.2013.12.027

Wang, H., Ma, C., & Zhou, L. (2009). A brief review of machine learning and its application. *International Conference on Information Engineering and Computer Science*.

Wang, S.-C. (2003). Artificial neural network. In *Interdisciplinary computing in java programming* (pp. 81–100). Springer. doi:10.1007/978-1-4615-0377-4_5

Weisberg, S. (2005). *Applied linear regression* (Vol. 528). John Wiley & Sons. doi:10.1002/0471704091

Yang, M.-S., Lai, C.-Y., & Lin, C.-Y. (2012). A robust EM clustering algorithm for Gaussian mixture models. *Pattern Recognition*, *45*(11), 3950–3961. doi:10.1016/j.patcog.2012.04.031

Yu, X., & Wang, H. (2014). Improvement of Eclat Algorithm Based on Support in Frequent Itemset Mining. *Journal of Computing*, *9*(9), 2116–2123. doi:10.4304/jcp.9.9.2116-2123

Zaremba, W., Sutskever, I., & Vinyals, O. (2014). Recurrent neural network regularization.

Zhang, X.-S. (2000). Introduction to artificial neural network. In *Neural Networks in Optimization* (pp. 83–93). Springer. doi:10.1007/978-1-4757-3167-5_5

Chapter 11
Securing Autonomous Robot Systems

Ambika N.
https://orcid.org/0000-0003-4452-5514
St. Francis College, India

ABSTRACT

The previous work is two versatile 3D methodologies for independent horticultural UAVs to find impact-free ideal ways in time with the interaction costs in various conditions, containing different obstructions. These strategies were created, roused by two metaheuristic calculations. This concentrate likewise involved an instrument for snagging the board. It observes the best way between the beginning and target stations of each UAV. In way arranging techniques, as a rule, either the robots haphazardly move, or expensive cycles are attempted in tracking down an advanced way. This interaction is performed steadily between the two stations. The UAV attempts to track down a method between two places. It enhances haphazardly made ways to observe the ideal direction. On account of this technique, robots can likewise effectively keep away from impediments. The amount of every one of the tuples' expenses is determined, and the expense of the way is acquired. The suggestion implements some security measures to make the system better.

INTRODUCTION

The internet of things-based (Ambika., 2019; Hassan, 2019; Lee, Bae, & Kim, 2017) innovations are utilized in agribusiness to recover data about the harvest field using sensors (Ambika., 2020; Nagaraj, 2021) for observing, intruding, and governing the arena region. The web of objects-based advances with the utilization of isolated detecting gadgets gives a method for detecting an item in a measured climate. It assists with diminishing the hole among the PC situated framework and this present reality. The benefit of presenting web of objects-based gadgets in farming is to amplify exactness and productivity with low human intercession. Horticulture 4.0 spotlights significantly on accuracy of agribusiness. Accuracy horticulture accomplishes in many ways. Some examples include refinement of development

DOI: 10.4018/978-1-6684-6413-7.ch011

Copyright © 2023, IGI Global. Copying or distributing in print or electronic forms without written permission of IGI Global is prohibited.

rehearses, selections of harvests, etc. It makes horticulture accurate through the web of objects Based gadgets in Farming.

The observation robot (Wang, 2011) practises its load to guide or hunt a region. The safety machine secures and shields a resource. The observation robot gathers information with its electronic shipment and transfers the crude information back to its control station or pre-processes the information with other data, for example, geographic data framework information, and transfers that data to a double-dealing element. The nature of surveillance is a component of the target range, how much objective mess, the sensor goal, and the likelihood of location of an objective. safety machine tasks comprise a versatile robot watching a characterized region, recognizing a dubious exercise, and starting a reaction. A safety machine can direct predefined or irregular watches and consequently perform reconnaissance undertakings, for example, checking for interlopers or surveying the situation with the edge safeguards. A reconnaissance robot can be more information serious when contrasted with the safety machine that frequently acts in an organized climate and has a smaller information burden to process.

The issue of learning planning between world states furthermore, activities empower a robot to choose an activity given its ongoing world state. It is a grouping of state-activity matches during the educator's show of the robot behavior (Laumond, 1998; Zhuang & Roth, 2018). The presence of robots inside society is turning out to be perpetually common space, robot soccer or a sporting robot for the home. It requires an effective independent robot activity that requires hearty control calculations.

The work (Kiani et al., 2022) is two versatile 3D methodologies for independent horticultural UAVs to find impact-free ideal ways with the interaction costs in various conditions containing different obstructions. These strategies were created, roused by two metaheuristic calculations. This concentrate likewise involved an instrument for snagging the board. It observes the best way between the beginning and target stations of each UAV. In way arranging techniques, as a rule, the robots haphazardly move, or expensive cycle tracks down an advanced way. This interaction is performed steadily between the two stations. The UAV attempts to track down a method between two places. It enhances haphazardly made ways to observe the ideal direction. On account of this technique, robots can likewise effectively avoid impediments. The number of the tuples' expenses is determined, and the payment of the way is acquired.

The suggestion implements some security measures to make the system better. The proposed work improves security in the network. The location information of the drones and target generates the encryption key. Using this key, the data to be communicated is encrypted and sent to the mainframe. The work secures the network by 10.17%.

LITERATURE SURVEY

The following segment details the various recommendations made by multiple writers toward the domain. The previous work (Kiani et al., 2022) uses two versatile 3D methodologies for independent horticultural UAVs to find impact-free ideal ways with the interaction costs in various conditions containing different obstructions. These strategies were created, roused by two metaheuristic calculations. This concentrate likewise involved an instrument for snagging the board. It observes the best way between the beginning and target stations of each UAV. In way arranging techniques, as a rule, the robots haphazardly move or have expensive cycles tracking down an advanced way. This interaction is performed steadily between the two stations. The UAV attempts to track down a method between two places. It enhances haphazardly made ways to observe the ideal direction. On account of this technique, robots can likewise effectively avoid impediments. The number of the tuples' expenses is determined, and the cost of the way is acquired.

The framework (Luo, Hsu, Lin, & Su, 2005) contains a super safety robot, remote manage PC, safety unit, and apparatus regulator component. The ISR has a fundamental regulator (IPC), secure location, RF interface, and GSM modem. The ISR can get the situation with the safety and machine control components utilizing a remote RF interface. The safety unit uses one-method correspondence with a portable robot and phone. The machine regulator component utilizes two-way correspondence with ISR and wireless. The ISR can speak with a GSM modem that has an RS232 interface. Wavecoll secludes the GSM. The measured is a self-limited E-GSM900/GSM1800 double band component. The RF connection point of the ISR can get data from the safety unit utilizing a remote. The remote oversee PC can cooperate with the ISR through Internet. The Web mainframe is typically put up on the nearby side together to diminish the web period interruption. The ISR utilizes an aluminum outline. The shape of the automaton is the chamber. The measurement is 50 cm, and the level is around 140 m. The ISR contains the chest area and lower body. The principal regulator of the safety machine is the PC with a Pentium-III 933 CPU and 256M RAM. The equipment has a contact shade, CCO, sensing devices and tactile circuits, driver framework, and others.

Complete Surveillance Manage (Kim, Bae, & Huh, 2010) is a clever observation and safety machine framework. It intends to check ways of behaving and exercises in a region using various reconnaissance gadgets. It includes machines, CCTV cameras, ground reconnaissance detectors, and sensors. Hosts can be assembled into the mainframe and widget the board systems. The fundamental mainframe merchant messages among consoles and oversees business rationales. The board mainframes control explicit gadgets sending and getting messages to and from the mainframe. It sends and gets to from the various control center. Gadget the executive's mainframes incorporate the automaton mainframe, versatile automaton mainframe, camera mainframe, detector mainframe, ANPR mainframe, and power the board mainframe. The control center geologically disperses. A guard automaton has cameras, sensing elements, and a concealment gadget intended for long-range reconnaissance and concealment. It involves two components, the following component and the recognition component. The identification component performs grouping visits that screen different pre-set regions for indicated time distinguishing targets. SGR-1 assumes a focal part in gadget interlocking, which helps administrators quickly and precisely sort out the circumstance of interruption caution. SGR-1 moves the following component to a target distinguished by insightful cameras or ground observation detectors in light of canny calculations. A deadly weapon has a rifle, a non-deadly acoustic gun, and a high-power incandescent light. With a novel single-turn configuration, a versatile watch automaton named STAR is driven by four freely controlled wheels, empowering shifting the course on the spot. The portable automaton intends for surveillance over a tremendous scope of fields with the direction of a GPS sensing element and an examining laser range finder. Ground observation detectors recognize moving items in an open and level region. It sends the distinguished objective data to the detector mainframe, which ships of the principal mainframe.

The proposition (Luo, Liao, Su, and Lin, 2005) is a cropping organiser approach. It is a compulsory supporting device for the safety machine restoring. The machine is utilized for a wellbeing watch and created by our research center. The equipment incorporates the focal regulator, detecting component circuits, servomotors, and a powerful framework. It has six subsystems in the robot - remote administrative, hindrance evasion, movement arranging, vision, and detecting component frameworks. It has two gatherings of re-energizing pins for the robot to dock. The superior component gives 12V/DC with a limit of 30A to reestablish PC and detect component circuits. The lower one gives 36V/DC with a limit of 15A to re-energize the batteries for engines and their drivers. They mount before the machine. The re-energizing connectors are within the openings. The opening is formed like a cone to help the docking flawlessly. Expansion runs take the re-energizing pins back to the middle after the robot separates. The section offset and point are permitted when the robot is in the docking system. The robot contacts the breaking point switch on the docking station. The auto-re-energizing gadget can give capacity to the machine within 30 seconds. The ongoing detecting component can recognize the constant worth to forestall the flawed status. The regulator is a chip and distinguishes the re-energizing current utilizing DC-type current detecting components. The information signals incorporate a breaking point switch and laser source. The result signals contain a wellbeing switch, laser light, and show. The safety switch might be removed or turned on by the re-energizing current. There is a camera mounted on top of the machine. The picture outline grabber catches the video signal. It is inside the focal regulator. The procured picture distinguishes the counterfeit milestone and finds the site of the docking station. Then the vision of the framework directs the movement to accomplish the docking.

The execution of the Robo-Guard (Birk & Kenn, 2002) depends on the supposed Cube System, a development pack for mechanical frameworks. It is designed at the VUB AI Lab. The VUB AI-lab has very some customs in creating adaptable automaton control equipment. Different trial stages have been developing from the mid-eighties up to this point. The focal point of the Cube System is the alleged Robo-Cube regulator equipment in light of the MC68332 processor. The conservative state of Robo-Cube is accomplished through a few methods. The board region is limited by utilizing SMD parts. Three sheets are stacked on one another, prompting a cubic plan. Robo-Cube has an open transport engineering that permits the addition of numerous sensing element/motor interfaces.

The safety automaton (Luo et al., 2007) utilizes an aluminum outline. The width is 50cm, and the level is 170cm. The framework construction of the safety automaton contains six subsystems -remote directing, movement, vision, tangible, automaton arms, and power assessment subsystems. The focal control unit is fundamental for handling the participation between subsystems. Every subsystem is running as an autonomous specialist. The astute portion arranges with the specialists between process correspondence. Two sorts of IPC have internetworking attachments. They offer memory and executes for communication. The primary regulator of the safety automaton is a PC with a Pentium(R) IV CPU and 1GB RAM. The lower body contains engine drivers, batteries, and two DC servomotors. There are sixteen Polaroid 6500 ultrasonic reach sensing elements appropriated around the versatile automaton for climate investigation. The terminating rate and succession constrain by a chip. The bar width for every sensing element is 22.5 degrees, and the reach a not entirely set in stone when of trip of an acoustic sign created by the transducer and reflected by the article. The automaton can utilize the sensing elements to identify the reach data from 30cm to 1500cm. The portable automaton uses sensing elements to recognize close-by objects.

The suggestion (Dieber et al., 2017) proposes safety engineering on the application layer. The ROS gets acknowledged at the application level. It is a committed Authentication Mainframe that keeps track of which ROS gateways might buy into or distribute to a topic. It deals with the validation of hubs and produces explicit encryption keys. The distribute or buy-in worldview reformulates as a telecom correspondence design. Numerous distributors of a specific point might be passing information to different supporters. The engineering depiction into stages indicates the lifecycle of a distributer and endorser. It portrays the significant cryptographic tasks per phase. A ROS hub firing up needs to confirm itself to the AS. A similar methodology finishes from the AS to the intersection. For a supporter, the AS sends the endorser a rundown of public mark check keys connected with distributers of the message subjects.

A automaton regulator (Quarta et al., 2017) is a complicated gadget. It is encased in at least one suspension, concealing several interconnected electrical and PC frameworks. A automaton regulator plans with an accentuation on effectiveness, complex movement depiction, nonlinear control, and collaboration with human operators. The regulator carries out a control framework to oversee the automaton's exercises, making it one of the most well-being and safety-basic parts: It epitomizes the rationale and usefulness expected to screen and pilot the mechanical pieces of the automaton and to speak with the climate. The control framework splits into three components and four progressive levels. The tangible handling components catch the condition of the framework. The human administrator is responsible for checking, beginning, and halting the activities. The developer is responsible for composing task programs. The regulator acknowledges task determinations composed utilizing a space explicit programming climate.

The Internet-based safety automaton framework (Liu, Wang, & Feng, 2005) utilizes a appearance acknowledgment method with consistent qualities. It includes face reply, lighting shift, and impeding structures.3-D model comprising three blocks. A focal specialist has to block the utilization of a clever identification plot on a solitary PC. An adjoining specialist has a block that permits the utilization of an element extraction plot in an equal figuring climate. A far-off specialist has to block the utilization of a remote chart matching plan in a conveyed figuring environment. Four specialists dwell at the focal regulator: the connection point specialist, the chroma handling specialist, the face-district location specialist, and the model matching specialist. The chroma diagram is focal to the programmed extraction of face-area districts. It finds picture locales where countenances might exist. The face-locale discovery specialist utilizes morphological administrators to separate different arched objects. The model-matching specialist presents a model face and a matching ability that underscores topological qualities. The matching capacity incorporates two-section - topological requirements and the mathematical limitations of every facial element. A two-venture matching plan completes mathematical-based coarse coordinating and dynamic-connect design-based fine coordinating. iBot stage model permits clients to remotely control an automaton because of living video pictures caught by the camera on the automaton. The iBot Mainframe interfaces the automaton and camera over a remote station.

The transfer hubs (Song et al., 2009) are dissipated in the rooms to frame a various-bounce network organizing topology. They speak with one another and course information to and from the household mainframe. The observation automaton upgrades the sensing element with velocity capabilities. The reconnaissance automaton can work in three modes according to client demands and errand assets. In the watching mode, the observation automaton meanders around in the rooms or follows predefined courses independently. In the person on-call mode, the reconnaissance automaton customizes to work in collaboration with other fixed observing gadgets. In controller mode, the reconnaissance automaton will be directed to the locale of interest and take care of a distant user. The projected automaton is 9cm in level and 250g in weight and can jump over hindrances multiple times its size. It relies upon the versatile components in a six-bar linkage leg framework to empower jumping motion. It can move openly on level floors and shift its bearings by the two-wheeled differential drive framework. It takes on the ZigBee convention for remote correspondence and subsequently adds to a ZigBee-based home control network as a portable video sensing element hub.

The framework (Luo et al., 2005) contains a safety automaton, a distant administrative PC, a safety component, and a machine control component. The ISR has the principal regulator, safety identification, RF interface, and GSM modem. The ISR can get the situation with the safety and machine control component utilizing a remote RF interface. The safety component has one-way correspondence with the safety automaton and wireless. The RF connection point of the ISR can get data from the safety component. The far-off administrative PC can cooperate with the ISR through the Internet. The Web mainframe is typically put on the nearby webpage to lessen the web time delay. The administrative PC and administrative PC impart in a similar neighborhood. The client can associate with the administrative PC to get safety data of the home or building, control the portable automaton to watch wherever in the home or building, and regulate the apparatus control component through the web. The client can obtain the safety data and manage the machine control component utilizing a mobile phone through GSM. The safety automaton uses an aluminum frame. The form of the automaton is round and hollow. The safety automaton contains a chest area and a lower body. The chest area of the safety automaton has the principal regulator. The lower body includes the drive framework, batteries, and two DC servomotors.

The framework (Kim, Kim, Lee, & Lee, 2006) can recognize safety caution made by sensing element hubs and transferred in the jump-by-bounce transmission. HSSN framework comprises Home Mainframe, Home Safety Mobile Automaton, and Sensing element Network designed by Cricket nodes. HS gets and disseminates home safety data utilizing RF inside and remote LAN outside. It recognizes the alert occasion data and communicates the occasion log message to the UI gadget. The safety pictures caught by HSMR in the home safety actuation method of HSSN sends through the remote internet. HSMR follows the ideal direction made by positioning produced by intertwining the neighborhood deterrent detecting and keeping away from data gained by the neighborhood infrared and ultrasonic sensing elements. The outright arranges data of every sensing element hub procured through the sensing element organization. HSMR catches the current setting mindful picture of the caution zone when it arrives at AZ and afterward sends the picture data to the client through HS. A client can perceive home safety through the alert occasion messages and pictures.

The automaton (Booth et al., 2017) piggybacks understudies into a college dormitory. It is set either outside or inside a residence secure-access door. If the member halted strolling, the automaton would rehash the same thing. The member overlooked the automaton and went on towards the entrance the automaton would request. If the member held the entry, the automaton would enter or exit. If the member didn't open the entryway, it would stay in its unique position. Because of the entrance shutting speed, the automaton couldn't follow the member inside the structure without the member either holding the entry open or squeezing the programmed entryway open button for wheelchair access.

In Safety Mode Transition Procedure (Kim et al., 2006), a CN recognizes a caution from a sensing element component and sends the setting mindful data to HS in bounce-by-jump through an RF signal. HS sends the occasion log message and illuminates home safety data to a far-off client. Then CNs travel their activity mode from the underlying ESM to CM since they need remote correspondence to engender what is going on toward the sensing element organization. Toward the finish of the SMTP, entire CNs transform into HSAM as indicated by HS. In this mode entire, CNs actuate RF and Ultrasonic sensors. In the HSSN Initiation Procedure, HSSNAP, HSMR plans an ideal direction for showing up at the objective trained by HS and starts the independent development. After showing up there quickly and precisely, HSMR takes the setting mindful pictures utilizing its CCD camera and communicating them to HS. At last, HS coordinates the information taken from all CNs and HSMR and sends them to the mortgage holder so that it makes the client accommodating to answer the dire.

The framework (Seeman et al., 2006) works with a coordinated effort between human administrators and automatons utilizing our variable independence control design, explicitly in the exhibition of distant examination undertakings. The common perceptual space we propose is a general instrument for a coordinated effort among people and automatons. The graphical UI with expanded computer-generated experience representation we present is an endeavor to mitigate the issues with control and discernment,

Figure 1. Previous method (Kiani et al., 2022)

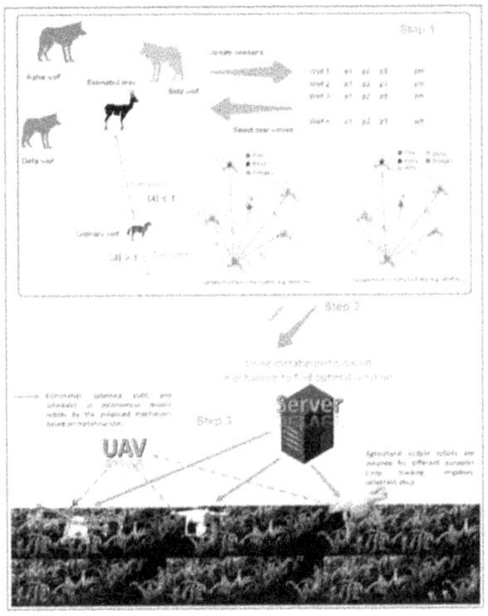

straightforwardly connected with a ground robot. Panda3D utilized has a straightforward mainframe design and dispersed object model.

Proposed Work

The previous work (Kiani et al., 2022) uses two versatile 3D methodologies for independent horticultural UAVs to find impact-free ideal ways with the interaction costs in various conditions containing different obstructions. These strategies were created, roused by two metaheuristic calculations. This concentrate likewise involved an instrument for snagging the board. It observes the best way between the beginning and target stations of each UAV. In way arranging techniques, as a rule, either the robots haphazardly move, or expensive cycles track down an advanced way. This interaction is performed steadily between the two stations. The UAV attempts to track down a method between two places. It enhances haphazardly made ways to observe the ideal direction. On account of this technique, robots can likewise effectively avoid impediments. The number of the tuples' expenses is determined, and the fee of the way is acquired.

Assumptions

- At the time of deployment, the network is free from attacks.
- The location of drones changes frequently (minor changes).
- The drones are liable to get compromised.
- The devices are capable of deriving hash code.

Representations used in the recommendation

Detailed Working of the System

The proposed work improves safety in the network. The location information of the drones and target generates the encryption key. Using this key, the data to be communicated is encrypted and sent to the server. In equation (1), the encryption key E_i is generated using the hash value of the location of the

Table 1. Notations used in the recommendation

Notations	Description
N	Network under consideration
E_i	Encryption key
D_L	Location information of drone under consideration
T_L	Location information of the target
$H(D_L)$	Hash value of Location information of drone
$H(T_L)$	Hash value of Location information of the target

target T_L and the hash value of the location of the drone D_L. Below is the algorithm used to generate the hash key.

$$E_i \neg H(D_L) \bullet H(T_L) \quad (1)$$

Algorithm to generate hash key
Step 1– Input information (16 bits)
Step 2– Insert 0 and 1 in alternate odd locations (0 in 1,5,9,13; 1 in 3,7,11,15)
(*Resultant has 24 bits*)
Step 3– Divide the resultant into two halves (12 bits each)
Step 4– Use right circular shift on first halve and left circular shift on second halve
Step 5– Combine both to obtain resultant bits
Step 6– Insert 8 1's in different location of the resultant (4,8,12,16,20,24)
Step 7– Divide the resultant into two (15 bits each)
Step 8– Xor one split part by another (resultant - 15 bits)
Step 9– Apply left circular shift on the resultant. (Hash key)

Analysis and Simulation of the Work

The work is simulated in Python. Table 2 jolts down the parameters used in the suggestion.

Security

The data transmitted from one end to another are liable to different kinds of attacks. Hence the data has to be protected. The previous suggestion (Kiani et al., 2022) is two versatile 3D methodologies for independent horticultural UAVs to find impact-free ideal ways with the interaction costs in various conditions containing different obstructions. These strategies were created, roused by two metaheuristic calculations. This concentrate likewise involved an instrument for snagging the board. It observes the best way between the beginning and target stations of each UAV. In way arranging techniques, as a rule, the robots haphazardly move, or expensive cycle tracks down an advanced way. This interaction is performed steadily between the two stations. The UAV attempts to track down a method between two places. It enhances haphazardly made ways to observe the ideal direction. On account of this technique,

Table 2. Parameters used in the suggestion

Parameters	Description
Considered area	200m * 200m
Number of targets used	5
Number of drones used	2
Length of location information	16 bits
Length of hash key	15 bits
Length of encryption key	15 bits + 15 bits
Time of simulation	60s

robots can likewise effectively avoid impediments. The number of the tuples' expenses is determined, and the cost of the way is acquired.

The proposed work improves security in the network. The location information of the drones and target generates the encryption key. The work secures the network by 10.17%. Figure 2 portrays the same.

Figure 2. Portrayal of security in the system

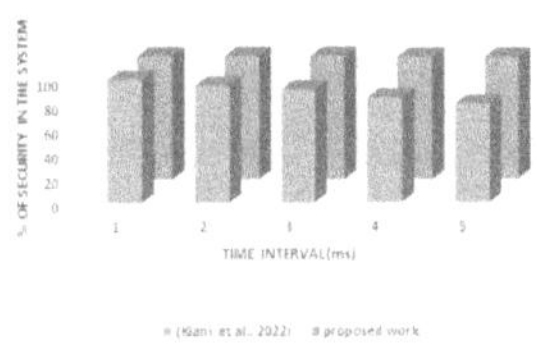

CONCLUSION

Independent framework plans and creates with expanded degrees of freedom and dynamic abilities. These limits will be productive in unsure circumstances. Human-robot trust is a significant part of Human-Robot Interaction. It is acquired expanding consideration among researchers in many disciplines. The previous suggestion is two versatile 3D methodologies for independent horticultural UAVs to find impact-free ideal ways with the interaction costs in various conditions containing different obstructions. These strategies were created, roused by two metaheuristic calculations. This concentrate likewise involved an instrument for snagging the board. It observes the best way between the beginning and target stations of each UAV. In way arranging techniques, as a rule, the robots haphazardly move, or expensive cycle tracks down an advanced way. This interaction is performed steadily between the two stations. The UAV attempts to track down a method between two places. It enhances haphazardly made ways to observe the ideal direction. On account of this technique, robots can likewise effectively avoid impediments. The number of the tuples' expenses is determined, and the cost of the way is acquired.

The proposed work improves security in the network. The location information of the drones and target generates the encryption key. The work secures the network by 10.17%.

REFERENCES

Ambika, N. (2019). Energy-Perceptive Authentication in Virtual Private Networks Using GPS Data. In M. Z. (eds), Security, privacy and trust in the IoT environment (pp. 25-38). Cham: Springer. doi:10.1007/978-3-030-18075-1_2

Ambika, N. (2020). SYSLOC: Hybrid Key Generation in Sensor Network. In S. P., B. B., P. M., K. N., & H. W. (eds), Handbook of Wireless Sensor Networks: Issues and Challenges in Current Scenario's. Advances in Intelligent Systems and Computing (Vol. 1132, pp. 325-347). Cham: Springer.

Birk, A., & Kenn, H. (2002). Roboguard, a teleoperated mobile security robot. *Control Engineering Practice*, *10*(11), 1259–1264. doi:10.1016/S0967-0661(02)00088-6

Booth, S., Tompkin, J., Pfister, H., Waldo, J., Gajos, K., & Nagpal, R. (2017). Piggybacking robots: Human-robot overtrust in university dormitory security. *ACM/IEEE International Conference on Human-Robot Interaction* (pp. 426-434). ACM. 10.1145/2909824.3020211

Dieber, B., Breiling, B., Taurer, S., Kacianka, S., Rass, S., & Schartner, P. (2017). Security for the robot operating system. *Robotics and Autonomous Systems*, *98*, 192–203. doi:10.1016/j.robot.2017.09.017

Hassan, W. H. (2019). Current research on Internet of Things (IoT) security: A survey. *Computer Networks*, *148*, 283–294. doi:10.1016/j.comnet.2018.11.025

Kiani, F., Seyyedabbasi, A., Nematzadeh, S., Candan, F., Çevik, T., Anka, F., Randazzo, G., Lanza, S., & Muzirafuti, A. (2022). Adaptive metaheuristic-based methods for autonomous robot path planning: Sustainable agricultural applications. *Applied Sciences (Basel, Switzerland)*, *12*(3), 943. doi:10.3390/app12030943

Kim, K., Bae, S., & Huh, K. (2010). Intelligent surveillance and security robot systems. *IEEE Workshop on Advanced Robotics and its Social Impacts* (pp. 70-73). IEEE.

Kim, Y. G., Kim, H. K., Lee, S. G., & Lee, K. D. (2006). Ubiquitous home security robot based on sensor network. *IEEE/WIC/ACM International Conference on Intelligent Agent Technology* (pp. 700-704). IEEE. 10.1109/IAT.2006.128

Kim, Y. G., Kim, H. K., Yoon, S. H., Lee, S. G., & Lee, K. D. (2006). Home security robot based on Sensor Network. *SICE-ICASE International Joint Conference* (pp. 5977-5982). IEEE.

Laumond, J. P. (1998). *Robot motion planning and control* (Vol. 229). Springer. doi:10.1007/BFb0036070

Lee, S. K., Bae, M., & Kim, H. (2017). Future of IoT networks: A survey. *Applied Sciences (Basel, Switzerland)*, *7*(10), 1072. doi:10.3390/app7101072

Liu, J. N., Wang, M., & Feng, B. (2005). iBotGuard: An Internet-based intelligent robot security system using invariant face recognition against intruder. *IEEE Transactions on Systems, Man and Cybernetics. Part C, Applications and Reviews*, *35*(1), 97–105. doi:10.1109/TSMCC.2004.840051

Luo, R. C., & Chou, Y. T. L. C., Lai, C. C., & Tsai, A. C. (2007). NCCU security warrior: An intelligent security robot system. *33rd Annual Conference of the IEEE Industrial Electronics Society* (pp. 2960-2965). IEEE.

Luo, R. C., Hsu, T. Y., Lin, T. Y., & Su, K. L. (2005). The development of intelligent home security robot. *International Conference on Mechatronics* (pp. 422-427). Taipei, Taiwan: IEEE. 10.1109/ICMECH.2005.1529294

Luo, R. C., Liao, C. T., Su, K. L., & Lin, K. C. (2005). Automatic docking and recharging system for autonomous security robot. *RSJ International Conference on Intelligent Robots and Systems* (pp. 2953-2958). IEEE. 10.1109/IROS.2005.1545197

Luo, R. C., Lin, T. Y., & Su, K. L. (2009). Multisensor based security robot system for intelligent building. *Robotics and Autonomous Systems*, *57*(3), 330–338. doi:10.1016/j.robot.2008.10.025

Nagaraj, A. (2021). Introduction to Sensors in IoT and Cloud Computing Applications. UAE: Bentham Science Publishers. doi:10.2174/97898114793591210101

Quarta, D., Pogliani, M., Polino, M., Maggi, F., Zanchettin, A. M., & Zanero, S. (2017). An experimental security analysis of an industrial robot controller. *IEEE Symposium on Security and Privacy (SP)* (pp. 268-286). IEEE. 10.1109/SP.2017.20

Seeman, M., Broxvall, M., Saffiotti, A., & Wide, P. (2006). An autonomous spherical robot for security tasks. *IEEE International Conference on Computational Intelligence for Homeland Security and Personal Safety* (pp. 51-55). IEEE. 10.1109/CIHSPS.2006.313312

Song, G., Yin, K., Zhou, Y., & Cheng, X. (2009). A surveillance robot with hopping capabilities for home security. *IEEE Transactions on Consumer Electronics*, *55*(4), 2034–2039. doi:10.1109/TCE.2009.5373766

Wang, X. (2011). Mixed reality and human-robot interaction (Vol. 47). Springer Science & Business Media.

Zhuang, H., & Roth, Z. S. (2018). *Camera-aided robot calibration*. CRC press. doi:10.1201/9781315138725

Chapter 12
Introduction to Machine Learning Methods With Application in Agriculture

Mohit Dayal
Dr. Akhilesh Das Gupta Institute of Technology and Management, India

Mehak Gupta
Dr. Akhilesh Das Gupta Institute of Technology and Management, India

Manvi Gupta
Dr. Akhilesh Das Gupta Institute of Technology and Management, India

Anant Rajee Bara
Central University of Haryana, India

Chitrangada Chaubey
Dronacharya Group of Institutions, India

ABSTRACT

Machine learning is the branch of artificial intelligence which makes the computers capable enough to learn from the input data using statistical models and algorithms to imitate the human learning process, gradually becoming more accurate at outcome prediction without being explicitly programmed to do so. Once the algorithm learns what task it has to do with the data, it works on it automatically. Deep learning, also referred to as deep neural network or deep neural learning, on the other hand, is the branch of machine learning, which is just a neural network with many layers to optimize and upgrade for accuracy.

INTRODUCTION

Machine learning is a subfield of Artificial Intelligence), its goal is to understand the structure and fit data in a model which is easy to understand and utilized. It's a field in Computer science, but differ from

DOI: 10.4018/978-1-6684-6413-7.ch012

Copyright © 2023, IGI Global. Copying or distributing in print or electronic forms without written permission of IGI Global is prohibited.

Figure 1. Machine Learning lifecycle

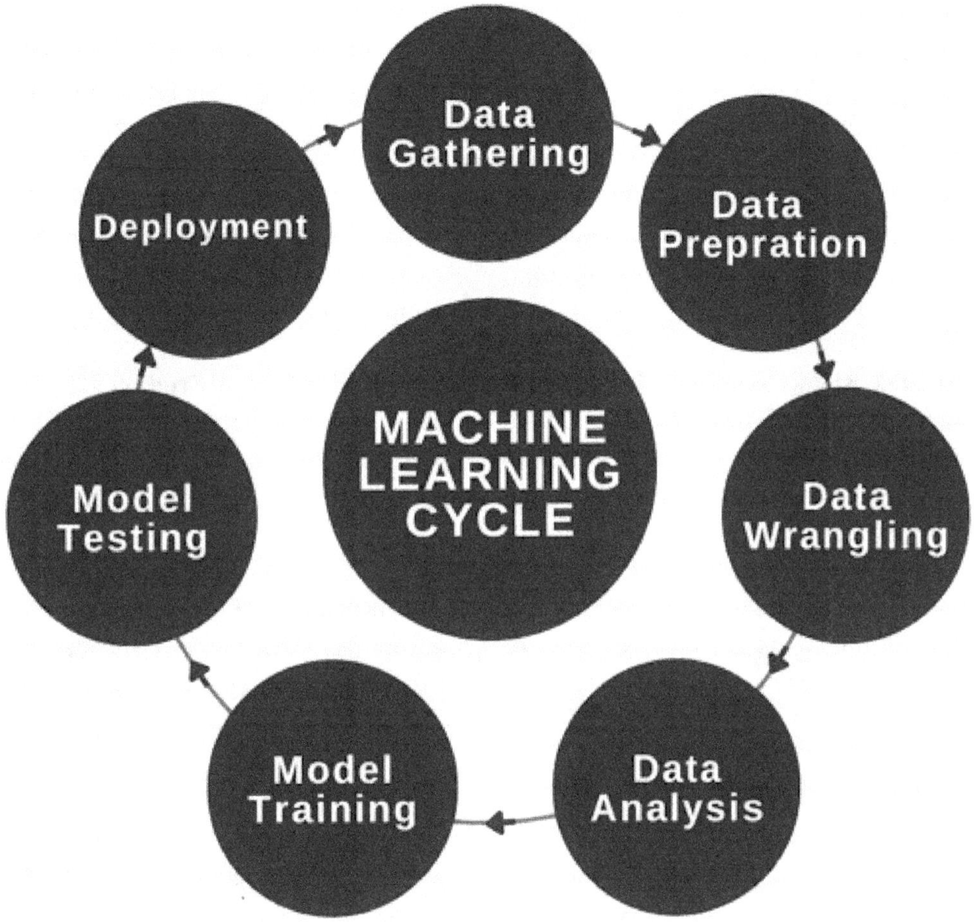

its traditional approaches, where algorithms are set of programmed instructions which is further used by the computer to execute on data inputs and use static analysis in the form to values of output which ranges in a specific range (Nishant et al., 2020). With the help of ML computers can built models from the data sample in the way to automate decision making process purely based on the data inputs.

LITERATURE REVIEW

1. **Evaluating machine learning models for engineering problems:** This article goal is to remedy this situation. First, several evaluation methods are discussed with their relative qualities. Second, these qualities are illustrated by using the methods to evaluate ANN performance in two engineering problems. Third, a systematic evaluation procedure for ML is discussed. (Reich & Barai, 1999)
2. **Coalescence of Neural Networks and Blockchain:** Brief introduction to neural networks then taking one step ahead with introduction to the terminologies related to neural networks and architecture of the neural network. Data propagation is the most important step in the neural network

where the model is trai7ned on basis of data propagated in forward and backward propagation. (Dayal, Chawla, & Khari, 2022)

3. **Applying Machine Learning to Extract New Knowledge in Precision Agriculture Applications:** By monitoring soil, crop and climate in a field and providing a decision support system that is able to learn, it is possible to deliver treatments, such as irrigation, fertilizer and pesticide application, for specific parts of a field in real time and proactively. In this context, we have applied machine learning techniques to automatically extract new knowledge in the form of generalized decision rules towards the best administration of natural resources like water.ne of the major problems faced in blockbhain technology is the large consumption of electricity. Application of artificial intelligence (AI) can help in reducing the unwanted hash operations performed and hence can help in saving electricity (Dimitriadis & Goumopoulos, 2008).
4. **NSL-BP: A Meta Classifier Model Based Prediction of Amazon Product Reviews:** In this research project, we introduce a novel approach to enhance the accuracy of rating prediction by machine learning methods by processing the reviewed text. We trained our model by using many methods, so we propose a combined model to predict the ratings of products corresponding to a given review content (Kumar et al., 2021).
5. **Combination of Convolutional Neural Networks and Blockchain in Handwritten Digit Recognition:** The chapter introduces the reader to the mechanism of the propagation of an image through various filters and how patterns are recognized from the image. The chapter concludes with a literature review and a brief overview of blockchain technology and the real-life application of blockchain technology and convolutional neural networks. (Dayal et al., 2021) .

Types of Machine Learning

Now as we have read about machine learning, let's go further into its types. It is further divided into four types:

1. Supervised learning
2. Unsupervised learning
3. Reinforcement learning
4. Semi-supervised learning

Supervised Learning

Here we use labeled datasets that means tagging the correct output with the input data for the training of algorithms in order to classify or predict the outcomes more precisely. One can call it as a part of cross validation process. The main goal of the algorithms which come under this category is to find the best function that describes the input data for mapping so as to map the input variable with the correct output variable accurately (Reich & Barai, 1999).

Unsupervised Learning

Here we use unlabeled datasets that means not tagging the correct output with the input data for the training of algorithms instead the algorithm itself studies the data and finds the patterns in it (Cunha

Figure 2. Reinforcement learning

et al., 2018). Here the algorithm tries to group the data and form clusters of similar types or sort it in a way to make it more organized. With the assessment of more data, its decision-making ability improves gradually and becomes more and more refined and precise.

Reinforcement Learning

It works on observations gathered from past experiences and then take the suitable actions in order to maximize the reward which obviously results in better performance. It is basically a cycle in which the agent observes the environment and then take actions on the basis of its environmental observations and past experiences and then analyze its action and take the feedback as reward for the learning purposes for its behavior so that it can adapt its response approach in any given situation for the generation of best results possible. The reward feedback is also called the reinforcement signal (Dayal et al., 2023).

Semi Supervised Learning

It is the intermediate ground between the other two machine learning types i.e. supervised and unsupervised learning. It was introduced in order to overcome the drawbacks of those two learnings which are high cost of processing and limited spectrum of application respectively (Katarya et al., 2020). During the training it trains upon the combination of both labeled and unlabeled data. It works on the following assumptions: continuity assumption, cluster assumption and manifold assumption.

KNOW ABOUT MACHINE LEARNING FRAMEWORKS

Machine Learning framework is an interface, a tool and a library which let the users to build their machine learning models efficiently, without knowing about the depth algorithms. Machine learning frameworks simply ease machine learning algorithms. There are a diverse machine learning framework, some are mentioned below.

There are some specific ways to select a perfect fit ML tool for you model:

- Know what exact your goal is: Machine Learning or Deep Learning
- To perform variety of tasks to evaluate data: Image, language and absolute data.

Popular Frameworks

TensorFlow

Tensor Flow is an acclaimed framework recently. It is an open-source library as well as an software library for arithmetical computation with data flow graphs (applying this), line data flow graphs, spot sets of data or tensors can also be set by series of also. Stated by graphs.

Theano

Theano is amazingly sealed over Keras, atypical state neural systems library, that runs virtually in co-ordinate with the Theano library. Keras "fundamental supportive position is a balanced Python library for deep discovering which can frequently running up Theano. It was created to make produce deep learning models as fast also simple and smooth for inventive work. Released under the tolerant MIT, it keeps working on Python 2.7 or 3.5 and can smoothly execute on GPUs and CPUs given the simple model (Kumar et al., 2021).

Sci-Kit Learn

Sci-Kit Learn is a most used ML library. It is mostly used in commanded and separately learning calculations. Precedents are implemented direct and calculated recurrence, bunching, and many more (Bhanu et al., 2020). This includes many calculations for AI and data mining assignments.

Caffe

Caffe is also very popular known architecture created with diction, speed, along with finite quality as absolute priority. It is created by BVLC and also by network donors. Googles DeepDream depends on it.

H20

H20 is also an open-source machine learning platform. It is an AI tool which is particularly a business-oriented and also helps in decision on data problem and permit the users to create insights. It is majorly used for auguring modelling, risks and fraud detection, insurance insights, advertising tech., health maintenance, and customer skill.

Amazon Machine Learning

Amazon Machine Learning provides decision tools which helps to know about the process of ML models without knowing deep algorithms of ML. A service which makes it readable and understandable for a developer of any skillset levels to implement ML technologies (McQueen et al., 1995). It a bridge to a data inbuilt in Amazon S3, Redshift, RDS, and can execute binary analysis, multi-class analysis, or relapse on the data to build model.

Torch

Torch serves most help for ML algorithms to GPUs firstly. It is easy to implement and dynamic because its easy and fast scripted language, **LuaJIT**. Its goal is to maximize affability and faster in creating scientific algorithms along with smooth and fast process.

Google Cloud ML Engine

Google Cloud ML Engine is a handled service that ease things for developers and data scientists in creating and executing exceptional ML models in management. It provides training and prophecy services that can be used in combine or individually.

Azure ML Studio

Azure ML Studio is a framework which provides user to build and lineup a model, then convert them in APIs which can also be engrossed by many services. You can connect Azure storage to service for large models (Kumar et al., 2015). To manage and use

Azure ML Studio, you need no account to test it. Can simply login from an anonymous account and start using Azure ML Studio for eight hrs.

Spark ML Lib

It's a part of Apache Spark's library. The goal of this framework is to create practical ML extensible and simple. It includes mostly used learning algorithms and utilities, with classification, clustering, and many more.

Libraries of Machine Learning

NumPy

NumPy is extremely popular and a library which deals with large multi dimension array and also with matrix process, with the help of large-scale mathematical functions. It is also useful for fundamental scientific computations.

Pandas

Pandas is the tool in ML which is used for cleaning and data analysis. It has many features which are used for exploring, transforming and visualizing from the data provided.

Matplotlib

Matplotlib is popular Python library used for data visualization. Some like Pandas, is not directly connected to ML. It particularly comes in handy only when a programmer needs to visualize the patterns.

Seaborn

Seaborn is a library used to creating statistical graphics in python. It created above matplotlib and integrates nearly with pandas data structures. Seaborn helps user to explore, understand their data. It is stating functions, operated on data frame and arrays which contains complete datasets and internal perform which are necessary semantic mapping and statistical aggregation to create informative plots.

POPULAR MACHINE LEARNING ALGORITHMS

Linear Regression

It is a regression algorithm and comes under supervised learning. Here, we find the best line also called the regression line which is given by the mathematical equation:

$Y = AX + B$ and fit it to the data by designing a linear relationship between the independent and dependent variables.

In the equation above:

Y = dependent variable
A = slope of the line
X = independent variable
B = intercept of the line

Figure 3. Linear Regression Graphical Representation

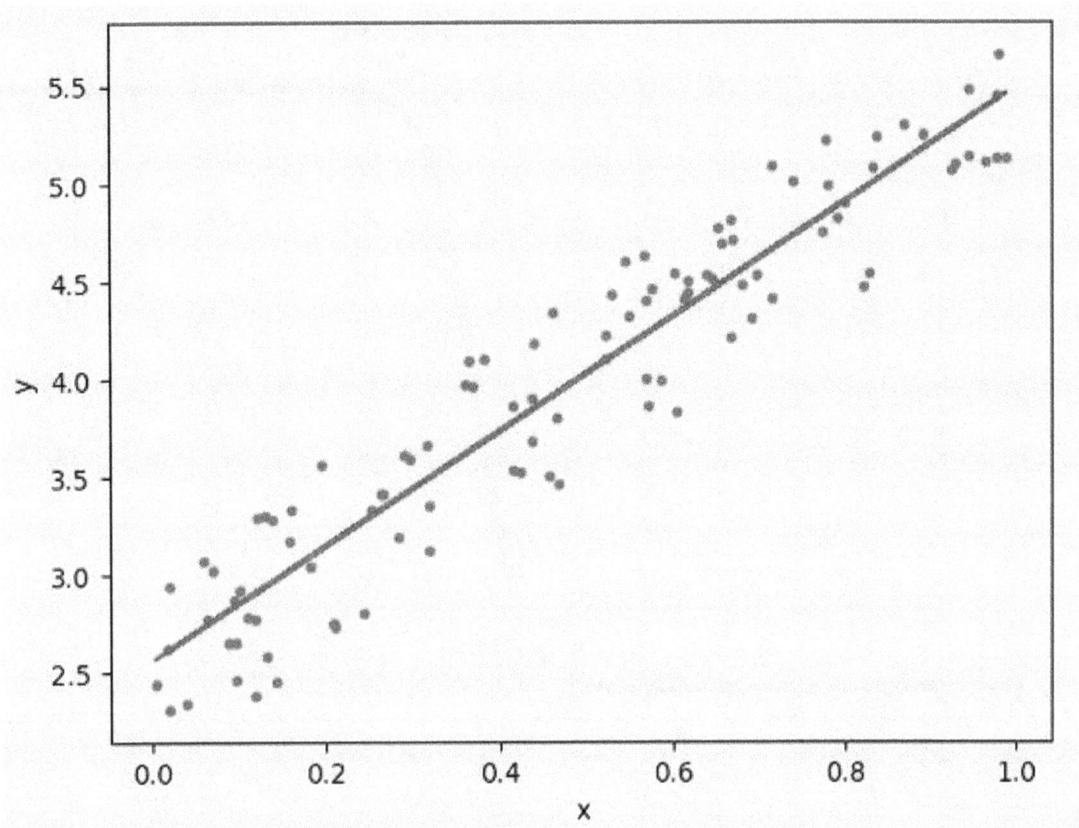

We can use any plot provided by matplotlib library as the tool for exploration of the relationship among the variables and then derive the coefficients A and B by ordinary least squares (OLE) technique in which we try to minimize the sum of the squared difference of distances between the linear regression line and the data points (Dayal et al., 2021).

Linear regression can be further break down into Simple and Multiple Linear Regression, characterized by one independent variable and multiple independent variables, respectively.

Logistic Regression

It is a classification algorithm which comes under supervised learning. Here the word "regression" can confuse us as it counters with "classification", but the highlight is on the word "logistic" which refers to logistic function which is the main doer of the classification task in this algorithm. It is used to find the discrete classification values (commonly binary values) from the set of given independent variables of the dataset. In other words, it predicts the probability of the event's occurrence by fitting the data to a logistic function, which is also referred as sigmoid function (Dayal, Chawla, & Khari, 2022).

Sigmoid Function

Figure 4. Logistic Function Graphical Representation

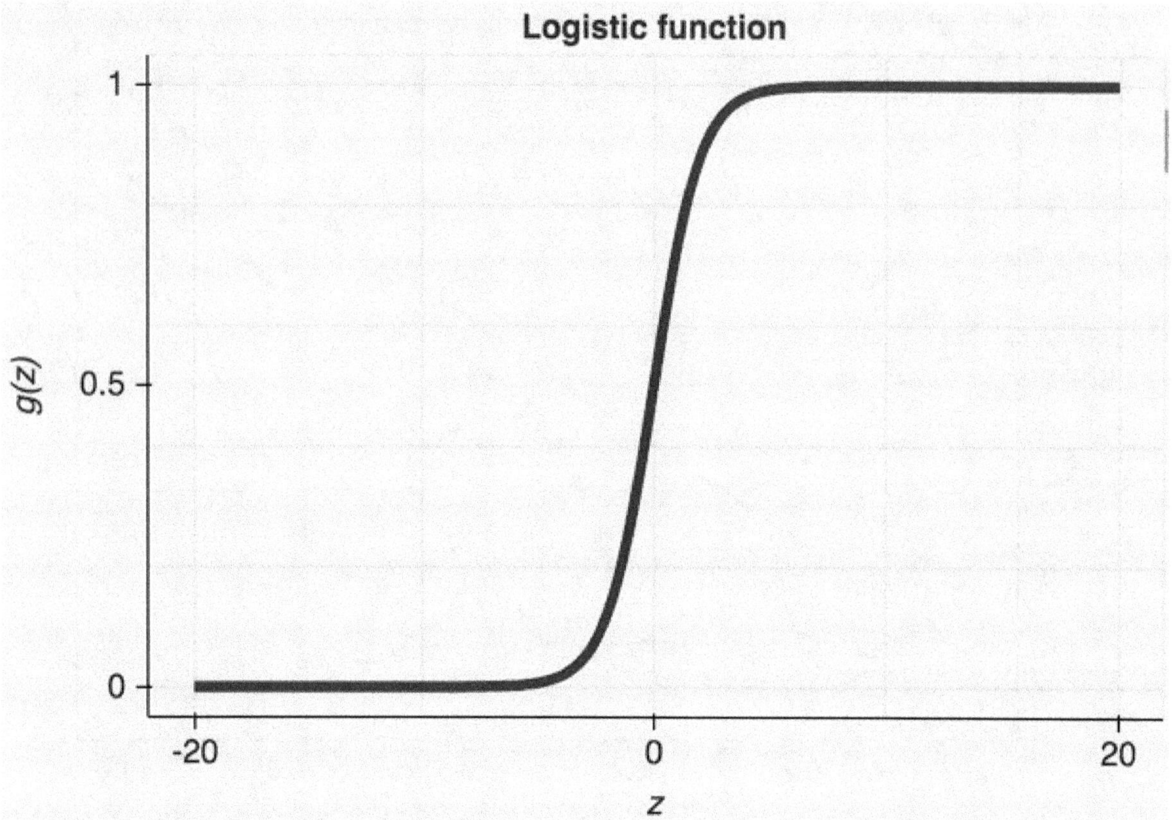

$$Sigmoid\ Function:\ y = \frac{1}{1+e^{-x}}$$

As we know, a logistic function is the inverse of logit function, that's why it is also termed as logit regression. It takes real valued numbers as input and gives the output values which lie between 0 and 1.

Decision Trees

A supervised learning algorithm which comes under both regression and classification but mostly used for classification problems. It works well with both categorical and continuous dependent variables. The algorithm usually does not need normalization of data or scaling of features.

It is a flowchart like tree structure that splits the data into two or more parallel sets. This is done on the basis of independent variables for making as distinguishable groups as possible. For splitting data into heterogeneous sets, techniques like Information Gain, entropy, Chi-square and Gini are used (Chlingaryan et al., 2018).

The main goal at each portioning is increasing the predictiveness as much as possible in order to keep gaining information about the dataset by the model. This algorithm is highly prone to overfitting so to generalize well, it needs to be ensemble well in order.

Figure 5. Decision Tree Representation

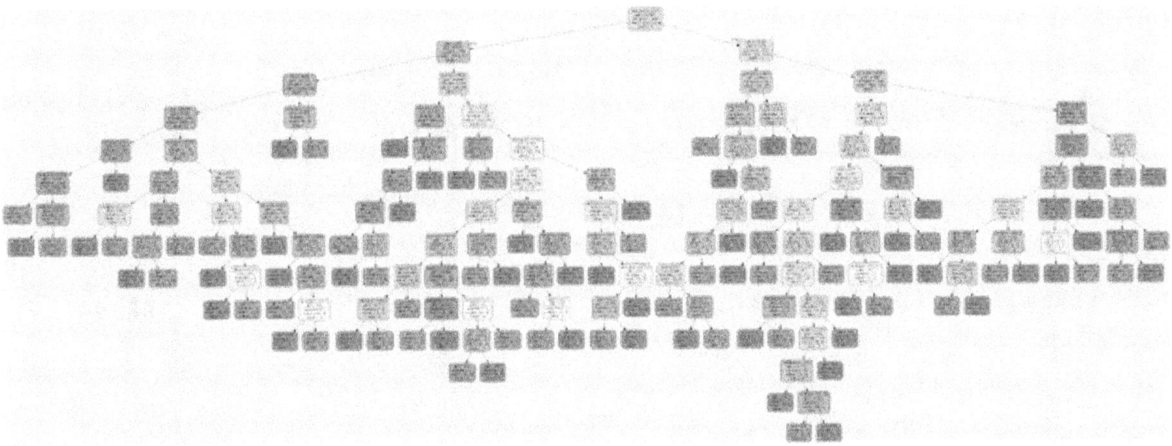

Random Forest

A supervised learning algorithm which comes under both regression and classification. Random forests or 'random decision forests' is a trademarked term for a collection of many decision trees, hence the name 'forest'. It is an ensemble learning method in which multiple algorithms are combined to result in better classification or regression. The forests are built by bagging method which uses decision trees as parallel estimators because individually it is weak but on combining, produces good results.

In case of classification problems, the result is derived as the majority of the vote (which is the classification given by the tree) of the products received from each individual decision tree whereas in case of regression problems, leaf node prediction is the mean value of different results received from each individual decision tree.(Dayal, Garg, Chaudhary et al, 2022)

For high success of this algorithm, random forests must have greater number of uncorrelated decision trees which is achieved by bootstrapping and feature randomness.

Naive Bayes Classifier Algorithm

A classification algorithm which comes under supervised learning. It is based on Bayes' Theorem and works on the assumption that each feature is independent of the presence of the other features and hence no correlation between features. Naïve Bayesian model is mainly used for large datasets and works really well despite its simplicity.

The Bayes' theorem is given in as:

$$P(A|B) = \frac{P(B|A)PA}{P(B)}$$

where:

p(A|B): Probability of event A given event B has already occurred

Figure 6. Random Forests Representation

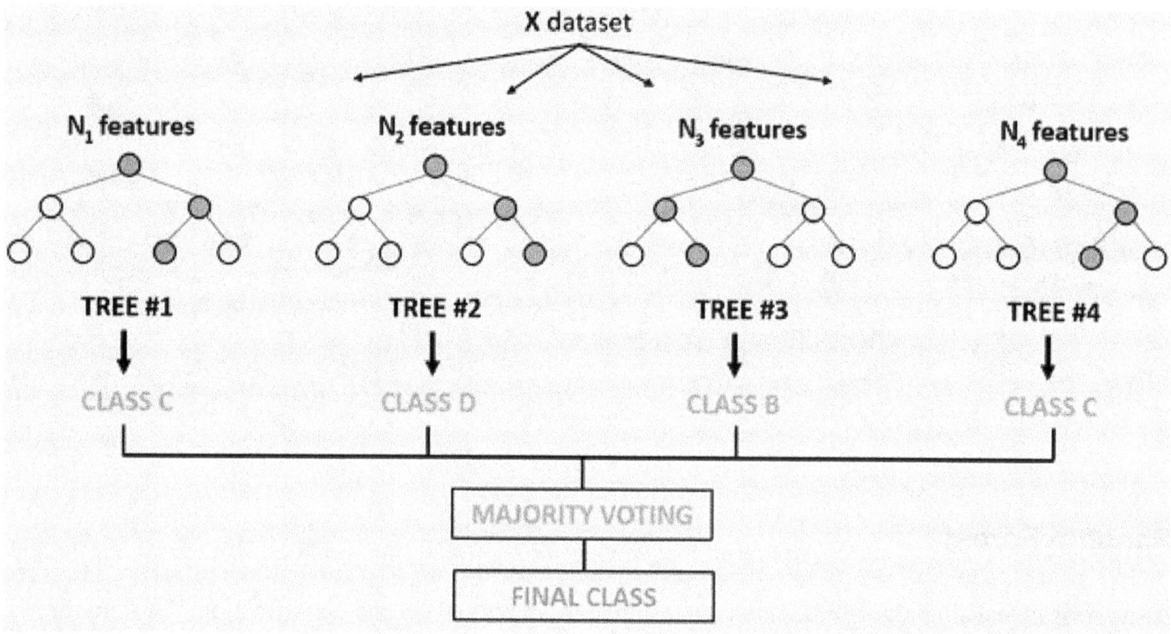

p(B|A): Probability of event B given event A has already occurred
p(A): Probability of event A
p(B): Probability of event B

In case of naive bayes classifier algorithm, the probability formula used for a given class of feature values is given by:

Likelihood, Prior, Posterior, Normalizing constant

$$\underbrace{P(A|B)}_{\text{Posterior}} = \frac{P(B|A)P(A)}{P(B)}$$

$$P(B) = \sum_{Y} P(B|A)P(A)$$

Here,

- P(A|B) - the posterior probability of class (target) given predictor (attribute).
- P(A) - the prior probability of class.

Figure 7. SVM Graphical Representation

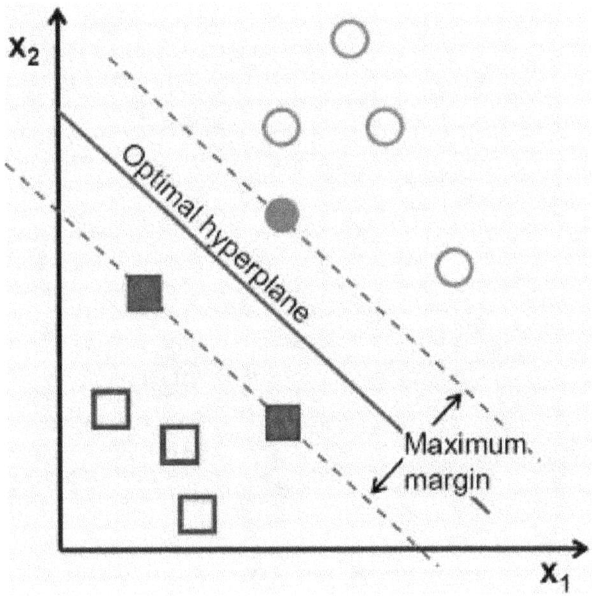

- P(B|A) - the likelihood which is the probability of predictor given class.
- P(B) - the prior probability of predictor (normalising constant).

SVM (Support Vector Machine) Algorithm

A supervised learning algorithm which comes under both regression and classification but mostly used for classification tasks. Here we plot the data in n-dimensional space considering n as the number of features of the data. The value of each feature is the value of that particular coordinate. Here we find an optimal hyperplane of n-1 dimensions which best separates the data in a way its distance from the nearest points is maximized. These coordinates or nearest points are termed as support vectors. This hyperplane is our classifier and helps in classification of new data (Nigam et al., 2019).

In case of non-linear data, it uses kernel trick in which the data points are projected in higher dimensional space to make the data linearly separable.

KNN (K – Nearest Neighbors) Algorithm

A supervised learning algorithm which comes under both regression and classification but mostly used for classification tasks. On the basis of input cases, it classifies the class or value of new cases by using the majority vote of its k nearest neighbor data points present around it. The data points must be normalized or else the variables of high range can bias it.

Figure 8. KNN Graphical Representation

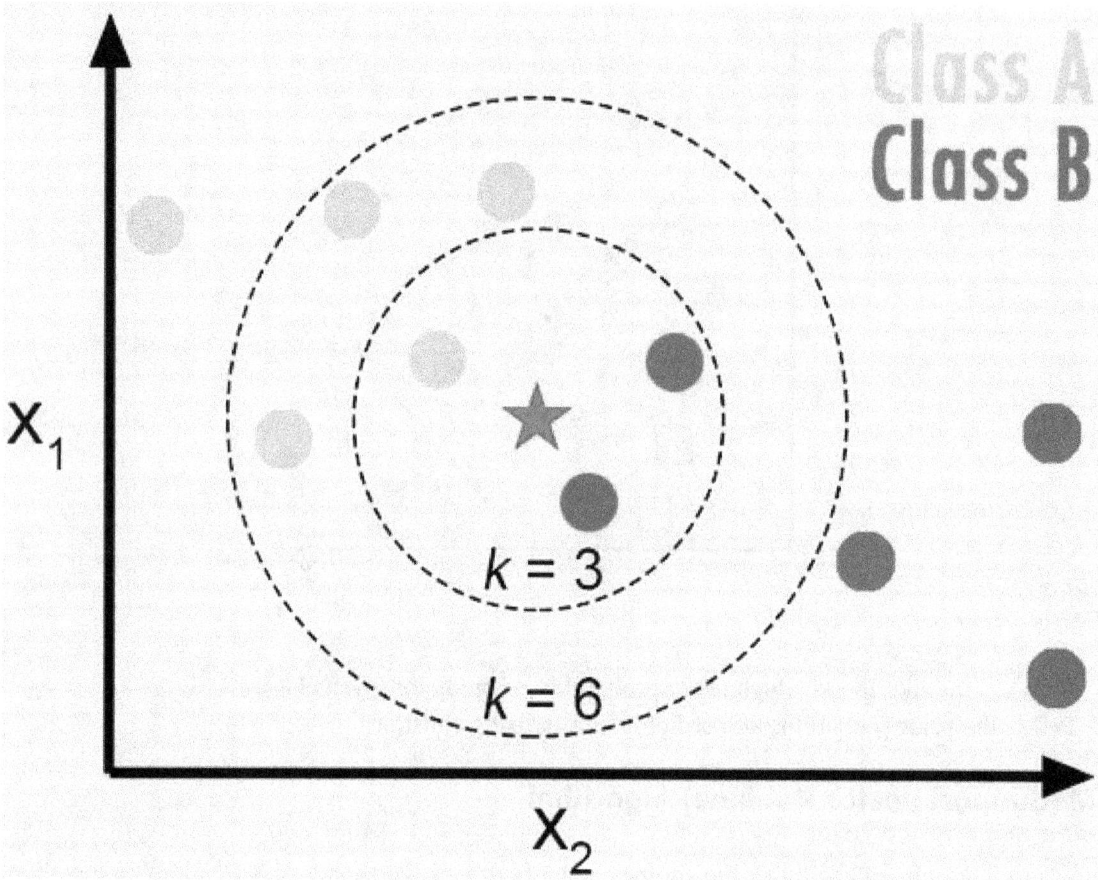

K Means Algorithm

It is a clustering algorithm which comes under unsupervised learning. Here the data is classified into sets of a particular number of clusters such that data points inside each cluster are homogeneous and are heterogeneous to the points outside in another clusters. Thus, it can also be referred as a partition-based clustering technique. The main idea is to minimize the distances within the cluster and maximize the distances among different clusters (Dimitriadis & Goumopoulos, 2008).

IMPLEMENTATION OF DIFFERENT MACHINE LEARNING ALGORITHMS USING PYTHON

Now let's apply few machine learning algorithms to the given dataset of India District-wise agriculture crop area and production statistics 1997-2020.

Applying various machine learning algorithms on the data set

Figure 9. K- Means Representation

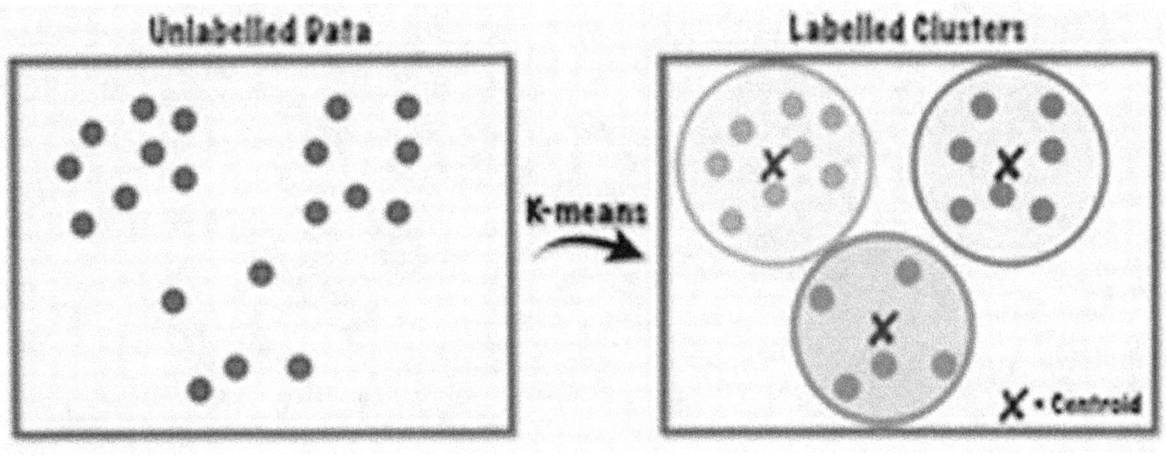

Figure 10. Initial steps of data preprocessing and importing required libraries.

```
#importing the libraries
import pandas as pd
import numpy as np
import matplotlib.pyplot as plt
import seaborn as sns

#reading the dataset from file saved locally
data = pd.read_csv('India Agriculture Crop Production.csv')

#this method prints information about a dataframe
data.info()

#this method generate Generate descriptive statistics
data.describe()

#dropping the duplicate data from the dataset
final_data = data.drop_duplicates()
final_data.info()
```

Figure 11. Preprocessing the data and splitting it into training and testing data.

```
import sklearn
```

```
from sklearn.preprocessing import LabelEncoder

# Creating a instance of label Encoder.
le = LabelEncoder()

# Using .fit_transform function to fit label
# encoder and return encoded label i.e. label encoding the categorical data
X['State'] = le.fit_transform(X['State'])
X['District'] = le.fit_transform(X['District'])
X['Season'] = le.fit_transform(X['Season'])
X['Crop'] = le.fit_transform(X['Crop'])
X['Year'] = le.fit_transform(X['Year'])
```

```
#converting the data from dataframes format to numpy arrays
X = X.values
y = y.values
```

```
#splitting the data into training data and testing data (4:1)
split = int(0.8*X.shape[0])
print(split)

X_train = X[:split,:]
y_train = y[:split]

X_test = X[split:,:]
y_test = y[split:]

split
```

Figure 12. Linear regression

```
#linear regression
from sklearn.linear_model import LinearRegression
```

```
reg = LinearRegression().fit(X_train, y_train)
reg.score(X_test, y_test)
```

```
0.06728154300815992
```

Figure 13. Neural networks

```
#neural networks
from sklearn.neural_network import MLPRegressor
regr = MLPRegressor(random_state=1, max_iter=500)
regr.fit(X_train, y_train)
```

```
regr.score(X_test, y_test)
```

```
0.16144897665210622
```

Figure 14. K nearest neighbours

```
#knn
from sklearn.neighbors import KNeighborsRegressor
neigh = KNeighborsRegressor(n_neighbors=5)
neigh.fit(X_train, y_train)
```

```
neigh.score(X_test,y_test)
```

```
0.5813571433772474
```

Figure 15. Decision tree

```
#decision tree
from sklearn.tree import DecisionTreeRegressor
regressor = DecisionTreeRegressor(random_state=0)
```

```
regressor.score(X_test,y_test)
```

```
0.9566171495357091
```

Figure 16. Random forests

```
import time
```

```
#random forests
from sklearn.ensemble import RandomForestRegressor
l1 = []
l2 = []
for i in range(2,15):
    start = time.time()
    regr = RandomForestRegressor(max_depth=i, random_state=0)
    regr.fit(X_train, y_train)
    end = time.time()
    l1.append(regr.score(X_test, y_test))
    l2.append(end-start)
```

```
l3 = []
for i in range(1,len(l1)):
    l3.append(l1[i] - l1[i-1])
plt.scatter(l2[1:], l3, color = 'red')
plt.plot(l2[1:],l3)
plt.xlabel('time')
plt.ylabel('difference in score')
plt.show()
```

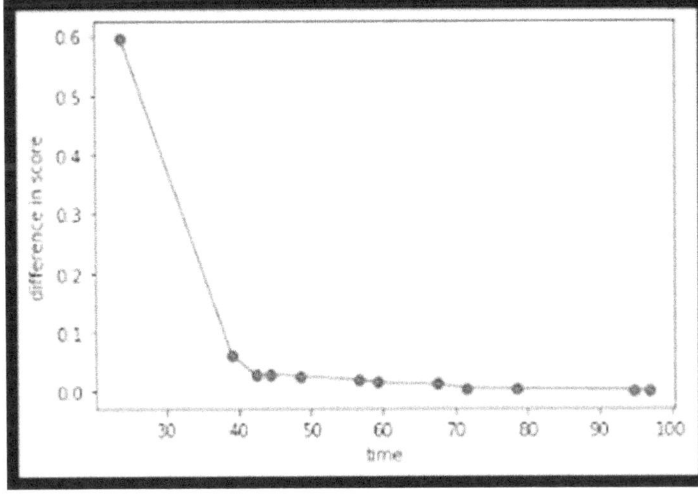

REFERENCES

Bhanu, K. N., Jasmine, H. J., & Mahadevaswamy, H. S. (2020, June). Machine learning implementation in IoT based intelligent system for agriculture. In *2020 International Conference for Emerging Technology (INCET)* (pp. 1-5). IEEE. 10.1109/INCET49848.2020.9153978

Chlingaryan, A., Sukkarieh, S., & Whelan, B. (2018). Machine learning approaches for crop yield prediction and nitrogen status estimation in precision agriculture: A review. *Computers and Electronics in Agriculture, 151*, 61–69. doi:10.1016/j.compag.2018.05.012

Cunha, R. L., Silva, B., & Netto, M. A. (2018, October). A scalable machine learning system for pre-season agriculture yield forecast. In *2018 IEEE 14th International Conference on e-Science (e-Science)* (pp. 423-430). IEEE. 10.1109/eScience.2018.00131

Dayal, M., Chawla, A., & Khari, M. (2022). Coalescence of Neural Networks and Blockchain. In *Handbook of Green Computing and Blockchain Technologies* (pp. 31–44). CRC Press.

Dayal, M., Chawla, A., Khari, M., & Mahajan, A. N. (2023). Artificial Intelligence-Based Smart Packet Filter. In *Proceedings of Third International Conference on Computing, Communications, and Cyber-Security* (pp. 791-801). Springer, Singapore.

Dayal, M., Chawla, A., Singh, D., & Garg, R. K. (2021). Combination of Convolutional Neural Networks and Blockchain in Handwritten Digit Recognition. In *Handbook of Green Computing and Blockchain Technologies* (pp. 55–67). CRC Press. doi:10.1201/9781003107507-5

Dayal, M., Garg, R. K., Chaudhary, S., Chaubey, C., & Chawla, A. (2022). Blockchain for Medicine and Healthcare. In *Handbook of Green Computing and Blockchain Technologies* (pp. 171–181). CRC Press.

Dimitriadis, S., & Goumopoulos, C. (2008, August). Applying machine learning to extract new knowledge in precision agriculture applications. In *2008 Panhellenic Conference on Informatics* (pp. 100-104). IEEE. 10.1109/PCI.2008.30

Katarya, R., Raturi, A., Mehndiratta, A., & Thapper, A. (2020, February). Impact of machine learning techniques in precision agriculture. In *2020 3rd International Conference on Emerging Technologies in Computer Engineering: Machine Learning and Internet of Things (ICETCE)* (pp. 1-6). IEEE. 10.1109/ICETCE48199.2020.9091741

Kumar, P., Dayal, M., Khari, M., Fenza, G., & Gallo, M. (2021). Nsl-bp: A meta classifier model based prediction of amazon product reviews.

Kumar, R., Singh, M. P., Kumar, P., & Singh, J. P. (2015, May). Crop Selection Method to maximize crop yield rate using machine learning technique. In 2015 international conference on smart technologies and management for computing, communication, controls, energy and materials (ICSTM) (pp. 138-145). IEEE. doi:10.1109/ICSTM.2015.7225403

McQueen, R. J., Garner, S. R., Nevill-Manning, C. G., & Witten, I. H. (1995). Applying machine learning to agricultural data. *Computers and Electronics in Agriculture, 12*(4), 275–293. doi:10.1016/0168-1699(95)98601-9

Nigam, A., Garg, S., Agrawal, A., & Agrawal, P. (2019). Crop Yield Prediction Using Machine Learning Algorithms. *Fifth International Conference on Image Information Processing (ICIIP)*, pp. 125-130, 2019. 10.1109/ICIIP47207.2019.8985951

Nishant, P. S., Venkat, P. S., Avinash, B. L., & Jabber, B. (2020, June). Crop yield prediction based on indian agriculture using machine learning. In *2020 International Conference for Emerging Technology (INCET)* (pp. 1-4). IEEE. 10.1109/INCET49848.2020.9154036

Reich, Y., & Barai, S. V. (1999). Evaluating machine learning models for engineering problems. *Artificial Intelligence in Engineering*, *13*(3), 257–272. doi:10.1016/S0954-1810(98)00021-1

Chapter 13
5G:
Rearchitecting the Edge Networks

Mohit Dayal
Dr. Akhilesh Das Gupta Institute of Technology and Management, India

Raghav Seth
Dr. Akhilesh Das Gupta Institute of Technology and Management, India

Deepti Singh
ABES Institute of Technology, India

Bhumika Pal
Dr. Akhilesh Das Gupta Institute of Technology and Management, India

Priya Paliwal
Dr. Akhilesh Das Gupta Institute of Technology and Management, India

Pravin Kumar
Indian Institute of Technology, Dhanbad, India

ABSTRACT

Cellular technology has undergone remarkable development over the last four decades. With the commercial introduction of 5G, humans are again going to take a considerable leap in the fields of IoT, augmented and virtual reality, and tactile internet. Like LTE, the 5G new radio technology is also based on orthogonal frequency-division multiplexing, but unlike the previous generations, it supports a large number of different use cases because of its ability to operate over both high and low-frequency bands.

INTRODUCTION

"5G" is the fifth generation of wireless telecommunications technology that will revolutionize many aspects of society. Virtual reality, high-res video streaming, and gaming have all contributed to the rapid

DOI: 10.4018/978-1-6684-6413-7.ch013

Copyright © 2023, IGI Global. Copying or distributing in print or electronic forms without written permission of IGI Global is prohibited.

growth of mobile networks over the last few years.

Compared to today's network, 5G offers much faster transmission speeds. Compared to 4G and 4G-LTE, 5G will be able to transmit data at speeds up to 10Gbps. With 5G, blockchain, big data, artificial intelligence (AI), the Internet of Things (IoT), and cloud computing can be combined with existing technologies to create innovative services. The 5G standard also offers lower latency in addition to increased speed. It has a response time of less than one millisecond (ms), making it close to real-world response times. Furthermore, 5G is expected to release a huge IoT network, unlike today's IoT services. Due to its high bandwidth, connectivity, and coverage, 5G can also be used for medical devices to provide real-time interaction. 5th-generation (5G) cellular technology has recently taken center stage in global competition because it is expected to be one of the most important sources of revenue for years to come.

In the energy sector, 5G will make the concept of smart virtual power plants more evident, as well as solve the most significant social problems associated with traffic congestion, disaster safety, and climate change (Tripathi and Prasad, 2018).

In addition, blockchain technology can be used to transform energy from producers to consumers in real time, and securely (Raj, Saini, & Surianarayanan, 2022).

Among the different types of 5G use cases, three stand out:

- Enhanced-Mobile-Broadband (eMBB): This refers to an evolution of the improvements made in the consumer experience in cell broadband, which encourages further improvements in customer service.
- Massive-Machine-type Connectivity (mMTC): Contains many devices, such as remote control, actuator, and tracking tools. These systems are primarily designed with a low system cost and relatively low computer power consumption, which allows for very long battery life. Most systems produce and absorb a limited volume of data, making large data volumes less important.
- Ultra-Reliable and Low-Latency Connectivity (URLLC): For use in industrial automation, automated driving, and virtual surgery, this type of connectivity provides low-latency reactive tools.

ARCHITECTURE

It is essential that, beyond 4G, some of the major objectives and demands are addressed, including increasing capacity, improving data rates, decreasing latency, and improving quality (Gupta and Jha, 2015).

The third-generation partnership project (3GPP) covers telecommunication technology, including radio access networks (RANs), service capabilities, and core transport networks. In comparison to earlier generations, the 3GPP has released entire system specifications for 5G.

Services are offered by a common framework to network services that need them. The 3GPP requirements for 5G network architecture also call for it to be modular, reusable, and self-contained.

As shown in above *figure 1*, the 5G architecture and its different components are explained briefly in the following sections.

Figure 1.

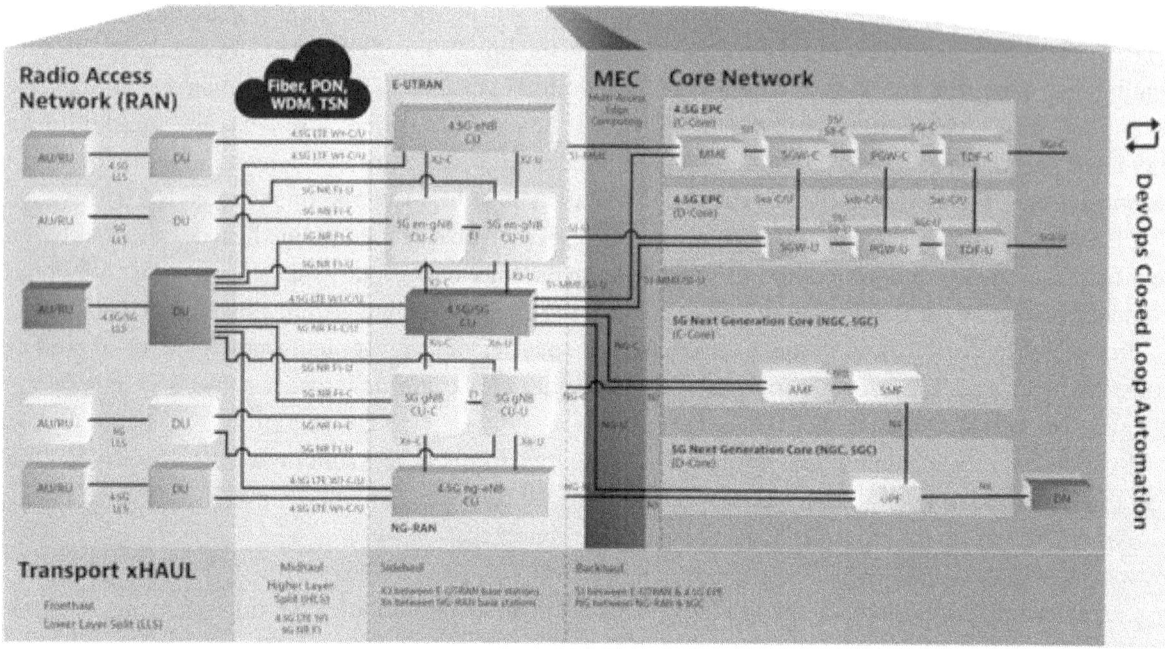

5G Spectrum and Frequency

There are now various frequency bands allocated to 5G new radio (NR). Millimeter wave refers to the range of wavelengths between 1-10 mm along with the frequency range of 30 GHz - 300 GHz. Worldwide, multiple regions are now allocating frequency ranges of 24 GHz and 100 GHz for 5G.

Mobile communications currently operate between 700 MHz and 3600 MHz. Data of this magnitude cannot be carried over this spectrum. A frequency band beyond 10 GHz is the most preferred band for 5G (Tripathi and Prasad, 2018).

In addition, UHF channels between 300 MHz and 3 GHz are being used for 5G. The variety of frequencies used may be customized to the specific applications since higher frequencies offer a wider bandwidth but a smaller operating range. The millimeter wave frequencies are useless for long-distance communications but work best in heavily populated areas. Within these high and low-frequency bands, each operator has begun separating off its own distinct, separate 5G spectrum sections.

MEC

Multi-access edge computing (MEC) is a crucial component of the 5G architecture. MEC allows for the distribution of programs from centralized data centers to the network edge, where end users and their appliances may access them. It shortens the network path between user and host, thus reducing the time it once took to deliver content.

5G networks built to 3GPP 5G standards are perfect for implementing MEC. Edge computing enablers are described as part of the 5G standard, enabling MEC and 5G to work together to transport traffic.

The MEC architecture's distributed processing capability will also improve the capabilities of the 5G rollout and the Internet of Things (IoT).

NFV and 5G

Different network functions, such as security systems, load balancers, and adapters, are decoupled from hardware through network function virtualization (NFV). The consumer will benefit from receiving revenue-generating services more quickly without having to buy a large number of costly hardware components as a result.

A 5G network is enabled through NFV by virtualizing appliances. It enables the simultaneous operation of multiple virtual networks using network slicing technology.

5G RAN Architecture

Network disaggregation allows NFV to reach the RAN through alliances like O-RAN. This flexibility and new prospects for competitiveness are made possible by offering open interfaces and open-source software development, enabling the widespread adoption of new features and technologies. The O-RAN ALLIANCE enables off-the-shelf deployment across several manufacturers to streamline and quicken interoperability. Network disaggregation makes it possible to virtualize network elements, facilitating scaling and enhancing user experience. Virtualizing RAN components has the advantage of being more

affordable from a hardware and software standpoint, notably in IoT operations where the number of connected devices is in the billions.

eCPRI

With new interfaces such as eCPRI, network dis-aggregation also provides other cost benefits. When a huge number of 5G carriers are tested, RF interfaces are not cost-effective because RF costs increase rapidly. With eCPRI interfaces, multiple 5G carriers can be tested with fewer interfaces, resulting in a more cost-effective solution. The eCPRI, like the DU used in the O-RAN front haul approach, is meant to be a common interface for 5G.

Network Slicing

Network slicing may be the critical component that enables 5G architecture to realize its full potential. As a result of this technology, multiple logical networks can run simultaneously over the same physical infrastructure in the NFV domain.

Network slicing is an efficient solution for IoT applications with many users, having little bandwidth demand. Due to the different requirements of each 5G vertical, network slicing becomes an important consideration when designing 5G networks. With this level of customization, costs, resource management, and network configuration flexibility can be optimized. As a result of network slicing, new 5G services can be evaluated faster and time-to-market is shortened.

Beamforming

5G technology is also based on beamforming, which is another breakthrough technology. Signals are broadcast by conventional base stations in multiple directions regardless of the location of the users or devices targeted. Multiple-input, multiple-output (MIMO) arrays, which include a large number of tiny antennas, can be used to pick the most effective transmission channel for individual packets to reach the final user in a preset order.

Due to 5G's millimeter wave frequencies, propagation loss and diffraction loss are significant due to the compact-sized antenna and the low wall penetration. The compact-sized antennas do, however, make it possible for considerably bigger arrays to fit into the same area. Massive beamforming for 5G is made feasible by the ability of several of these compact-sized antennas to change their beam orientation several times every millisecond. With massive MIMO, smaller beams may be produced in the same area, increasing throughput and improving user monitoring.

5G Core Architecture

To meet the demands of new services across a wide range of industries, the 3GPP standards development organization is developing a new 5G system architecture (5GS), including a new 5G New Radio access network (5NR) and a new 5G core network (5GC). This new core network is fundamental for the commercial success of 5G since it will enable new service types and benefits from cloud economics (Brown, 2017).

5G's core network architecture enables its increased throughput demand and is at the core of its new specification. The new 5G core, as described by 3GPP, includes authentication, security, session management, and traffic aggregation from end devices. It makes use of a service-based architecture (SBA). NFV is also emphasized as part of the 5G core, enabling virtualized software functions to be deployed via MEC technology, the infrastructure at the core of 5G's architecture.

A 5G core architecture consists of the following components (see *figure 2*)

- User plane Function (UPF)
- Data network (DN)
- Core Access and Mobility Management Function (AMF)
- Authentication Server Function (AUSF)
- Session Management Function (SMF)
- Network Slice Selection Function (NSSF)
- Network Exposure Function (NEF)
- NF Repository Function (NRF)
- Policy Control function (PCF)
- Unified Data Management (UDM)
- Application Function (AF)

Here is how the 5G core architecture works:

1. A 5G User Equipment (UE) like a 5G smartphone or 5G cellular device connects to a 5G core, then to a Data Network (DN), such as the internet, over the 5G New Radio Access Network.

Figure 2

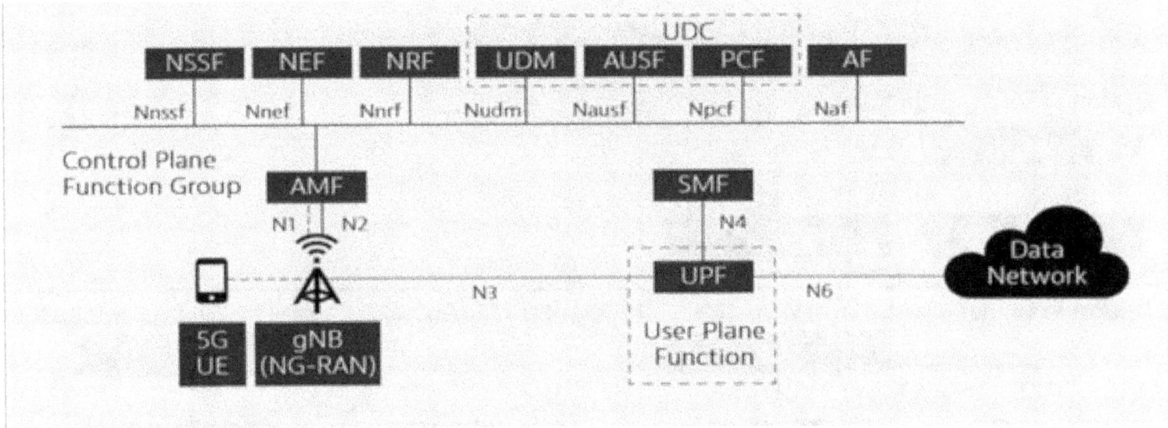

2. UE connections are managed by the Access and Mobility Management Function (AMF).
3. The AMF selects the relevant Session Management Function (SMF) for managing the user session based on the service requested by the UE.
4. IP data traffic (user plane) is transported between the User Equipment (UE) and the external networks by the User Plane Function (UPF).
5. As part of the Authentication Server Function (AUSF), the AMF authenticates the UE and gives it access to 5G core services.
6. A policy control framework is provided by functions such as Session Management Functions (SMFs), Policy Control Functions (PCFs), Application Functions (AFs), and Unified Data Management (UDMs). These functions apply policy decisions and access subscription information to govern network behavior.

5G Architecture Options

A well-thought-out approach and modest measures are required to bridge the gap between 4G and 5G. As a symbol of this shift, 5G architecture options will gradually move from non-standalone to standalone mode. With the inclusion of a 5G component carrier, LTE RAN and core networks serve as an anchor to establish the standard for 5G non-standalone in late 2017. The non-standalone mode will enhance bandwidth while employing millimeter-wave frequencies, as seen in figure 3.

Figure 4 shows that the full deployment of all 5G hardware, capabilities, and functionality is referred to as 5G standalone mode. Careful planning and execution are essential when non-standalone mode eventually gives way to new 5G mobile phone network architecture installations to make this transition painless for the user base.

Security in 5G Architecture

By utilizing a lot of cloud-based services, virtualization, network slicing, and other cutting-edge technologies, 5G will deliver tremendous performance benefits. Consequently, 5G security architectures are exposed to new security risks and additional attack surfaces.

Figure 3.

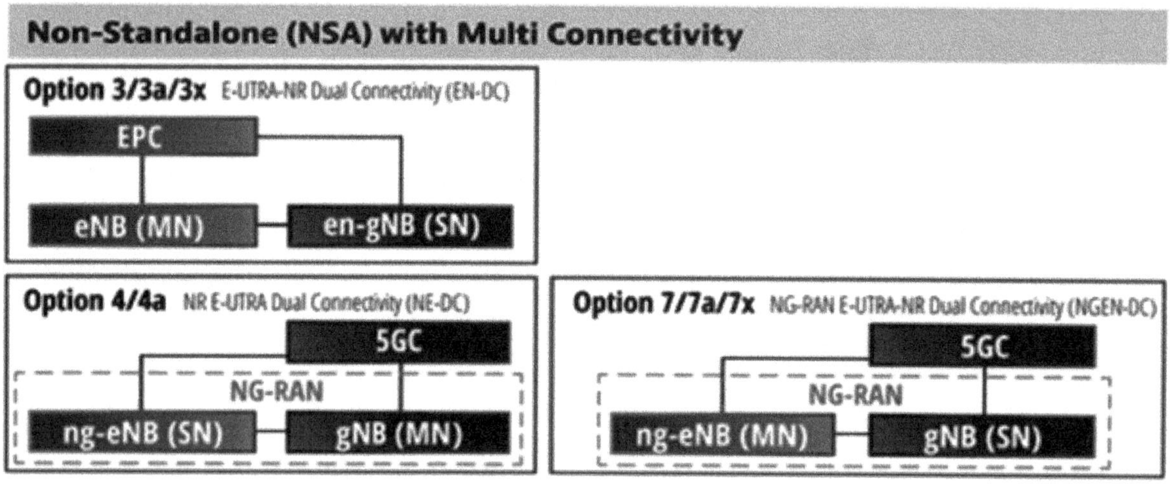

5G presents a typical use case for wireless networking and communication convergence, incorporating functions such as Software Defined Networks (SDNs), Network Functions Virtualization (NFVs), and edge cloud. But this convergence extends many of the security challenges and opportunities applicable to SDN/NFV and cloud to 5G networks (Dutta & Hammad, 2020).

The 3GPP standards for 5G security outline several benefits, such as unified authentication that decouples authentication from access points and extensible authentication protocols that support mul-

Figure 4.

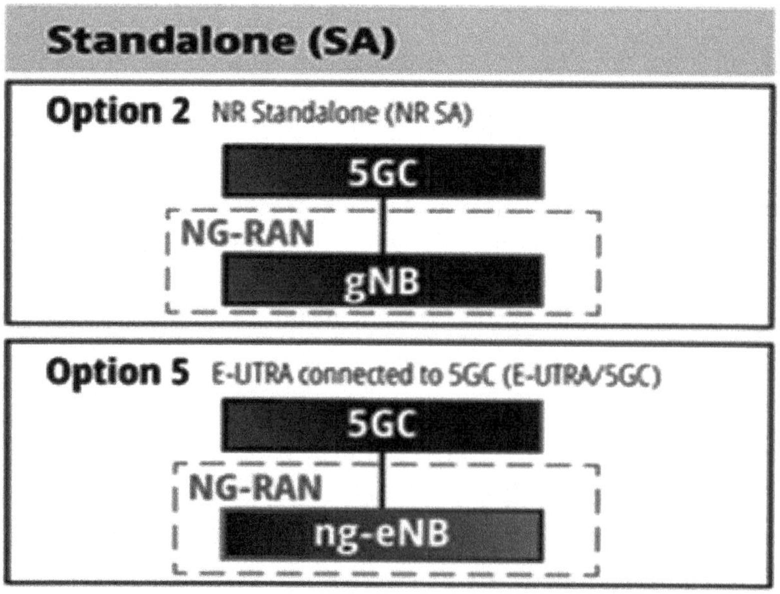

tifactor authentication, versatile security protocols to support new use cases, and subscriber permanent identifiers (SUPI) that guarantee network confidentiality.

Operators will need to regularly monitor and evaluate security performance as a result of the rollout of 5G and the growing virtualization of important nodes. To adhere to best practices, system architecture, devices, and applications must be monitored end-to-end for network security.

USE CASES

Fast Data Rate

High-speed broadband speeds and smart grids define 5G networks. It takes about 8 minutes to download a 4G movie, however, 5G will enable them to finish it in under 10 seconds. Networking speeds can accommodate technology like:

- social networking sites
- multimedia TVs
- the high definition of 3D content
- augmented reality, robotics
- driverless cars
- advanced manufacturing advances.

If tons of computers are connected, then there is no need of transferring d data simultaneously. One system might need immediate linking to the network, so others can transfer their data during off-peak hours. For a 5G advanced world, we need networks that can effectively handle data traffic while also making split-second choices. The engineers are currently making a lot of development in making these wireless networks. The core of the next 5G system is anticipated to be a high-speed connection and minimal latency. You may watch 8K in only one-second thanks to 5G's output speed, which is 10 times quicker than 4G.

Augmented Reality on 5G

The popularity of augmented and virtual reality has grown recently because of the increased capabilities of video streaming and mobile networks. However, researchers are being bombarded with various machine and deep learning algorithms, which are categorized as a part and parcel of the enigmatic AI discipline (Saini and Raj, 2021). High-fidelity telepresence, interactive apps, and augmented reality are not yet possible due to bandwidth and delay restrictions, although some of these activities can still be completed. The fact that developers and architects are aware of these problems and are putting 5G networks in place to help us migrate to this new era of software applications is fortunate.

Current technology cannot support such applications because of the strict end-to-end latency requirement, which should not exceed 20 ms to avoid motion sickness and nausea. Enable immersive interactive AR/VR, which will require 5G ultra-reliable low-latency communications (URLLC) and tactile Internet of Things (IoT) (Saini, Mishra, & Srivastava, 2021).

Although, 5G should have a wide variety of opportunities for wireless connectivity due to the two major demand influences of wireless Internet and IoT. It is expected that 5G will have a wide range of uses, including augmented and virtual reality, wireless computing, eHealth, smart cars, etc.

Despite the network requirements of emerging technology sectors like AR and VR, there is a lot of enthusiasm and anticipation surrounding the launch of 5G network technologies.

A hyper-connected society where mobile devices are playing a more significant role is made possible by the 5G standard, which includes a wide variety of applications including higher connection (such as high-speed trains) and considerable improvements in energy efficiency and network capacity (Torres Vega et al., 2020).

The main source of network traffic is video streaming services. In many instances, live streaming demands from internet users compensate for the data traffic generated by popular web services that use asynchronous information reuse. One of the first killer apps of 5G was expected to be smartphone Augmented Reality and Virtual Reality (AR / VR).

By enabling device-to-device connectivity and IoT, 5G infrastructure aims to enable a variety of conventional as well as new technologies. The purpose of this study is to examine the capabilities and features of 5G that could aid the deployment of advanced e-learning systems utilizing virtual reality and augmented reality.

Additionally, it's crucial to provide impromptu classes everywhere by using user equipment. To find out if these technologies can be included in e-learning systems, we are assessing current 5G experiments, especially in Europe.

Smart Mobility Using 5G

The whole world nowadays widely uses the current trend called Industry 4.0. Manufacturing Industry also uses it for smarter decisions as it makes the techniques of gathering information and processing data easier (Kumari & Sainin, 2021). And the 5G mobility technology spectrum consists of road preparation technologies (connected vehicles) and new automated driving systems (connected vehicles). There are many advantages to smart mobility, including road management, safe navigation, avoiding incidents, fuel conservation, lowering prices, and reducing emissions.

Smart Cities Using 5G

As 5G technology advances, it will connect every part of the world in an ever-online manner, from the biggest megacities to the smallest IoT devices. A connected architecture like this will combine smart cities, smart homes, and IoT into one cohesive system.

Research on smart cities and self-organizing networking techniques has been done in the context of 5G wireless networks. Smart cities require 5G for huge data transfers, but the network itself is unsure of the data that streams over it. Stronger cooperation between the two will result in a shared collaboration as the sensors gather an enormous quantity of data, improving connectivity performance with the information produced.

Additionally, 5G intends to integrate a variety of access technologies, resulting in a much more efficient communication network and easier information exchange between heterogeneous systems and services.

In smart cities, the tactile internet will calculate, manage, track, and scale smart devices in actual or virtual reality. Among the key characteristics of the tactile internet would be its low latency, stability, and quality of access.

CONCLUSION

This chapter has covered a variety of topics related to the impending 5G network, including segments necessary for its deployment. Everything will be connected in 5G networks, from users to the internet, basic sensors to automated automobiles, integrated sensors in all sorts of hardware to automated cars, aviation to smart cities, and 5G networks will connect everything. As compared to the current network, the 5G network has a much higher capacity, lower latency, and much higher bandwidth. As a result, 5G will contribute to what could be the biggest technical revolution in human history. Human lives can be changed and preserved by improving emergency response care. For 5G to be commercially successful, network capacity and flexibility must be continually improved to accommodate a variety of use cases and business models. Keeping an eye on the energy and cost efficiency of 5G technology is also important. This chapter presents 5G's energy efficiency areas, various efficient antennas, architectures, and wide uses, as well as various other aspects of 5G technology.

REFERENCES

Brown, G. (2017). Service-based architecture for 5G core networks. *Huawei White Paper, 1.*

Dutta, A., & Hammad, E. (2020, September). 5G security challenges and opportunities: a system approach. In *2020 IEEE 3rd 5G World Forum (5GWF)* (pp. 109-114). IEEE.

Gupta, A., & Jha, R. K. (2015). A survey of 5G network: Architecture and emerging technologies. *IEEE Access: Practical Innovations, Open Solutions*, *3*, 1206–1232. doi:10.1109/ACCESS.2015.2461602

Kumari, R., & Saini, K. (2021, March). Advanced Automobile Manufacturing: An Industry 4.0. In *International Conference on Computing for Sustainable Global Development (INDIACom)* (pp. 899-904). IEEE.

Raj, P., Saini, K., & Surianarayanan, C. (2022). *Edge/Fog Computing Paradigm: The Concept, Platforms and Applications*. Academic Press. .

Saini, K., Mishra, I., & Srivastava, S. (2021, December). Farmer's E-mart: An E-Commerce Store For Crops. In *3rd International Conference on Advances in Computing, Communication Control and Networking (ICAC3N)* (pp. 346-350). IEEE.

Saini, K., & Raj, P. (Eds.). (2021). *Advancing Smarter and More Secure Industrial Applications Using AI, IoT, and Blockchain Technology*. IGI Global.

Torres Vega, M., Liaskos, C., Abadal, S., Papapetrou, E., Jain, A., Mouhouche, B., Kalem, G., Ergüt, S., Mach, M., Sabol, T., Cabellos-Aparicio, A., Grimm, C., De Turck, F., & Famaey, J. (2020). Immersive interconnected virtual and augmented reality: A 5G and IoT perspective. *Journal of Network and Systems Management*, *28*(4), 796–826. doi:10.100710922-020-09545-w

Tripathi, P. S., & Prasad, R. (2018). Spectrum for 5G services. *Wireless Personal Communications*, *100*(2), 539–555. doi:10.100711277-017-5217-9

Chapter 14
Edge AI for Real-Time and Intelligent Agriculture

Jigna Bhupendra Prajapati
Acharya Motibhai Patel Institute of Computer Studies, Ganpat University, India

Akash Kumar
https://orcid.org/0000-0001-7969-5666
Center for Health and Applied Sciences, Ganpat University, India

Jhilam Pramanik
Center for Health and Applied Sciences, Ganpat University, India

Bhupendra G. Prajapati
https://orcid.org/0000-0001-8242-4541
Shree S. K. Patel College of Pharmaceutical Education and Research, Ganpat University, India

Kavita Saini
https://orcid.org/0000-0003-4058-8622
Galgotias University, India

ABSTRACT

Advancements of the last decade in edge computing, edge IoT, and edge artificial intelligence now allow for autonomous, efficient, and intelligent systems to be proposed for various industrial applications. Intelligence agricultural solutions allow farmers to achieve more with less while improving quality and providing a rapid go-to-market approach for produce. Using AI is an effective technique to detect any crop health concerns or nutrient inadequacies in the field. Plant diseases affect the food system, economy, and environment. This chapter covers intelligent agriculture & challenges in front of technology. It focuses AI application using machine learning, artificial neural network (ANN), and deep learning. The various AI applications in agriculture for land monitoring, crop and varietal selection, smart irrigation or automation of irrigation, monitoring of crop health, crop disease detection, predictive analytics, weed control, precision agriculture, harvesting, yield estimation and phenotyping, supply chain management, and food quality.

DOI: 10.4018/978-1-6684-6413-7.ch014

Copyright © 2023, IGI Global. Copying or distributing in print or electronic forms without written permission of IGI Global is prohibited.

Figure 1. Elementary Agriculture

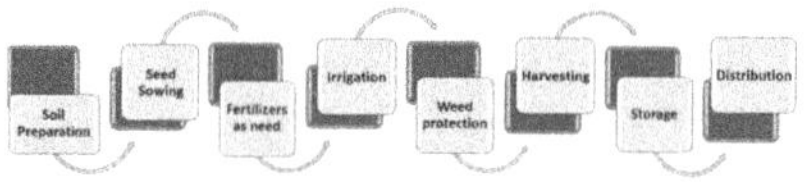

INTRODUCTION

Agriculture

Agriculture is one of the oldest economic sectors for thousands of years ago (Mazoyer & Roudart, 2006). The farming began manually with the help of a few animals & with a few manual devices. Agriculture is an important sector, of the overall economy anywhere (Alston & Pardey, 2014). The various factor dependable on each other for overall growth. Agriculture has a wide range of subsections such as farming, food industries, animal centers, food trails & errors, services sector, agricultural research, and many more. Agriculture can cover different types of farming as Subsistence, Intensive Subsistence, Mediterranean, Arable, Shifting Cultivation, etc (Komarek, De Pinto, & Smith, 2020). Today, there is an increment in the human population that needs more agricultural production with concern to a varied range of food products. There is a large volume of farmers associated with middle or low income of agriculture which is accounting for national income. To maintain the routine of the Environment cycle agriculture is helpful in many nodes as reduces erosion & soil moisture, Soil carbon sequestration (Palm et al., 2014). Another side some pesticides used in farming create environmental pollution to a certain extent (Tudi et al., 2021). There can be various impacts on CO2 by renewable energy consumption (Waheed et al., 2018). Agriculture affects directly and indirectly many other industries for prospects.

Agriculture methods have been changing dynamically. The conventional way of agriculture production involved the preparation of soil, sowing of seeds, Adding Fertilizers, Irrigation, weed protection, Harvesting, Storage, and distribution as described in Figure 1. Each phase is modernized as time passes. The new technology assists profoundly in enhancing the performance of Agribusiness.

Intelligent Agriculture

Agriculture with AI-aided machinery is becoming popular as intelligent Agriculture using IOT support. Intelligent agriculture utilizes various types of robots as intelligent machines for different task execution. Automation surely benefited in almost all fields but in agriculture, it is a big revolution as it is the prime source of income for a large volume of people. The smart devices enabled AI support pushes agriculture towards intelligent agriculture with human intervention. The traditional agricultural tasks such as harvesting, weed identification, and pest examination are also done with autonomous robots which are AI aided. The usage of intelligent machines in agriculture will enhance the speed of work compared to human laborers (Ahmad & Nabi, 2021). There is a higher number of technical integrations with agriculture by various smart devices & mechanisms for sensing, identification, transmission, monitoring, analysis, and prediction. Such IoT devices will save time & cost & improves crop yields (Chen & Yang, 2019). The remote sensors, GIS, GPS, drones, and many others are actively playing an important role in Intelligent Agriculture.

AGRICULTURE CHALLENGES IN FRONT OF TECHNOLOGY

Many factors affect traditional farming as rainfall, temperature, and humidity. An imbalance in climatic changes may cause deforestation and pollution which create a hurdle for farmers to take decisions to prepare the soil, sow seeds, and harvest (Harris et al., 2001). The nutrient deficit can result in poor crop quality as every crop requires a certain type of soil nourishment (Nair, 2019). Weed protection is also crucial to increasing the production of nutrients (Ibáñez & Blázquez, 2019). Agriculture is not an isolated sector; it is an integral part of a sound ecosystem. There are many challenges in agriculture as horticulture, animal centers, dairy, and other allied sectors (McCalla & Economics, 2001). Technology is widely integrated into multiple stages of the agriculture process. The internal & external resources can be utilized with a better prospect using current trends of technologies. The broadly technological challenges can be classified as

- Techno savvy Device usage
- Food nutrition identification
- Prediction of soil, weather, weed
- Prediction of post-harvest losses & wastage
- Production & Import analysis of specific crop
- Planning Market & distribution linkages

ARTIFICIAL INTELLIGENCE

Artificial Intelligence is part of science and technology which prepare intelligent machines, program, and computer system to perform several tasks similar to human. AI can take a decision as human takes (Bhbosale, Pujari, & Multani, 2020). AI is the approach to intelligent behaviour toward building intelligent machines which can be able to understand, behave & think like human thinks (Ahmad et al., 2021). These are high computing machines for executing various simulations for the integration of different modeling approaches. The machinery is working with the intelligence in AI-enabled environment (Daugherty & Wilson, 2018). It enabled machines to learn by themselves and perform decision-making as per need. It can correct the action if any error is found. AI is used for complex problem solving and step-by-step reasoning. AI has many subsets such as Machine learning, Deep learning, Neuron network, natural language processing, robotics, expert system, speech recognition, etc. Each subset of AI is providing solutions in its expert domain.

Machine Learning

The machine which takes action and reactions like human intelligence, the machine which learns itself to take a decision is popular as Machine Learning (ML). ML is a subset of AI. Machine learning is popular to solve complex problems as humans solved (Carbonell, Michalski, & Mitchell, 1983). Machine learning is working with a large dataset and the shortest time. Machine Learning algorithms are broadly categorized as supervised learning (SL), unsupervised learning (UL), and semi-supervised learning (SSL) (Bonaccorso, 2017). Supervised learning is functioning with the labeled data set. The label data set is used to have desired outcomes & prediction of results (Schwenker & Trentin, 2014). Semi-supervised

Figure 2. Basic ANN Approach

learning (SSL) uses a labeled and unlabelled dataset. The labeled data is to learn pattern formation & unlabelled to refine the decision limit (Hady & Schwenker, 2013; Sawant & Prabukumar, 2020). There is a certain difficult domain where a combination of labeled and unlabelled needs to construct a reliable classifier. In some situations where labeled data is very time taking and difficult to manage. In such a scenario, SSL is one of the best practices (van Engelen & Hoos, 2020). Unsupervised learning uses no label on the training data set. It works on decision boundaries by recognizing the elements towards result-oriented approach (Barlow, 1989; Li, Shepperd, & Guo, 2020).

Artificial Neural Network (ANN)

Artificial neural networks are computer programs designed based on biological simulation. ANNs are processed as the human brain processes information. The data and patterns are two important elements from which ANNs build their knowledge. ANN learns through experience (Agatonovic-Kustrin & Beresford, 2000). An ANN is made of artificial neurons connected with other coefficients in various layers. The connected neural in-network creates sturdy neural computations (Zou, Han, & So, 2008). Each neuron has input weight associated with output using the learning rule. Biological neurons send & receive signals from connected neurons. One sends a signal to the other, another passes the signal to the next one. Such a cycle continues until the completion of tasks (Kukreja et al., 2016). ANN is parameterized with adjustable weights. The activation of neurons is based on the weighted sum of inputs that passes the signal to function resulting in a single output. It is done with non-linearity using interconnections for optimized error predictions & accuracy. The predicted output can be achieved from trained and tested transfer function with learning rule & connection formula.

Figure 2 presents the Input 1 to N to the neuron. W0 to Wn are the weights. Signals are forwarded through connected nodes. The signal strength depends on the product of weight and input. The single output results from various inputs from different sources (Livingstone, 2008; Mishra & Srivastava, 2014).

Deep Learning

Deep learning (DL) is working on multiple processing layers to learn from data. Deep learning learns from complicated large data sets using the backpropagation algorithm (LeCun, Bengio, & Hinton, 2015). It can manage internal parameters to compute from various layers as shown in Figure 3. DL also learns itself through multiple levels of representations with the distribution of the data by various models for solving complex problems (Lauzon, 2012). Deep learning covers the spectrum of processing images, video, speech, and audio, whereas recurrent nets have shone a light on sequential data such as text and speech (Nikolenko, Kadurin, & Arkhangelskaya, 2018). DL can learn supervised & unsupervised sets of elements for better accuracy and outcomes. The idea behind deep learning is similar to the human brain identifying, recognizing and filtering data by image, sound, text, numeric or alphanumeric. DL

Figure 3. Deep learning Approach

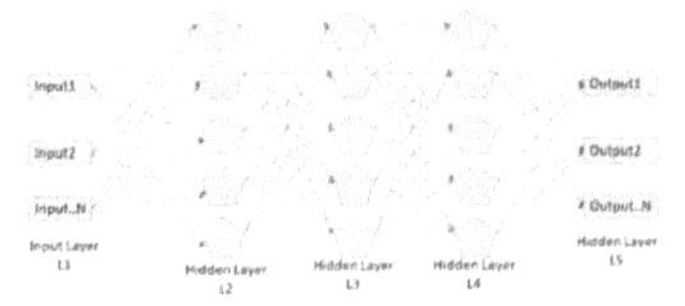

algorithms correlate predefined learned data for accurate results. As there is an increment in the volume of data, DL performance can be closer to real machine intelligence. DL also claims for more reliable and concise results (dos Santos Ferreira et al., 2019). DL is very popular in image intelligent scanning, voice scanning, computer vision, natural language processing, predictive models, etc (Shinde & Shah, 2018).

ARTIFICIAL INTELLIGENCE IN AGRICULTURE

Artificial intelligence become popular in every sector for analysis, decision making, prediction and enhancing the better performance of any conventional system. AI works like a man-made action-reaction. The technology is being used in Agriculture for AI sensor base farming, food processing, food control, food quality testing and many others. Technology integrated with agriculture will assure better prospects. AI systems can help in many ways for improving yield production along with quality measures as shown Figure 4. AI and its subset can help to analyse diseases, pests and poor nutrition, and detect or predict relevant crop-related matters. AI sensors can detect weather humidity, rain, and weeds and suggest action accordingly. Precision farming using emerging technologies supports many advantages. AI-enabled agriculture has proven to increase production, productivity, and profitability (O'Sullivan et al., 2019).

Machine Learning in agriculture

Jagtap et al. have discussed the classification of objects and stated that k-Nearest Neighbor classification is a simple and effective technique. They elaborated on classification's effectiveness and how to apply various machine learning techniques in agriculture (Jagtap et al., 2022). Pallathadka et al. have discussed about optimization of water irrigation & role of AI in better crop production (Pallathadka et al., 2021). Aiello et al. has discussed the operator's activities and safety with ML classifier. They have

Figure 4. Artificial Intelligence in Agriculture

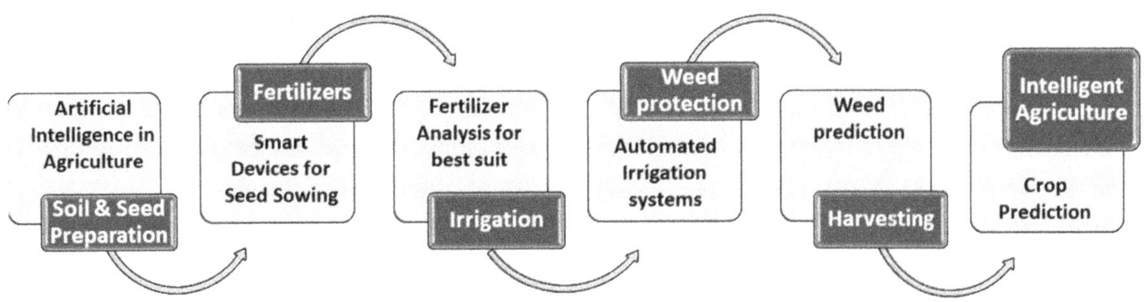

also approached the wearable machine for monitoring purposes. They have tried to validate the process in real conditions through experimental tests (Aiello et al., 2022). Ting Li et al. have discussed about agriculture satellite images dataset usage with machine learning for food production & water usage planning (Li, Johansen, & McCabe, 2022). Pant et al. has devolved a trained model to identify the patterns among data and it is used for Maize, Potatoes, Rice and wheat prediction. (Pant et al., 2021)

ANN in Agriculture

Saldana-Robles et al. have worked on agricultural backhoe structural design using Theoretical Analysis, Finite Element Analysis, Response Surface Methodology and Artificial Neural Network which carry multiple assignments in agriculture and livestock (Saldaña-Robles et al., 2020). Monteiro et al. has discussed ANN-NLP is effective in defining weed control (Monteiro et al., 2021). Have proposed to find the water table imprints by judicious application of artificial neural networks (Chattopadhyay & Rangarajan, 2014). Li and Chao (2020) have discussed that in precision farming plant disease detection & pests' control are very important for further decisions. The prosed ANN-based continual classification method for better identification of diseases and pests.

Deep Learning in Agriculture

Fanyu Bu & Xi Wang have designed a smart agriculture system using edge-cloud computing, they have presented many representative deep reinforcement learning models with challenges & applications in smart agriculture (Bu & Wang, 2019). Karar et al. have studies on pests causes & crop production. They stated that deep learning methods outperform in concern with their study (Karar et al., 2021). Cheng et al. (2017) have used deep learning with insect pest images in complex farmland backgrounds. Kasinathan, Singaraju, & Uyyala (2021) have studied deep learning with a better solution for achieving accurate identification of tomato pests. Yadav et al. have processed the images with a deep learning model for the identification of bacteriosis from peach leaves (Yadav et al., 2021).

AI APPLICATIONS IN AGRICULTURE

AI Applications in Agriculture for Land Monitoring

The land is a reservoir of nutrients for plants, and microbes, and holds the water required for plant growth. The physical qualities of soil impact how much water it can store and make available to plants. This quantity controls how long a crop can live without even being watered. It specifies the periodicity of irrigation as well as the size of the irrigation system required to maintain continuous crop development (Soil –Plant – Water – Relationships, 2018)

In terms of plant development, soil water is the most important element of the land. Crops may rapidly absorb land water if appropriate water is available in the soil for plant development. The importance of soil water is shown in Figure 5 (Soil Moisture: Importance of Soil Water, 2017)

In agriculture, the soil is a critical element. Crop yield and quality degradation result from poor irrigation and soil management. Bralts et al. developed a rule-based intelligent framework for monitoring the performance of micro irrigation systems (Bralts et al., 1993). Sicat et al. designed a fuzzy control

Figure 5. Importance of soil water (Data is retrieved from Soil Moisture: Importance of Soil Water, 2017)

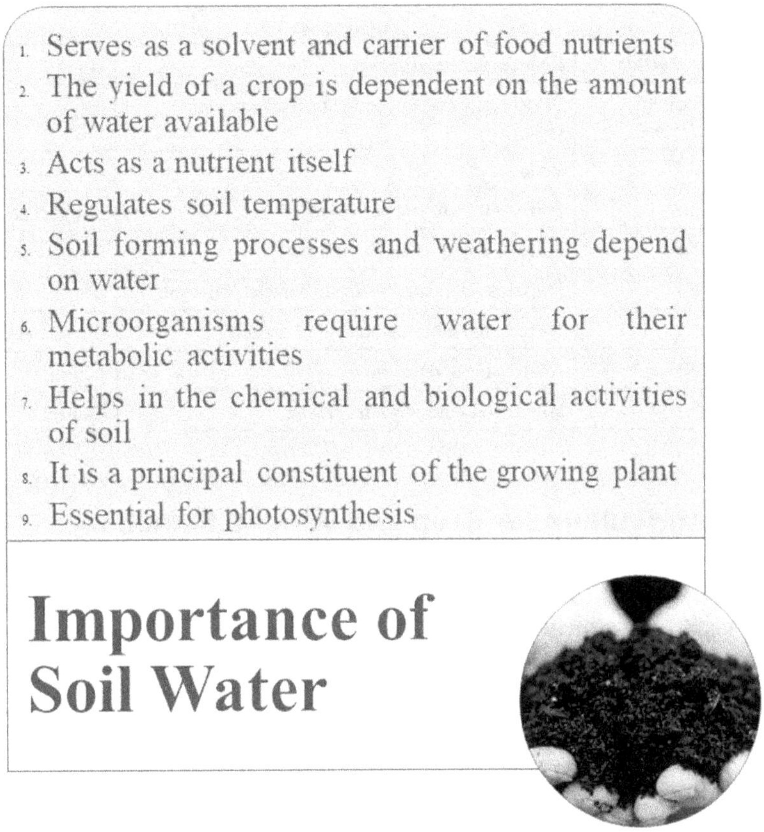

system to select plant varieties based on land compatibility (Sicat et al., 2005). Valdes-Vela et al. (2015) estimated a plant's stem water potential using a Takagi Sugeno Kang fuzzy inference system based on the water content of soil and the meteorology. Arif et al. (2012b) created an artificial neural network-based approach for estimating soil moisture. "Soilsens," a prospective AI-based system, evaluate soil health (Busch, 2012). The farmer gets help from this tool to select the appropriate amount and type of inputs required. This innovation was developed by Proximal Soilsens Technologies Pvt. A soil moisture sensor, temperature sensor, a humidity sensor, and an air temperature sensor are present in this device. Based on observed data, the smartphone application recommends the farmers the quantities of fertilizer and the method of irrigation that should be used. The Soilsens technology assists producers in avoiding over-irrigation, protecting the crops against insects' infestation, preventing nutrients depletion from the soil, and providing voluntary recommendations and advice (Yeasin et al., 2021). Soil retains moisture, nitrogen, phosphate, and potassium which are essential for efficient crop production. Condition of soil may be addressed via manure and fertilizer, which promote soil properties, as well as an alternate plowing strategy, which prevents soil deterioration. Negative variables like soil contaminants could be reduced by soil management (Abawi et al., 2000). The other application of AI is to create soil maps, which aid in the display of soil-landscape interactions as well as the various layers of soil (Syers, 1997). Table 1 represents the different model and their application in soil analysis.

Table 1. Artificial intelligence in Soil Management

Model	Application	References
Management-Oriented Modeling (MOM)	Reduces nitrate leaching	(Li et al., 2000)
Fuzzy logic Soil Risk Characterization Decision Support System (*SRC-DSS*)	Classification of soil on the basis of associated risks.	(López et al., 2008)
Decision Support System (DSS)	Reduces soil erosion.	(Montas et al., 1992)
Artificial Neural Networks (ANN)	Forecast the enzymatic activity of soil.	(Tajik et al., 2012)
Artificial Neural Networks (ANN)	Forecast the temperature of the soil	(Bilgili M.., 2011)
Artificial Neural Networks (ANN)	Forecasts the texture moisture content of the soil	(Zhao et al., 2009)
Artificial Neural Networks (ANN)	Forecast the moisture content of the soil.	(Elshorbagy et al., 2008)
Artificial Neural Networks (ANN)	Analyze the soil texture.	(Chang D. et al., 2000)
Artificial Neural Networks (ANN)	Predict soil nutrients after erosion	(Kim et al., 2008)

AI Applications in Agriculture for Crop and Varietal Selection

In agriculture, suggestions must be precise since inaccuracies can result in significant material and capital loss. Many different types of studies are being conducted in order to develop an efficient and accurate crop prediction model. Satish Babu proposed a model for Kerala State, where the holding size is significantly lesser than the rest of India. The overall aim is to provide direct advising services to even the smallest producer at the level of her or his smallest crop plot, utilizing accessible tools such as text and email. As a result, with minor alterations, this model may be positioned elsewhere in India (Babu, 2013). Kumar et al. discussed the importance of produces selection and the factors that can influence produces selection, such as production rate, government policy, and market price. Their work provides a Crop Selection Method (CSM), which solves the produce selection problem and increases the crop's net harvest rate. This method recommends an order of produces to be planted throughout the season, factors are taken into consideration such as weather, soil and crop type, and water density (Kumar et al., 2015). The study of Khedr et al. (2015) addresses the issue of food insecurity in Egypt. It gives a framework for forecasting output and imports for that particular year. For the forecasting, WEKA utilizes Multi-layer Perceptrons. This tool forecasts the yearly required quantities of the CROPS.

AI Applications in Agriculture for Smart Irrigation or Automation of Irrigation

A smart irrigation system is required to address the problem of water scarcity. Through this method, more land can be irrigated with the help of a small quantity of water. Jha et al. (2019) provide a detailed overview of several smart irrigation approaches. There are several systems in which a low quantity of water can irrigate more area but in those systems for the operation human being are required. There is an opportunity to design smart irrigation systems by implementing AI in the current approaches. This smart system constantly analyses the amount of water required by crops and compares it with the amount of water available on the land. This system automatically activates sprinklers or drips based on the crop's water requirements. Arif et al. (2012a) introduced ANN-based soil moisture prediction models.

Hinnell et al. (2010) introduced a Neural Network-based irrigation system. Nawandar and Satpute (2019) have introduced an IoT-based system. Information gathering is required for smart irrigation.

Major data such as soil water content, water required for plants, temperature, humidity level, etc. Soil moisture sensors, humidity sensors, temperature sensors, and other devices can be used to gather data. These devices are linked to low-cost Arduino-based systems for storing collected data and operating analysis techniques to predict water requirements for irrigation. Efficient water utilization is a difficult task, especially in areas where water availability is the main consideration. To reach the highest yields, the ground must be irrigated when necessary. The term "smart" refers to the devices that can detect water demand for crops. This capability is attained by combining several technologies such as knowledge, and sensors. Even these models are not expensive, many growers continue to use traditional irrigation methods due to a lack of awareness, resulting in inadequate water utilization for irrigation and a low yield of crops. Smart irrigation is accomplished through the use of smart integrated devices. Intelligent irrigation frameworks save a tremendous quantity of water (Sinwar et al., 2020). Smart irrigation systems make sure that water reaches all the plants in the required quantity which provides good to crops and keeps away diseases. Improper methods of irrigates cause disease in plants, which could occur because of the adaptation of traditional irrigation systems. To avoid any kind of failures, Repeated checking of sprinklers/ drip modules is required at a particular interval of time. These devices may require recalibration from time to time owing to performance deterioration or damage caused through environment and animals. In this case, parts of the field are heavily watered while others are left unirrigated. The nonuniformity of land and crop kinds is the fundamental problem in building a smart irrigation system (Sinwar et al., 2020). Suresh and Umasankar designed an intelligent irrigation system that runs on solar electricity. By employing solar power, the system can reduce electricity bills. It is made up of an automatic water flow system that detects the moisture level in the soil. The goal of this system's creation is to conserve both power and water in the Indian context. The system is powered by a solar panel, which is utilized to activate the stepper motor, which feeds water from the well to the storage tank based on moisture detected (Harishankar, S. et al., 2014). As per Sinwar et al the water release valve on the storage tank is automatically get started by a moisture sensor. The sensors are installed within the farm where the crop is being grown. Devices may transform moisture level to voltage, and compared with a reference voltage. The variance level in the voltage, is utilized to calculate the requirement of water. Their trial findings demonstrated that the system can irrigate the area while using the least amount of water and energy (Sinwar et al., 2020). Kehui et al. (2010) created an approach that uses sensors to monitor the level of water in a farm. The device was put through its paces in real-time and found to be suitable for rice field irrigation. Vellidis et al. (2008) created a prototype for scheduling cotton field irrigation based on temperature and soil moisture. This approach is based on the concept of smart nodes, which are made up of a sensor array, an RFID tag for transmitting data to the central computer, and electrical circuitry. Nasiakou et al. (2016) provided a technology solution for decreasing irrigation costs by integrating the smartness of renewable energy sources with irrigation. Navarro-Hellin et al. introduced a smart irrigation decision support system (SIDSS) that uses sensors and machine learning (ML) algorithms to estimate weekly watering requirements (Navarro-Hellín et al., 2016). A solution based on IoT is developed by Goap et al. (2018) The system is fully functioning and utilizing cloud integration through web-based services. The technology works based on sensed data. The suggested system is smart enough to provide real-time irrigation decisions. Gu et al. (2017) proposed Root Zone Water Quality Model (RZWQM), a software-based method for scheduling irrigation requirements. Because it appears to be more difficult to assess this stress in real-time, the system can identify irrigation requirements based on crop estimated based on agricultural models. Irrigation timings are advised based on model water stress (WS) estimates, whilst irrigation depth is computed based on soil moisture threshold. The model incorporates weather inputs from on-site sensors or from the database of weather forecast websites (historical and four days ahead of data). Nawandar and Satpute (2019) presented a low-cost automated irrigation system internet of Things (IoT) and neural networks. Internet of Things is utilized to link various equipment, while Neural Networks are employed to make irrigation decisions. Through the message queue telemetry transport (MQTT) protocol, the system transmits crop-related information such as remote data monitoring, zone-wise irrigation alarms, soil statistics, and estimated irrigation requirements.

Many researchers have developed several smart irrigation methods to overcome the limitations of traditional irrigation systems that require continual field analysis. The majority of smart irrigation approaches are capable of automating agricultural field irrigation based on sensor data. Sensor nodes are the most important component of these approaches (Sinwar et al., 2020). Other critical components include the controller unit, wireless media, GSM module, sprinkler/drip modules, and so on. The distinction between these systems originates from two factors: sensor node variety/capability and arrangement with additional components Temperature sensors measure the amount of heat necessary for a certain crop, whereas moisture sensors measure the amount of moisture in the soil. The values from these sensors must be obtained on a continuous or periodic basis by the system designer. Some systems were identified to be energy saving in the manner that wireless nodes were calibrated to sense and broadcast information at user-specified interval thresholds rather than continuously. A controlled device continuously checks these sensed values and permits water pump operations (On and Off). drip/ Sprinklers modules should be installed in a manner that is particular to the land structure and crop/plant. There is no generic irrigation system that can fulfil the irrigation needs of all types of crops/fields since the structure and irrigation demands of various crops are not the same (Sinwar et al., 2020). Given this field/crop versatility, intelligent irrigation systems would be able to make judgments based on real-time sensed information as well as assessing past knowledge. Following the examination of certain systems of irrigation in the preceding section, an architecture of a smart irrigation system known as cloud and solar power enabled hybrid smart irrigation system (CS-HYSIS) has been presented (Sinwar et al., 2020). The system includes smart sensor nodes that are powered by solar panels. The type of solar panels used may vary depending on the weather (if the system is to be put in a cloudy/rainy location, advanced solar panel controllers must be fitted). The sensor nodes are made up of three sensor units, which are moisture and temperature sensor in the soil and humidity sensor (Sinwar et al., 2020)

Applications in Agriculture for Monitoring of Crop Health and Disease Detection

Traditional crop health monitoring methods are inefficient and time-consuming, particularly for big areas. Many scientists have established various AI-based architectural systems to address the difficulties found in traditional agriculture. Some architectural designs are discussed by Kamilaris and Prenafeta-Bold (2018). AI-based approaches that include image analysis, deep learning, and data analysis make prediction simple and effective about the diseases and health conditions of the crop. Crop photos are captured by high-definition (HD) camera-equipped drones, or satellite photography. The collected pictures are used to train Convolution Neural Networks (CNNs). The CNNs system identifies relevant information from supplied photos and predicts crop disease(s). The device is effective at continually monitoring plant health and offers better methods for deciding the amount of pesticide to spray. Foliar pictures are a valuable and comprehensive source of data on plant disease and morphology; thus, this data must be retrieved and studied extensively. Picture processing is critical in the identification and study of leaf diseases (Devaraj et al., 2019). The capture of photos is the first stage in recognizing illnesses. Images may usually be retrieved from a digital camera or an imaging device. Because raw pictures include noise, these contaminants must be removed. As a result, the second stage is known as picture pre-processing, and it includes removing undesired distortions as well as enhancing contrast to clarify and brighten the image characteristics (Mojjada et al., 2020). The third phase is image segmentation, in which the picture is separated from its backdrop and the region of interest (ROI) is divided to focus on the main characteristics (Iqbal & Talukder, 2020). The fourth phase is feature extraction, which reveals an image's information and details. Leaf characteristics such as form, texture, and color are used to identify crops. As a result, the selected characteristics create an input feature vector that is subsequently given to the classifier (Srivastava & Venkatesan, 2020). This vector can be used to distinguish one type of item from another. The final stage is categorization. It should be noted that the selection of an appropriate classifier is dependent on the particular problem. The lassifier's goal is to recognize photos by categorizing them into many predetermined groups based on the feature vector acquired in the fourth phase. The categorization task is divided into two parts for this purpose: training and testing. The classifier is trained on a training dataset during the training process; consequently, the more training sets used, the better the accuracy attained (Whitmire et al., 2021).

Deep learning (Ferentinos, 2018; Fuentes et al., 2017) and transfer learning (Chen et al., 2020) systems in agriculture have acquired broad popularity and yielded extremely promising results. The application of transfer learning is a technique that becomes extremely popular and is frequently employed by researchers (Ramcharan et al., 2017; Too et al., 2019). Mohanty et al. demonstrated that the fine-tuning technique outperforms a CNN model. Neural Network (NN) is widely used and recommended for analyzing hyperspectral data for disease identification. Its core function was motivated by the nervous system of human, and it has particular features like learning and generalization that help in crop disease identification (Mohanty et al., 2016). Similarly, Zhu et al. used machine variable selection approach and the learning classifiers to investigate the potential of hyperspectral imaging as a non-invasive tool for the early identification of TMV illness (Zhu et al., 2017). Cui et al. presented a review on several approaches, highlighting their benefits and limitations, stating that the smart nose is a rapid way for crop disease identification (Cui et al., 2018). Ma et al. developed a system based on a deep convolutional neural network (DCNN) to find out the diseases in four different types of cucumber and the accuracy of this system is up to 93.41% (Ma et al., 2018). Tran et al. proposed a system to monitor the development of tomatoes and also yield improvement. This system was able to categorize nutritional deficits and illnesses. As a consequence, agricultural specialists use the information to analyze the symptoms and protect tomato crops (Tian et al., 2019). The hyperspectral photography approach has been widely employed to assess abiotic and biotic stressors in agricultural plants over the last twenty years (Bioucas-Dias et al., 2013; Mananze et al., 2018). Hyperspectral imaging is a technology that combines spectroscopy with photography to get spatial and spectral information about an item at the same time. Changes in plant biophysical and biochemical parameters, such as pigment content, water content, tissue structure, and transpiration rate, are caused by disease infection. Plant spectral characteristics, intercellular space, and water content can all be affected by these changes (Rumpf et al., 2010).

AI Applications in Agriculture for Predictive Analytics

Machine learning is a key decision-making tool for agricultural produce yield prediction, including advising on which crops to cultivate and what to do during the crop's growing season. Temperature, rainfall, and soil type are the most commonly utilized characteristics in these models, while the most commonly used technique is Artificial Neural Networks. Convolutional Neural Networks (CNN) is the most extensively utilized deep learning algorithms in these investigations, with Deep Neural Networks and Long-Short Term Memory (LSTM) also being widely employed (DNN) (van Klompenburg et al., 2020). Charoen-Ung and Mittrapiyanuruk estimated a farmer plot's sugarcane output with the help of the RF approach (Charoen-Ung & Mittrapiyanuruk, 2019). Precise crop yield forecast in developing nations before harvest is crucial for the reduction of famine, enhancement of food security, and ensuring agriculture's long-term growth. Existing methods are costly and difficult to expand since they rely on locally acquired survey data. Approaches based on remotely sensed data, such as satellite images, have the potential to be a low-cost, high-impact alternative (Wang et al., 2018). Predicting agricultural yields from natural photographs is a difficult issue that machine learning (ML) can assist. Convolutional Neural Networks have improved various ML applications, such as natural language processing, speech recognition, and computer vision. The study utilized convolution neural networks to develop models capable of estimating a grapes' weight from a photograph. The system achieves a cross-validation yield estimation accuracy of 87 percent after being trained and evaluated with a dataset of 60 photos of grapevines (Monga, 2018). Bangladesh's economy is heavily reliant on agriculture, therefore predicting crop yields is critical. According to Osman et al. a model that analyses weather and climate before crop planting is an important tool. It enables to delivery of information regarding crop state, quantity, and quality, as well as early alerts so that timely interventions may be carried out (Osman et al., 2017). Wheat yield prediction is a significant issue in Turkey. Cakir et al. developed an artificial neural network model that can accurately estimate wheat production using environmental information such as rainfall records and temperature (Cakir et al., 2014).

AI Applications in Agriculture for Precision Weed Control

"A History of Weed Science in the United States," mentioned the research of Thomas K. Pavlychenko, who researched plants' competitiveness (Research & 2010, 2014). After doing extensive studies on this subject, found that the struggle for water and nutrients starts when the roots of plants overlap with each other within the soil and for water competition weeds are the fiercest competitor. Water requirements for different types of crops may vary to reach maturity. Sunlight is also a major aspect of the development of plants. Weeds that become tall typically obstruct the flow of sunlight to the crops. Redroot pigweed and green foxtail are shade-intolerant weeds, whereas Arkansas rose, spotted spurge, common milkweed, and field bindweed, are shade-tolerant weeds.

Weeds removal is critical; it may not only take farmland but also hurt the development of many other crops (Bak et al., 2004). Lie Tang et al. proposed a vision-based weed identification method. It was made by applying hereditary computation to identify a region in Hue-Saturation-Intensity (HSI) shading space (GAHSI) (Tang et al., 2000). According to several researchers, before the development of an automated weed management program, we must first distinguish between crop seedlings and weeds (C. Chang et al., 2018; Talaviya et al., 2020). Aitkenhead et al. distinguished ryegrass and carrot seedlings by simple morphological feature evaluation of leaf shape. This approach has a varied efficacy of 52 to 75 percent. Another weeding method was developed using digital pictures. This concept made use of a self-organizing neural network. However, this approach did not produce the intended results for commercial use (Aitkenhead et al., 2003). Weeding, according to Nrremark and Griepentrog (2004), is dependent on the position and number of weeds. Agricultural tines have been used to promote weed

Table 2. AI models in weed detection.

No.	Tool for Weed Detection	Reference
1.	MVA	(Partel et al., 2019)
2.	MVA	(Bakker et al., 2006)
3	CNN	(Ngo et al., 2019)
4.	MVA	(Åstrand & Baerveldt, 2002)
5.	MVA	(Blasco et al., 2002)
6	MVA	(Lamm et al., 2002)

witling. However, it was not a recommended practice since tillage might destroy the contact of both the soil and the crop (Nørremark & Griepentrog, 2004). Heisel et al. introduced micro spraying and laser treatments and these methods are no contact methods (Heisel et al., 2001). Nakai and Yamada described the approach of using farming robots for weed control and establishing ways of regulating robot postures in irregular lands in rice production (Nakai et al., 2014). It employed the Laser Range Fielder (LRF) technology to control the weeds. Astrand and Baerveldt (2002) demonstrated a weed-controlling robot. The robot was outfitted with several sensing technologies (Åstrand & Baerveldt, 2002). Fennimore et al. utilized the sight technologies to guide the robots along with the row structure to eradicate weeds (Fennimore et al., 2016). Table2 represents the AI tools used in weed detection.

AI Applications in Agriculture for Precision Agriculture (Pa)

The former is described as "an information-led management concept in both plant and animal production that is based on a wide range of technologies". This notion is focused on the digital processing of particular information to aid decision-making processes. The latter relates to a "knowledge-based approach in which machines can at least take partially autonomous decisions in collaboration with management systems. The decisions taken by machines are based on autonomously obtained and processed information in real-time, even though farmers always can correct them". (Munz et al., 2020). PA concept is a "farming management concept based upon observing, measuring and responding to inter and intra-field variability in crops or aspects of animal rearing" (Daheim et al., 2019). PA causes a shift in land use by fostering "whole-farm management strategies using information technology, highlighting the potential improvements on production while reducing environmental impacts". Yost et al. (2017) simplified the definition of PA as a "suite of Information Technologies that focus on producing immediate benefits by being conscious of the environment". Lezoche et al. (2020) have defined the term "Agriculture 4.0" as "the provision of advanced technologies to farmers to meet their production challenges".

Liakos et al. (2018) published a thorough assessment of research on machine learning (ML) uses in agricultural systems. Crop management, water management, and soil management were identified as the three most important sectors of application. They concluded that agriculture will undeniably gain from ML technologies, which will enable Farm Management Systems to transform into real-time decision support systems (DSS) through the use of AI algorithms. Van Klompenburg, Kassahun, and Catal (2020) conducted a systematic literature review on the Machine learning algorithms and elements utilized in agricultural production prediction research. The results demonstrate that no definite conclusion can be reached about which model is the best, but they do show that several ML models, such as random forests, neural networks, linear regression, and gradient boosting trees, are utilized more than others. Noon et al. provided a thorough assessment of the deep learning (DL) approach for plant leaf stress recognition. Based on stress type, dataset size, training/test size, and the deep network utilized, the methodologies studied were classified into vegetables, fruits, and other crops. Several studies need to be done, including the recognition and categorization of all stress severities to detect infectious illnesses, as well as the applicability of algorithms to crop varieties (Noon et al., 2020). Finally, AI and robots are playing an important role in assisting or replacing manual involvement in the agriculture industry, which includes a wide range of applications. R robotics have played a significant role in crop production, with robots a variety of independent tasks such as weeding, irrigating, and monitoring crops can be carried out. Zhang et al. present a complete overview of current robotic grippers, grasping, and sensor-based control approaches, as well as their applications in robotic agricultural works. When relative to industrialized grasping, agricultural grasping has greater requirements although most agriculture and food goods are fragile, bruised, and sticky (Zhang et al., 2020). On the other hand, Drones are being used in farming to assess crop health, irrigation equipment, and weed detection. Roshanianfard et al. addressed several elements of autonomous agricultural vehicles produced at Hokkaido University. Each unmanned aerial vehicle (UAV development) technique and attributes were compared and reviewed. As realistic instances and representative development issues, case studies of autonomous cars created and the development process has been provided (Roshanianfard et al., 2020).

In recent years, the EU has been actively engaged in R&D initiatives linked to the application of AI for agricultural automation. Concerning the former, they intend to emphasize the growing interest in certain research efforts as well as the recognition of the underlying technologies (Bacco et al., 2019).

AI Applications In Agriculture For Harvesting

After the crops ripen, harvesting is the last horticultural practice conducted on the field. Crop output projections give farmers vital information for planning the best practices during harvesting and post-harvesting operations (Seok Ahn et al., 2018). Machine learning algorithms are focused on the prediction of agricultural yield with the help of remote sensing technology (Haghverdi et al., 2018). Kouadio et al. evaluated soil characteristics by ensemble learning approach and used it for forecasting coffee yield during the harvesting (Kouadio et al., 2018).

Deep neural networks (Mehra et al., 2018) and data mining approaches (Mucherino et al., 2009) are utilized for the prediction of yield during harvesting. Support Vector Machine and machine vision are employed for plant segmentation by detecting foliage (Rico-Fernández et al., 2019) Machine learning algorithms are utilized to forecast crop color changes throughout the different harvesting stages (Sadgrove et al., 2018).

AI Applications in Agriculture for Yield Estimation & Phenotyping

For many decades, scientists have been interested in yield prediction. Yield prediction necessitates the use of yield mapping instruments, which are currently not widely available to farmers (Nevavuori et al., 2019). CNN can handle the challenge of yield prediction cost-effectively and simply. Many researchers have created models to forecast crop yields using AI-based approaches that employ RGB/NDVI pictures (Nevavuori et al., 2019; Pantazi et al., 2016). Their experimental findings on data sets demonstrate the utility of their models. A comparison of CNN-based models with AI-based models clearly shows that CNNs outperform standard Machine Learning (ML) and AI-based methodologies (You et al., 2017).

Manual record-keeping and monitoring of agricultural activities and yields is a time-consuming and labor-intensive operation. Farmers may not be accessible all of the time to keep track of all agricultural operations. As a result, inconsistent and incomplete data is collected. Artificial Intelligence-based solutions are useful in dealing with such circumstances. It is essential to inspire and teach farmers about the use of new technologies and procedures. This is very desirable to decrease manual labor, enhance crop quality, and fulfill the increasing food demand for a growing population. Precision agriculture offers an innovative solution to this issue experienced in traditional agriculture. Farmers are becoming more aware and educated about precision agriculture, which is prompting them to consider adopting it. Farmers are eager to use new precision agricultural tactics, yet exact yield prediction remains a difficulty owing to form inconsistencies, lighting conditions, and so on. Precision agriculture benefits from continuous monitoring of all agricultural activities, from seeding through harvesting. Nevavuori et al. show how to establish a schedule for agricultural operations based on prior experience and acquired data. Farmers are creating a timeline and keeping track of the following elements to make better predictions. Soil fertility and fertility enhancement through the use of manures, compost, organic and inorganic fertilizers (Nevavuori et al., 2019). Liakos et al. explained the different machine learning models to solve real-world problems. Most commonly used ML models are Ensemble Learning (EL), Bayesian Models (BM), Support Vector Machine (SVM), Artificial Neural Networks (ANNs), Deep Learning (DL), Dimensionality Reduction (DR), and Decision Trees (DT). There are many machine learning techniques available such as; Convolution Neural Networks (CNNs, Hopfield Network (HN), Deep Boltzmann Machine (DBM), Ensemble Neural Networks (ENN), Generative Adversarial Networks (GAN), and Multiple Linear Regression (MLR) (Liakos et al., 2018). Wang et al. created a computer vision-based system that can automatically evaluate the quantity of apples in an orchard. The technology is both precise and rapid in its operation. This system includes a vehicle for moving around in the field. Both the vehicle and the cameras are precisely calibrated using coordinate frames. Pixel characteristics are used to identify Apple regions. Their approach identified specular reflections using hue, saturation, and value. The results of their experiments and analyses reveal that their model can determine the number of apples properly and quickly (Wang et al., 2013). Ramos et al. developed a technique for determining the number of coffee fruits. Their machine vision system (MVS) can count and categorize coffee fruit. This work classified fruits as harvestable or non-harvestable. They used a mobile-based system to acquire photos, which were then processed using a linear estimating model to compute the quantity of coffee fruits on a branch (Ramos et al., 2017). Amatya et al. created a MV system for autonomous harvesting of Cherry fruits since human harvesting is time-consuming and costs more than half of the production cost. The algorithm can recognize branches with an efficiency of almost 90% (Amatya et al., 2016). Sengupta and Lee established an SVM-based model for detecting immature green Citrus fruit under different illumination conditions (Sengupta & Lee, 2014). Liakos et al. conducted a thorough study of several Machine Learning approach for agricultural applications. These approaches have been categorized into many classes, such as weed monitoring, disease detection, and yield prediction (Liakos et al., 2018).

AI Applications in Agriculture for Supply Chain Management

The distribution and commercialization phases connect crop production and processing to final consumption, completing the farm-to-fork cycle (R. Accorsi et al. 2018). The processed farming product is delivered to distribution channels and warehouses during the delivery process. Before reaching the end customer, the majority of the product passes via a distribution chain (Beaman and Johnson, 2006). In the distribution of goods, transportation plays an important role. A genetic algorithm was used in many studies and these studies are generally focused on traveling distance, protecting the quality of the product, vehicle routing, and lessening product's damage (Qiang & Jiuping, 2008; X. P. Wang et al., 2018). While some studies used memetic algorithms for distribution scheduling problems and handling integrated production (Devapriya et al., 2017). Also identified the storage location assignment problem (Hui et al., 2016). For freight estimation, machine learning algorithms can be used (Krisztin, 2018), the shelf life of the product (Larsen et al., 2009), payment delays, and inventory management (Shukla & Jharkharia, 2014), predicting the risks of the supply chain (Luangkesorn et al., 2016).

AI Applications in Agriculture for Food Quality

During manual handling of foods, there are chances of contamination that's why robots are highly used due to their sterility. This is a major factor in reducing the cases of foodborne diseases. Numerous foods do not need refrigerating conditions for storage and are placed in the prone area of contamination. The Food Safety Modernization Act (FSMA) has established advanced hygienic necessities that should apply to the supply chain (Fedorova et al., 2020.; Misra et al., 2020). Technavio reported that the utilization of robots in the food industry would expand by thirty percent (I. Kumar et al., 2021). AI-based automated systems are quicker and more accurate than humans. Various types of smells can be identified via sensors and these systems are able to sense the surrounding odor and then transferred the sensed data to the data center where machine learning systems access these data (Fedorova et al., 2020). According to the report of the Department of Agriculture, USA, "In the United States, food waste is estimated at between 30 and 40% of the food supply. This estimate, based on estimates from the USDA's Economic Research Service of 31 percent food loss at the retail and consumer levels, corresponded to approximately 133 billion pounds and $161 billion worth of food in 2010. This amount of waste has far-reaching impacts on the society" (I. Kumar et al., 2021). According to McKinsey, artificial intelligence may resolve the issue of food waste by 2030 (Filimonau et al., 2020). Various AI-based systems have to be used in smart farming for the gathering and examination of the data. After the proper analysis of data appropriate decisions are taken. For the execution of such plans, all the organizations and food industries have to work together. Figure 6 represents the methods that should be utilized for the reduction of food wastage via AI (I. Kumar et al., 2021).

One of the time-consuming and labor-intensive operations for the food industries is the packing and ordering of foodstuffs. Therefore, time taking activity may be completed via AI approaches, resulting in reducing the possibility of mistakes and substantially increasing the industry's output rate. Because of the variations in the dimensions, and colors of vegetables and fruits, developing an AI system is a tough task. A tremendous quantity of data is required when an AI-based approach is developed for sorting and packing (Dewi et al., 2020; Tripathi et al., 2020). AI's precise decision-making process utilizes various techniques such as X-ray-based systems, laser-technology-based systems, high-resolution cameras, and IR spectroscopy. Such techniques are generally utilized to assess every feature of foodstuffs. Traditional techniques could only differentiate between excellent and poor products based on their appearances. In potatoes detaching problem is reduced by 5–10% by using the AI system (Bünger, L., 2021; MO. Onishchuk, 2020). Food companies' distribution networks utilize to examine each process (Kakani et al.,

Figure 6 Methods to reduce food wastage. (Data is retrieved from Kumar et al., 2021).

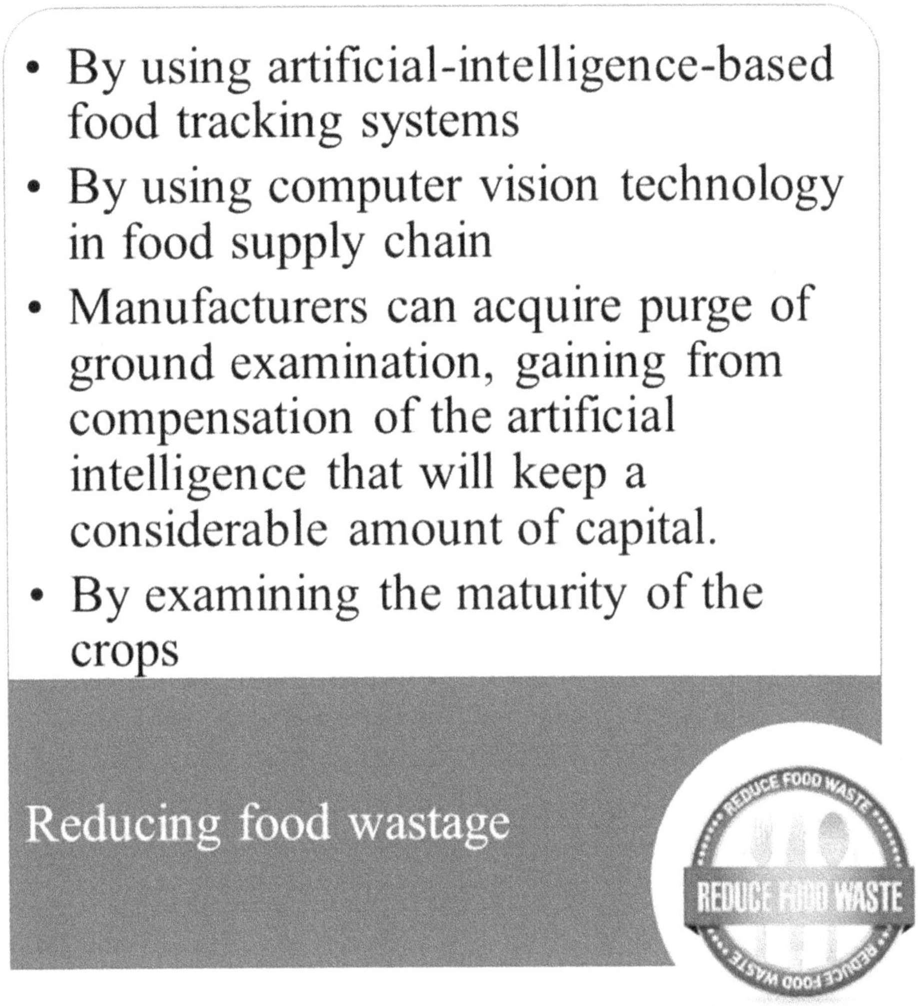

2020). It handles almost everything from pricing to inventory systems. In retailing, AI helps in various operations such as booking, billing, and inventory control (Bhattacharyya et al., 2021).

CHALLENGES IN ADOPTING AI IN AGRICULTURE

The next agricultural revolution will be driven by artificial intelligence. However, many countries, have still yet to implement artificial intelligence. In general, adopting and disseminating new technical breakthroughs in agriculture is quite challenging. There are several barriers to adopting new techniques. AI-based advancements in farming such as; AI-based tools, sensors, etc. but these tools are too expensive for small farmers. Repair and maintenance costs were likewise exorbitant. As a result, one of the greatest barriers to AI adoption in agriculture is a lack of funds and awareness. Internet connectivity is one of the most significant impediments to adopting new technologies in developing or underdeveloped countries. The insecurity of India's IT industry poses a significant hurdle to AI in agriculture.

CONCLUSION

This chapter has covered Agriculture & Intelligent Agriculture in depth. The challenges in front of technology for agriculture has been elaborated carefully. The artificial intelligence is discussed with working of machine learning, artificial neural network (ANN) & deep learning. The role & action of Machine Learning in Agriculture, ANN in Agriculture & Deep Learning in Agriculture is given with suitable examples. The various Artificial Intelligence Applications in agriculture is discussed as AI Applications in Agriculture for land monitoring, crop and varietal selection, smart irrigation or automation of irrigation, monitoring of crop health, crop disease detection, predictive analytics, weed control, precision agriculture, harvesting, yield estimation and phenotyping, supply chain management & food quality. The current challenges in adopting AI in Agriculture covers with all concern.

REFERENCES

Abawi, G. S., & Widmer, T. L. (2000). Impact of soil health management practices on soilborne pathogens, nematodes and root diseases of vegetable crops. *Applied Soil Ecology*, *15*(1), 37–47. doi:10.1016/S0929-1393(00)00070-6

Accorsi, R., Cholette, S., Manzini, R., & Tufano, A. (2018). A hierarchical data architecture for sustainable food supply chain management and planning. *Journal of Cleaner Production*, *203*, 1039–1054. doi:10.1016/j.jclepro.2018.08.275

Agatonovic-Kustrin, S., & Beresford, R. (2000). Basic concepts of artificial neural network (ANN) modeling and its application in pharmaceutical research. *Journal of Pharmaceutical and Biomedical Analysis*, *22*(5), 717–727. doi:10.1016/S0731-7085(99)00272-1 PMID:10815714

Agriinfo. (2018). Soil –Plant – Water – Relationships. *Agriinfo.* https://agriinfo.in/soil-plant-water-relationships-2206/

Ahmad, I., Shahabuddin, S., Sauter, T., Harjula, E., Kumar, T., Meisel, M., Juntti, M., & Ylianttila, M. (2021). The Challenges of Artificial Intelligence in Wireless Networks for the Internet of Things: Exploring Opportunities for Growth. *IEEE Industrial Electronics Magazine*, *15*(1), 16–29. doi:10.1109/MIE.2020.2979272

Ahmad, L., & Nabi, F. (2021). *Agriculture 5.0: Artificial Intelligence, IoT and Machine Learning*. CRC Press.

Ahn, H.S., Dayoub, F., Popovic, M., MacDonald, B.A., Siegwart, R.Y., & Sa, I. (2018). An Overview of Perception Methods for Horticultural Robots: From Pollination to Harvest. ArXiv, abs/1807.03124.

Aiello, G., Catania, P., Vallone, M., & Venticinque, M. (2022). Worker safety in agriculture 4.0: A new approach for mapping operator's vibration risk through Machine Learning activity recognition. *Computers and Electronics in Agriculture*, *193*, 106637. doi:10.1016/j.compag.2021.106637

Aitkenhead, M. J., Mcdonald, A. J. S., Dawson, J. J., Couper, G., Smart, R. P., Billett, M., Hope, D., & Palmer, S. (2003). A novel method for training neural networks for time-series prediction in environmental systems. *Elsevier, 162*, 87–95. doi:10.1016/S0304-3800(02)00401-5

Alston, J. M., & Pardey, P. G. (2014). Agriculture in the Global Economy %J. *The Journal of Economic Perspectives*, *28*(1), 121–146. doi:10.1257/jep.28.1.121

Amatya, S., Karkee, M., Gongal, A., Zhang, Q., & Whiting, M. D. (2016). Detection of cherry tree branches with full foliage in planar architecture for automated sweet-cherry harvesting. *Biosystems Engineering*, *146*, 3–15. doi:10.1016/j.biosystemseng.2015.10.003

Arif, C., Indra Setiawan, B., Mizoguchi, M., & Doi, R. (2012a). Estimation of soil moisture in paddy field using Artificial Neural Networks. *International Journal of Advanced Research in Artificial Intelligence*, *1*(1). doi:10.14569/IJARAI.2012.010104

Arif, C., Mizoguchi, M., Mizoguchi, M., & Doi, R. (2012b). Estimation of soil moisture in paddy field using Artificial Neural Networks. *International Journal of Advanced Research in Artificial Intelligence*, *1*(1). doi:10.14569/IJARAI.2012.010104

Åstrand, B., & Baerveldt, A. J. (2002). An agricultural mobile robot with vision-based perception for mechanical weed control. *Autonomous Robots*, *13*(1), 21–35. doi:10.1023/A:1015674004201

Babu, S. (2013). A software model for precision agriculture for small and marginal farmers. *Global Humanitarian Technology Conference: South Asia Satellite,* 352–355. IEEE. 10.1109/GHTC-SAS.2013.6629944

Bacco, M., Barsocchi, P., Ferro, E., Gotta, A., & Ruggeri, M. (2019). The Digitisation of Agriculture: A Survey of Research Activities on Smart Farming. *Array*, *3–4*, 100009. doi:10.1016/j.array.2019.100009

Bak, T., & Jakobsen, H. (2004). Agricultural robotic platform with four wheel steering for weed detection. *Biosystems Engineering*, *87*(2), 125–136. doi:10.1016/j.biosystemseng.2003.10.009

Bakker, T., van Asselt, K., Bontsema, J., Müller, J., & van Straten, G. (2006). An autonomous weeding robot for organic farming. *Springer Tracts in Advanced Robotics*, *25*, 579–590. doi:10.1007/978-3-540-33453-8_48

Barlow, H. B. (1989). Unsupervised Learning. *Neural Computation*, *1*(3), 295–311. doi:10.1162/neco.1989.1.3.295

Beaman, J. A., & Johnson, A. J. (2006). Food distribution channel overview: a guide for new manufacturers. *OSU.* https://ir.library.oregonstate.edu/concern/open_educational_resources/8623hz116

Bhattacharyya, S. S., Maitra, D., & Deb, S. (2021). Study of adoption and absorption of emerging technologies for smart supply chain management: A dynamic capabilities perspective. [IJAL]. *International Journal of Applied Logistics*, *11*(2), 14–54.

Bhbosale, S., Pujari, V., & Multani, Z. J. A. I. I. R. J. (2020). Advantages And Disadvantages Of Artificial Intellegence. 227-230.

Bilgili, M. (2011). The use of artificial neural networks for forecasting the monthly mean soil temperatures in Adana, Turkey. *Turkish Journal of Agriculture and Forestry*, *35*(1), 83–93. https://dergipark.org.tr/en/pub/tbtkagriculture/issue/11593/138197

Bioucas-Dias, J. M., Plaza, A., Camps-Valls, G., Scheunders, P., Nasrabadi, N. M., & Chanussot, J. (2013). Hyperspectral remote sensing data analysis and future challenges. *Geoscience and Remote Sensing Magazine, 1*(2), 6–36. IEEE . doi:10.1109/MGRS.2013.2244672

Blasco, J., Aleixos, N., Roger, J. M., Rabatel, G., & Moltó, E. (2002). AE—Automation and emerging technologies: Robotic weed control using machine vision. *Biosystems Engineering*, *83*(2), 149–157.

Bonaccorso, G. (2017). *Machine learning algorithms*. Packt Publishing Ltd.

Bralts, V. F., Driscoll, M. A., Shayya, W. H., & Cao, L. (1993). An expert system for the hydraulic analysis of microirrigation systems. *Computers and Electronics in Agriculture*, *9*(4), 275–287.

Bu, F., & Wang, X. (2019). A smart agriculture IoT system based on deep reinforcement learning. [doi:https://doi.org/10.1016/j.future.2019.04.041]. *Future Generation Computer Systems*, *99*, 500–507.

Bünger, L. (2021). *Robotic waste sorting*. Worcester Polytechnic Institute.

Busch, G. (2012). GIS-based Tools for Regional Assessments and Planning Processes Regarding Potential Environmental Effects of Poplar SRC. *BioEnergy Research* *5*(3), 584–605. doi:10.1007/S12155-012-9224-0

Cakir, Y., Kirci, M., & Gunes, E. O. (2014). Yield prediction of wheat in south-east region of Turkey by using artificial neural networks. *The 3rd International Conference on Agro-Geoinformatics.* Agro-Geoinformatics . doi:10.1109/AGRO-GEOINFORMATICS.2014.6910609

Carbonell, J. G., Michalski, R. S., & Mitchell, T. M. (1983). 1 - AN OVERVIEW OF MACHINE LEARNING. In R. S. Michalski, J. G. Carbonell, & T. M. Mitchell (Eds.), *Machine Learning* (pp. 3–23). Morgan Kaufmann.

Chang, D. H., & Islam, S. (2000). Estimation of soil physical properties using remote sensing and artificial neural network. *Remote Sensing of Environment*, *74*(3), 534–544.

Chang, C. & Robotics, K. L. (2018). Smart agricultural machine with a computer vision-based weeding and variable-rate irrigation scheme. *Mdpi.Com.* doi:10.3390/robotics7030038

Charoen-Ung, P., & Mittrapiyanuruk, P. (2019). Sugarcane Yield Grade Prediction Using Random Forest with Forward Feature Selection and Hyper-parameter Tuning. *Advances in Intelligent Systems and Computing, 769*, 33–42. doi:10.1007/978-3-319-93692-5_4

Chattopadhyay, P. B., & Rangarajan, R. (2014). Application of ANN in sketching spatial nonlinearity of unconfined aquifer in agricultural basin. [doi:https://doi.org/10.1016/j.agwat.2013.11.007]. *Agricultural Water Management*, *133*, 81–91.

Chen, J., Chen, J., Zhang, D., Sun, Y., & Nanehkaran, Y. A. (2020). Using deep transfer learning for image-based plant disease identification. *Computers and Electronics in Agriculture*, *173*, 105393. https://doi.org/10.1016/J.COMPAG.2020.105393

Chen, J., & Yang, A. (2019). Intelligent Agriculture and Its Key Technologies Based on Internet of Things Architecture. *IEEE Access: Practical Innovations, Open Solutions*, *7*, 77134–77141. doi:10.1109/ACCESS.2019.2921391

Cheng, X., Zhang, Y., Chen, Y., Wu, Y., & Yue, Y. (2017). Pest identification via deep residual learning in complex background. [doi:https://doi.org/10.1016/j.compag.2017.08.005]. *Computers and Electronics in Agriculture*, *141*, 351–356.

Cui, S., Ling, P., Zhu, H., & Keener, H. M. (2018). Plant Pest Detection Using an Artificial Nose System: A Review. *Sensors 18*(2), 378. doi:10.3390/S18020378

Daheim, C., Poppe, K., & Schrijver, R. (2019). *Precision agriculture and the future of farming in Europe: scientific foresight study*. European Parliament. https://data.europa.eu/doi/10.2861/175493

Daugherty, P. R., & Wilson, H. J. (2018). *Human+ machine: Reimagining work in the age of AI*. Harvard Business Press.

Devapriya, P., Ferrell, W., & Geismar, N. (2017). Integrated production and distribution scheduling with a perishable product. *European Journal of Operational Research*, *259*(3), 906–916. https://doi.org/10.1016/J.EJOR.2016.09.019

Devaraj, A., Rathan, K., Jaahnavi, S., & Indira, K. (2019). Identification of plant disease using image processing technique. *Proceedings of the 2019 International Conference on Communication and Signal Processing*, 749–753. IEEE. doi:10.1109/ICCSP.2019.8698056

Dewi, T., Risma, P., & Oktarina, Y. (2020). Fruit sorting robot based on color and size for an agricultural product packaging system. *Bulletin of Electrical Engineering and Informatics*, *9*(4), 1438–1445.

dos Santos Ferreira, A., Freitas, D. M., da Silva, G. G., Pistori, H., & Folhes, M. T. (2019). Unsupervised deep learning and semi-automatic data labeling in weed discrimination. [doi:https://doi.org/10.1016/j.compag.2019.104963]. *Computers and Electronics in Agriculture*, *165*, 104963.

Elshorbagy, A., & Parasuraman, K. (2008). On the relevance of using artificial neural networks for estimating soil moisture content. *Journal of Hydrology (Amsterdam)*, *362*(1-2), 1–18.

Fedorova, E., Darbasov, V., & Okhlopkov, M. (2020). The role of agricultural economists in study on problems related to regional food safety. In *E3S Web of Conferences* (Vol. 176, p. 05011). EDP Sciences.

Fennimore, S. A., Slaughter, D. C., Siemens, M. C., Leon, R. G., & Saber, M. N. (2016). Technology for automation of weed control in specialty crops. *Cambridge.Org, 30*, 823–837. doi:10.1614/WT-D-16-00070.1

Ferentinos, K. P. (2018). Deep learning models for plant disease detection and diagnosis. *Computers and Electronics in Agriculture*, *145*, 311–318. https://doi.org/10.1016/J.COMPAG.2018.01.009

Filimonau, V., Todorova, E., Mzembe, A., Sauer, L., & Yankholmes, A. (2020). A comparative study of food waste management in full service restaurants of the United Kingdom and the Netherlands. *Journal of Cleaner Production*, *258*, 120775.

Fuentes, A., Yoon, S., Kim, S. C., & Park, D. S. (2017). A Robust Deep-Learning-Based Detector for Real-Time Tomato Plant Diseases and Pests Recognition. *Sensors, 17*(9), 2022. doi:10.3390/S17092022

Goap, A., Sharma, D., Shukla, A. K., & Rama Krishna, C. (2018). An IoT based smart irrigation management system using Machine learning and open source technologies. *Computers and Electronics in Agriculture*, *155*, 41–49. https://doi.org/10.1016/J.COMPAG.2018.09.040

Gu, Z., Qi, Z., Ma, L., Gui, D., Xu, J., Fang, Q., Yuan, S., & Feng, G. (2017). Development of an irrigation scheduling software based on model predicted crop water stress. *Computers and Electronics in Agriculture*, *143*, 208–221. https://doi.org/10.1016/J.COMPAG.2017.10.023

Hady, M. F. A., & Schwenker, F. (2013). Semi-supervised Learning. In M. Bianchini, M. Maggini, & L. C. Jain (Eds.), *Handbook on Neural Information Processing* (pp. 215–239). Springer Berlin Heidelberg.

Haghverdi, A., & Washington-Allen, R. (2018). Prediction of cotton lint yield from phenology of crop indices using artificial neural networks. *Elsevier*. doi:10.1016/j.compag.2018.07.021

Harishankar, S., Kumar, R. S., Sudharsan, K. P., Vignesh, U., & Viveknath, T. (2014). Solar powered smart irrigation system. *Advance in electronic and electric engineering, 4*(4), 341-346.

Harris, D., Pathan, A. K., Gothkar, P., Joshi, A., Chivasa, W., & Nyamudeza, P. (2001). On-farm seed priming: Using participatory methods to revive and refine a key technology. [doi:https://doi.org/10.1016/S0308-521X(01)00023-3]. *Agricultural Systems*, *69*(1), 151–164.

Heisel, T., Schou, J., Christensen, S., & Andreasen, C. (2001). Cutting weeds with a CO 2 laser. *Weed Research*, *41*(1), 19–29. https://doi.org/10.1046/J.1365-3180.2001.00212.X

Hinnell, A., Lazarovitch, N., & Furman, A. (2010). Neuro-Drip: estimation of subsurface wetting patterns for drip irrigation using neural networks. *Springer, 28*(6), 535–544. doi:10.1007/s00271-010-0214-8

Hui, Y. Y., Choy, K. L., Ho, G. T. S., Leung, K. H., & Lam, H. Y. (2016). A cloud-based location assignment system for packaged food allocation in e-fulfillment warehouse. *Sage Journals, 8*, 1–15. doi:10.1177/1847979016684832

Ibáñez, M. D., & Blázquez, M. A. (2019). Ginger and Turmeric Essential Oils for Weed Control and Food Crop Protection. *Plants*, *8*(3). Advance online publication. doi:10.3390/plants8030059

Iqbal, M. A., & Talukder, K. H. (2020). Detection of Potato Disease Using Image Segmentation and Machine Learning. *International Conference on Wireless Communications, Signal Processing and Networking,* 43–47. doi:10.1109/WISPNET48689.2020.9198563

Jagtap, S. T., Phasinam, K., Kassanuk, T., Jha, S. S., Ghosh, T., & Thakar, C. M. (2022). Towards application of various machine learning techniques in agriculture. *Materials Today: Proceedings, 51,* 793-797. doi:10.1016/j.matpr.2021.06.236

Jha, K., Doshi, A., Patel, P., & Shah, M. (2019). A comprehensive review on automation in agriculture using artificial intelligence. *Artificial Intelligence in Agriculture*, *2*, 1–12.

Kakani, V., Nguyen, V. H., Kumar, B. P., Kim, H., & Pasupuleti, V. R. (2020). A critical review on computer vision and artificial intelligence in food industry. *Journal of Agriculture and Food Research*, *2*, 100033. https://doi.org/10.1016/J.JAFR.2020.100033

Kamilaris, A., & Prenafeta-Boldú, F. X. (2018). Deep learning in agriculture: A survey. *Computers and Electronics in Agriculture*, *147*, 70–90. https://doi.org/10.1016/J.COMPAG.2018.02.016

Karar, M. E., Alsunaydi, F., Albusaymi, S., & Alotaibi, S. (2021). A new mobile application of agricultural pests recognition using deep learning in cloud computing system. *Alexandria Engineering Journal*, *60*(5), 4423-4432. doi:10.1016/j.aej.2021.03.009

Kasinathan, T., Singaraju, D., & Uyyala, S. R. (2021). Insect classification and detection in field crops using modern machine learning techniques. [doi:https://doi.org/10.1016/j.inpa.2020.09.006]. *Information Processing in Agriculture*, *8*(3), 446–457.

Kehui, X., Deqin, X., & Xiwen, L. (2010). Smart water-saving irrigation system in precision agriculture based on wireless sensor network. *Nongye Gongcheng Xuebao (Beijing)*, *26*(11), 170–175. https://doi.org/10.3969/j.issn.1002-6819.2010.11.030

Khedr, A. E., Kadry, M., & Walid, G. (2015). Proposed Framework for Implementing Data Mining Techniques to Enhance Decisions in Agriculture Sector Applied Case on Food Security Information Center Ministry of Agriculture, Egypt. *Procedia Computer Science*, *65*, 633–642. https://doi.org/10.1016/J.PROCS.2015.09.007

Kim, M. agriculture, J. G.-C. and electronics in, & 2008, undefined. (2008). Artificial Neural Network estimation of soil erosion and nutrient concentrations in runoff from land application areas. *Elsevier*. doi:10.1016/j.compag.2008.05.021

Komarek, A. M., De Pinto, A., & Smith, V. H. J. A. S. (2020). A review of types of risks in agriculture: What we know and what we need to know. *Agricultural Systems, 178*, 102738.

Kouadio, L., Deo, R. C., Byrareddy, V., Adamowski, J. F., & Mushtaq, S. (2018). Artificial intelligence approach for the prediction of Robusta coffee yield using soil fertility properties. *Computers and Electronics in Agriculture*, *155*, 324–338.

Krisztin, T. (2018). Semi-parametric spatial autoregressive models in freight generation modeling. *Transportation Research Part E, Logistics and Transportation Review*, *114*, 121–143. https://doi.org/10.1016/J.TRE.2018.03.003

Kukreja, H., Bharath, N., Siddesh, C., & Kuldeep, S. J. I. J. A. R. I. I. E. (2016). An introduction to artificial neural network. 1, 27-30.

Kumar, I., Rawat, J., Mohd, N., & Husain, S. (2021). Opportunities of Artificial Intelligence and Machine Learning in the Food Industry. *Journal of Food Quality*. doi:10.1155/2021/4535567

Kumar, R., Singh, M. P., Kumar, P., & Singh, J. P. (2015). Crop Selection Method to maximize crop yield rate using machine learning technique. *International Conference on Smart Technologies and Management for Computing, Communication, Controls, Energy and Materials*, 138–145. doi:10.1109/ICSTM.2015.7225403

Lamm, R. D., Slaughter, D. C., & Giles, D. K. (2002). Precision weed control system for cotton. *Transactions of the ASAE. American Society of Agricultural Engineers*, *45*(1), 231.

Larsen, R. A., Schaalje, G. B., & Lawson, J. S. (2009). Food shelf life: estimation and optimal design. *Journal of Statistical Computation and Simulation, 80*(2), 143–157. doi:10.1080/00949650802549135

Lauzon, F. Q. (2012, 2-5 July 2012). An introduction to deep learning. Paper presented at the *International Conference on Information Science, Signal Processing and their Applications* (ISSPA).

LeCun, Y., Bengio, Y., & Hinton, G. (2015). Deep learning. *Nature*, *521*(7553), 436–444. doi:10.1038/nature14539

Lezoche, M., Hernandez, J. E., Díaz, M. D. M. E. A., Panetto, H., & Kacprzyk, J. (2020). Agri-food 4.0: A survey of the supply chains and technologies for the future agriculture. *Computers in Industry*, *117*, 103187.

Li, M., & Yost, R. S. (2000). Management-oriented modeling: Optimizing nitrogen management with artificial intelligence. *Agricultural Systems*, *65*(1), 1–27.

Li, N., Shepperd, M., & Guo, Y. (2020). A systematic review of unsupervised learning techniques for software defect prediction. [doi:https://doi.org/10.1016/j.infsof.2020.106287]. *Information and Software Technology*, *122*, 106287.

Li, T., Johansen, K., & McCabe, M. F. (2022). A machine learning approach for identifying and delineating agricultural fields and their multi-temporal dynamics using three decades of Landsat data. [doi:https://doi.org/10.1016/j.isprsjprs.2022.02.002]. *ISPRS Journal of Photogrammetry and Remote Sensing*, *186*, 83–101.

Li, Y., & Chao, X. (2020). ANN-Based Continual Classification in Agriculture. *Agriculture*, *10*(5). doi:10.3390/agriculture10050178

Liakos, K. G., Busato, P., Moshou, D., Pearson, S., & Bochtis, D. (2018). Machine Learning in Agriculture: A Review. *Sensors (Basel)*, *18*(8), 2674. https://doi.org/10.3390/s18082674

Livingstone, D. J. (Ed.). (2008). *Artificial neural networks: methods and applications* (pp. 185–202). Humana Press.

López, E. M., García, M., Schuhmacher, M., & Domingo, J. L. (2008). A fuzzy expert system for soil characterization. *Environment International*, *34*(7), 950–958.

Luangkesorn, K. L., Klein, G., & Bidanda, B. (2016). Analysis of production systems with potential for severe disruptions. *International Journal of Production Economics*, *171*, 478–486. https://doi.org/10.1016/J.IJPE.2015.09.014

Ma, J., Du, K., Zheng, F., Zhang, L., Gong, Z., & Sun, Z. (2018). A recognition method for cucumber diseases using leaf symptom images based on deep convolutional neural network. *Computers and Electronics in Agriculture, 154*, 18–24. https://doi.org/10.1016/J.COMPAG.2018.08.048

Mananze, S., Pôças, I., & Cunha, M. (2018). Retrieval of Maize Leaf Area Index Using Hyperspectral and Multispectral Data. *Remote Sensing, 10*(12), 1942. doi:10.3390/RS10121942

Mazoyer, M., & Roudart, L. (2006). *A history of world agriculture: from the neolithic age to the current crisis*. NYU Press.

McCalla, A. F. J. U. A., & Economics, R. (2001). Challenges to world agriculture in the 21st century. *Agricultural and Resource Economics, 4*(3), 1-2.

Mehra, M., Saxena, S., Sankaranarayanan, S., Tom, R. J., & Veeramanikandan, M. (2018). IoT based hydroponics system using Deep Neural Networks. *Computers and Electronics in Agriculture, 155*, 473–486.

Mishra, M., & Srivastava, M. (2014). A view of artificial neural network. Paper presented at the *International Conference on Advances in Engineering & Technology Research (ICAETR).*

Misra, N. N., Dixit, Y., Al-Mallahi, A., Bhullar, M. S., Upadhyay, R., & Martynenko, A. (2020). *IoT, big data and artificial intelligence in agriculture and food industry*. IEEE Internet of Things Journal.

Mohanty, S. P., Hughes, D. P., & Salathé, M. (2016). Using deep learning for image-based plant disease detection. *Frontiers in Plant Science, 7*(September), 1419. https://doi.org/10.3389/FPLS.2016.01419/BIBTEX

Moisture, S. (2017). Importance of Soil Water. *Agriinfo*. https://agriinfo.in/soil-moisture-importance-of-soil-water-263/

Mojjada, R. K., Kiran Kumar, K., Yadav, A., & Satya Vara Prasad, B. V. V. (2020). Detection of plant leaf disease using digital image processing. *Materials Today: Proceedings*. doi:10.1016/J.MATPR.2020.11.115

Monga, T. (2018). Estimating Vineyard Grape Yield from Images. Lecture Notes in Computer Science (Including Subseries Lecture Notes in Artificial Intelligence and Lecture Notes in Bioinformatics). LNAI, 339–343. doi:10.1007/978-3-319-89656-4_37

Montas, H., & Madramootoo, C. A. (1992). A decision support system for soil conservation planning. *Computers and Electronics in Agriculture, 7*(3), 187–202.

Monteiro, A. L., Freitas Souza, M. d., Lins, H. A., Teófilo, T. M. S., Barros Júnior, A. P., Silva, D. V., & Mendonça, V. (2021). A new alternative to determine weed control in agricultural systems based on artificial neural networks (ANNs). [doi:https://doi.org/10.1016/j.fcr.2021.108075]. *Field Crops Research, 263*, 108075.

Mucherino, A., & Papajorgji, P. Research, P. P.-O., & 2009, undefined. (2009). A survey of data mining techniques applied to agriculture. *Springer, 9*(2), 121–140. doi:10.1007/s12351-009-0054-6

Munz, J. M., Gindele, N., & Doluschitz, R. (2020). Exploring the characteristics and utilisation of Farm Management Information Systems (FMIS) in Germany. *Computers and Electronics in Agriculture*, *170*, 105246.

Nair, K. P. (2019). Soil Fertility and Nutrient Management. In K. P. Nair (Ed.), *Intelligent Soil Management for Sustainable Agriculture: The Nutrient Buffer Power Concept* (pp. 165–189). Springer International Publishing.

Nakai, S., & Yamada, Y. (2014). Development of a weed suppression robot for rice cultivation: Weed suppression and posture control. *Int J Electr Comput Electron Commun Eng*, *8*, 1736–1740.

Nasiakou, A., Vavalis, M., & Zimeris, D. (2016). Smart energy for smart irrigation. *Computers and Electronics in Agriculture*, *129*, 74–83. https://doi.org/10.1016/J.COMPAG.2016.09.008

Navarro-Hellín, H., Martínez-del-Rincon, J., Domingo-Miguel, R., Soto-Valles, F., & Torres-Sánchez, R. (2016). A decision support system for managing irrigation in agriculture. *Computers and Electronics in Agriculture*, *124*, 121–131. https://doi.org/10.1016/J.COMPAG.2016.04.003

Nawandar, N. K., & Satpute, V. R. (2019). IoT based low cost and intelligent module for smart irrigation system. *Computers and Electronics in Agriculture*, *162*, 979–990. https://doi.org/10.1016/J.COMPAG.2019.05.027

Nevavuori, P., Narra, N., & Lipping, T. (2019). Crop yield prediction with deep convolutional neural networks. *Computers and Electronics in Agriculture*, *163*, 104859. https://doi.org/10.1016/J.COMPAG.2019.104859

Ngo, H. C., Hashim, U. R., Sek, Y. W., Kumar, Y. J., & Ke, W. S. (2019). Weeds detection in agricultural fields using convolutional neural network. *International Journal of Innovative Technology and Exploring Engineering*, *8*(11), 292–296. https://doi.org/10.35940/IJITEE.K1327.0981119

Nikolenko, S., Kadurin, A., & Arkhangelskaya, E. J. S. P. (2018). Deep learning. *Journal of the American Heart Association.*

Noon, S. K., Amjad, M., Qureshi, M. A., & Mannan, A. (2020). Use of deep learning techniques for identification of plant leaf stresses: A review. *Sustain. Comput. Informatics Syst.*, *28*, 100443.

Nørremark, M., & Griepentrog, H. (2004). Analysis and definition of the close-to-crop area in relation to robotic weeding. *Orgprints.* https://orgprints.org/4834/

O'Sullivan, C. A., Bonnett, G. D., McIntyre, C. L., Hochman, Z., & Wasson, A. P. (2019). Strategies to improve the productivity, product diversity and profitability of urban agriculture. [doi:https://doi.org/10.1016/j.agsy.2019.05.007]. *Agricultural Systems*, *174*, 133–144.

Onishchuk, M. O. (2020). Opto-mechanical sorting of municipal solid waste [Doctoral dissertation, BHTY, USA].

Osman, T., Psyche, S. S., Kamal, M. R., Tamanna, F., Haque, F., & Rahman, R. M. (2017). Predicting Early Crop Production by Analysing Prior Environment Factors. *Advances in Intelligent Systems and Computing,* 470–479. doi:10.1007/978-3-319-49073-1_51

Pallathadka, H., Mustafa, M., Sanchez, D. T., Sekhar Sajja, G., Gour, S., & Naved, M. (2021). IMPACT OF MACHINE learning ON Management, healthcare AND AGRICULTURE. *Materials Today: Proceedings*. doi:10.1016/j.matpr.2021.07.042

Palm, C., Blanco-Canqui, H., DeClerck, F., Gatere, L., & Grace, P. (2014). Conservation agriculture and ecosystem services: An overview. [doi:https://doi.org/10.1016/j.agee.2013.10.010]. *Agriculture, Ecosystems & Environment*, *187*, 87–105.

Pant, J., Pant, R. P., Kumar Singh, M., Pratap Singh, D., & Pant, H. (2021). Analysis of agricultural crop yield prediction using statistical techniques of machine learning. *Materials Today: Proceedings*, *46*, 10922-10926. doi:10.1016/j.matpr.2021.01.948

Pantazi, X. E., Moshou, D., Alexandridis, T., Whetton, R. L., & Mouazen, A. M. (2016). Wheat yield prediction using machine learning and advanced sensing techniques. *Computers and Electronics in Agriculture*, *121*, 57–65. https://doi.org/10.1016/J.COMPAG.2015.11.018

Partel, V., Kakarla, S. C., & Ampatzidis, Y. (2019). Development and evaluation of a low-cost and smart technology for precision weed management utilizing artificial intelligence. *Computers and Electronics in Agriculture*, *157*, 339–350.

Qiang, L., & Jiuping, X. (2008). A Study on Vehicle Routing Problem in the Delivery of Fresh Agricultural Products under Random Fuzzy Environment. *International Journal of Information and Management Sciences*, *19*(4), 673–690.

Ramcharan, A., Baranowski, K., McCloskey, P., Ahmed, B., Legg, J., & Hughes, D. P. (2017). Deep learning for image-based cassava disease detection. *Frontiers in Plant Science*, *8*, 1852. https://doi.org/10.3389/FPLS.2017.01852/BIBTEX

Ramos, P. J., Prieto, F. A., Montoya, E. C., & Oliveros, C. E. (2017). Automatic fruit count on coffee branches using computer vision. *Computers and Electronics in Agriculture*, *137*, 9–22. https://doi.org/10.1016/J.COMPAG.2017.03.010

Research, R. Z.-P. J. of W. S., & 2010, undefined. (2014). Ethics for weed science. *Researchgate.Net*. doi:10.1017/S0043174500089633

Rico-Fernández, M. P., Rios-Cabrera, R., Castelán, M., Guerrero-Reyes, H. I., & Juarez-Maldonado, A. (2019). A contextualized approach for segmentation of foliage in different crop species. *Computers and Electronics in Agriculture*, *156*, 378–386.

Roshanianfard, A., Noguchi, N., Okamoto, H., & Ishii, K. (2020). A review of autonomous agricultural vehicles (The experience of Hokkaido University). *Journal of Terramechanics*, *91*, 155–183.

Rumpf, T., Mahlein, A. K., Steiner, U., Oerke, E. C., Dehne, H. W., & Plümer, L. (2010). Early detection and classification of plant diseases with Support Vector Machines based on hyperspectral reflectance. *Computers and Electronics in Agriculture*, *74*(1), 91–99. https://doi.org/10.1016/J.COMPAG.2010.06.009

Sadgrove, E. J., Falzon, G., Miron, D., & Lamb, D. W. (2018). Real-time object detection in agricultural/remote environments using the multiple-expert colour feature extreme learning machine (MEC-ELM). *Computers in Industry*, *98*, 183–191.

Saldaña-Robles, A. L., Bustos-Gaytán, A., Diosdado-De la Peña, J. A., Saldaña-Robles, A., Alcántar-Camarena, V., Balvantín-García, A., & Saldaña-Robles, N. (2020). Structural design of an agricultural backhoe using TA, FEA, RSM and ANN. *Computers and Electronics in Agriculture*, *172*, 105278.

Sawant, S. S., & Prabukumar, M. (2020). A review on graph-based semi-supervised learning methods for hyperspectral image classification. [doi:https://doi.org/10.1016/j.ejrs.2018.11.001]. *The Egyptian Journal of Remote Sensing and Space Sciences*, *23*(2), 243–248.

Schwenker, F., & Trentin, E. (2014). Pattern classification and clustering: A review of partially supervised learning approaches. [doi:https://doi.org/10.1016/j.patrec.2013.10.017]. *Pattern Recognition Letters*, *37*, 4–14.

Sengupta, S., & Lee, W. S. (2014). Identification and determination of the number of immature green citrus fruit in a canopy under different ambient light conditions. *Biosystems Engineering*, *117*(C), 51–61. https://doi.org/10.1016/J.BIOSYSTEMSENG.2013.07.007

Shinde, P. P., & Shah, S. (2018, 16-18 Aug. 2018). A Review of Machine Learning and Deep Learning Applications. Paper presented at the *Fourth International Conference on Computing Communication Control and Automation (ICCUBEA)*.

Shukla, M., & Jharkharia, S. (2014). An inventory model for continuously deteriorating agri-fresh produce: An artificial immune system-based solution approach. *International Journal of Integrated Supply Management*, *9*(1–2), 110–135. https://doi.org/10.1504/IJISM.2014.064362

Sicat, R. S., Carranza, E. J., & Nidumolu, U. B. (2005). Fuzzy modeling of farmers' knowledge for land suitability classification. *Agricultural Systems*, *83*, 49–75.

Sinwar, D., Dhaka, V. S., Sharma, M. K., & Rani, G. (2020). AI-Based Yield Prediction and Smart Irrigation, 155–180. doi:10.1007/978-981-15-0663-5_8

Srivastava, A. R., & Venkatesan, M. (2020). Tea leaf disease prediction using texture-based image processing. *Advances in Intelligent Systems and Computing, 1054,* 17–25. doi:10.1007/978-981-15-0135-7_3

Syers, J. K. (1997). Managing soils for long-term productivity. *Philosophical Transactions of the Royal Society of London. Series B, Biological Sciences*, *352*(1356), 1011–1021. https://doi.org/10.1098/RSTB.1997.0079

Tajik, S., Ayoubi, S., & Nourbakhsh, F. (2012). Prediction of soil enzymes activity by digital terrain analysis: Comparing artificial neural network and multiple linear regression models. *Environmental Engineering Science*, *29*(8), 798–806. https://doi.org/10.1089/EES.2011.0313

Tanha Talaviya, Dhara Shah, D. Shah, Nivedita Patel, N. Patel, Hiteshri Yagnik, H. Yagnik, & Manan Shah, M. Shah. (2020). Implementation of artificial intelligence in agriculture for optimisation of irrigation and application of pesticides and herbicides. *Artificial intelligence in agriculture, 4,* 58-73. doi:10.1016/j.aiia.2020.04.002

Tang, L., & Tian, L. ASAE, B. S.-T. of the, & 2000, undefined. (2000). Color image segmentation with genetic algorithm for in-field weed sensing. *Elibrary.Asabe.Org.* https://elibrary.asabe.org/abstract.asp?aid=2970

Tian, Y., Yang, G., Wang, Z., Wang, H., Li, E., & Liang, Z. (2019). Apple detection during different growth stages in orchards using the improved YOLO-V3 model. *Computers and Electronics in Agriculture, 157*, 417–426. https://doi.org/10.1016/J.COMPAG.2019.01.012

Too, E. C., Yujian, L., Njuki, S., & Yingchun, L. (2019). A comparative study of fine-tuning deep learning models for plant disease identification. *Computers and Electronics in Agriculture, 161*, 272–279. https://doi.org/10.1016/J.COMPAG.2018.03.032

Tripathi, S., Shukla, S., Attrey, S., Agrawal, A., & Bhadoria, V. S. (2020). Smart Industrial Packaging and Sorting System. 245–254. doi:10.1007/978-981-15-3647-2_18

Tudi, M., Daniel Ruan, H., Wang, L., Lyu, J., Sadler, R., Connell, D., & Phung, D. T. (2021). Agriculture Development, Pesticide Application and Its Impact on the Environment. *International Journal of Environmental Research and Public Health, 18*(3). doi:10.3390/ijerph18031112

Valdés-Vela, M., Abrisqueta, I., Conejero, W., Vera, J., & Ruiz-Sánchez, M. C. (2015). Soft computing applied to stem water potential estimation: A fuzzy rule based approach. *Computers and Electronics in Agriculture, 115*, 150–160. https://doi.org/10.1016/J.COMPAG.2015.05.019

van Engelen, J. E., & Hoos, H. H. (2020). A survey on semi-supervised learning. *Machine Learning, 109*(2), 373–440. doi:10.100710994-019-05855-6

van Klompenburg, T., Kassahun, A., & Catal, C. (2020). Crop yield prediction using machine learning: A systematic literature review. *Computers and Electronics in Agriculture, 177*, 105709. https://doi.org/10.1016/J.COMPAG.2020.105709

Vellidis, G., Tucker, M., Perry, C., Kvien, C., & Bednarz, C. (2008). A real-time wireless smart sensor array for scheduling irrigation. *Computers and Electronics in Agriculture, 61*(1), 44–50. https://doi.org/10.1016/J.COMPAG.2007.05.009

Waheed, R., Chang, D., Sarwar, S., & Chen, W. (2018). Forest, agriculture, renewable energy, and CO2 emission. [doi:https://doi.org/10.1016/j.jclepro.2017.10.287]. *Journal of Cleaner Production, 172*, 4231–4238.

Wang, A. X., Tran, C., Desai, N., Lobell, D., & Ermon, S. (2018). Deep transfer learning for crop yield prediction with remote sensing data. *Proceedings of the 1st ACM SIGCAS Conference on Computing and Sustainable Societies,* COMPASS. doi:10.1145/3209811.3212707

Wang, Q., Nuske, S., Bergerman, M., & Singh, S. (2013). Automated Crop Yield Estimation for Apple Orchards. 745–758. doi:10.1007/978-3-319-00065-7_50

Wang, X. P., Wang, M., Ruan, J. H., & Li, Y. (2018). Multi-objective optimization for delivering perishable products with mixed time windows. *Advances in Production Engineering & Management, 13*(3), 321–332. https://doi.org/10.14743/APEM2018.3.293

Whitmire, C. D., Vance, J. M., Rasheed, H. K., Missaoui, A., Rasheed, K. M., & Maier, F. W. (2021). Using Machine Learning and Feature Selection for Alfalfa Yield Prediction. *AI, 2*(1), 71–88. doi:10.3390/AI2010006

Yadav, S., Sengar, N., Singh, A., Singh, A., & Dutta, M. K. (2021). Identification of disease using deep learning and evaluation of bacteriosis in peach leaf. [doi:https://doi.org/10.1016/j.ecoinf.2021.101247]. *Ecological Informatics*, *61*, 101247.

Yeasin, M., Dhandapani, A., & Ravichandran, S. (2021). Artificial Intelligence in Agriculture. In Ch. Srinivasarao (Ed.), *Agricultural Research, Technology and Policy: Innovations and Advances, ICAR-National Academy of Agricultural Research Management (NAARM), Hyderabad* (pp. 291–306).

Yost, M. A., Kitchen, N. R., Sudduth, K. A., Sadler, E. J., Drummond, S. T., Volkmann, M. R., & Yost MattYost, M. A. (2017). Long-term impact of a precision agriculture system on grain crop production. *Springer, 18*(5), 823–842. doi:10.1007/s11119-016-9490-5

You, J., Li, X., Low, M., Lobell, D., & Ermon, S. (2017). Deep Gaussian Process for Crop Yield Prediction Based on Remote Sensing Data. *Proceedings of the AAAI Conference on Artificial Intelligence*, *31*(1). https://ojs.aaai.org/index.php/AAAI/article/view/11172

Zhang, B., Xie, Y., Zhou, J., Wang, K., & Zhang, Z. (2020). State-of-the-art robotic grippers, grasping and control strategies, as well as their applications in agricultural robots: A review. *Computers and Electronics in Agriculture*, *177*, 105694. https://doi.org/10.1016/J.COMPAG.2020.105694

Zhao, Z., Chow, T. L., Rees, H. W., Yang, Q., Xing, Z., & Meng, F. R. (2009). Predict soil texture distributions using an artificial neural network model. *Computers and Electronics in Agriculture*, *65*(1), 36–48.

Zhu, H., Chu, B., Zhang, C., Liu, F., Jiang, L., & He, Y. (2017). Hyperspectral Imaging for Presymptomatic Detection of Tobacco Disease with Successive Projections Algorithm and Machine-learning Classifiers. *Scientific Reports, 7*(1), 1–12. doi:10.1038/s41598-017-04501-2

Zou, J., Han, Y., & So, S. S. (2008). Overview of artificial neural networks. *Artificial Neural Networks*, 14-22.

Chapter 15
Smart Systems and Services by Artificial Intelligence Algorithms

Esha Saxena
Dr. Akhilesh Das Gupta Institute of Technology and Management, India

Preety Shoran
Dr. Akhilesh Das Gupta Institute of Technology and Management, India

Meenakshi Yadav
Dr. Akhilesh Das Gupta Institute of Technology and Management, India

ABSTRACT

Artificial intelligence is facilitating the future of smart systems and their associated services. The world is witnessing the rapid growth in smart systems with artificial intelligence (AI) algorithms. Services must include a set of features to handle future trends, intelligent decisions, security, adaptivity, scalability, user-friendliness, time-saving ability, and adaptability to new architectural models. To satisfy these objectives, smart systems and services need advancements in working models of IoTs and applications of cloud computing. The AI and IoT create a great combination in various fields; health systems, agriculture, weather forecasting, manufacturing units, prediction, etc. Some of its applications include smart homes, smart cities, and smart businesses. This chapter also discuss about the role of IoT and cloud computing in smart systems, followed by smart city development, education system, environment protection, healthcare, water reservation, security, etc.

INTRODUCTION

This chapter concentrate on Smart systems and AI algorithms. Because of market rivalry, items' quality and dependability must always be improved. Artificial intelligence helps in analyzing the data and develops the corresponding smart and automated application. Smart systems and services facilitate growing urban populations. The rising demand for automation, security and comfort leads to completely new application domains. Various real-world application fields, such as cybersecurity systems, smart cities,

DOI: 10.4018/978-1-6684-6413-7.ch015

Copyright © 2023, IGI Global. Copying or distributing in print or electronic forms without written permission of IGI Global is prohibited.

healthcare, e-commerce, agriculture are the most popular piece of work. The change from product to service oriented services is another factor that plays an important role, which involve a huge role of Internet of Things (IoT) and applications of cloud computing. A wide range of household appliances, including those used for entertainment, energy management, security, health, and wellness, will have some kind of IT capability created to support smart services. Some of these gadgets will perform complementary tasks, supporting the fusion of smart home services. The research for Smart Systems came into action as with the advancements in AI algorithms. Through the analysis of explicit definitions of smartness and smart systems published in scientific journals, we pay particular attention to defining the properties of a smart system. This chapter will focus on explaining different AI based algorithms and how it helps in creating a smart system. Smart system and devices demand a need for storage, data transfer channels and self-decision making devices. The data can be structured, semi-structured, or unstructured, thus create a need for data management tools or platform. AI algorithms that can be applied to enhance the intelligence and the capabilities of smart devices. The types differentiate as classification analysis, regression, clustering, and dimensionality reduction or reinforcement learning techniques, which effectively works on building data-driven systems. The selection of appropriate algorithm depends on selected target system.

AI Algorithms

Artificial Intelligence is an area that deals with the progressive development of science and technology. AI is a decision-making and prediction process, it reduces the time, cost and increase the accuracy level of problem detection. AI refers to the systems that works on intelligent behaviour and their predictive environment (Jin, 2020). AI includes a wide variety of applications and algorithms which train them to automatically learn and improve from experience. Currently, the most common AI applications are applied on healthcare system, hybrid intelligent systems, robotics, e-commerce, navigation, agriculture, gaming and building smart cities/homes. The steps involved to build an AI model are listed below with expected training and test data.

Supervised Learning

In supervised learning, the machine is trained with labelled data, in means to provide it with the expected output. The term “supervised learning method” refers to the process of setting appropriate learning objectives prior to learning. During the initial training of the machine, the machine relies on information to learn. Machines must learn the content in a controlled setting in order to gather basic data (Mahesh, 2020). After the system learning is finished, it can help the machine to solve some classification or regression problems. Currently, widely accepted techniques of learning include random forest, SVN, KNN, etc. Because the entire learning process has purpose, using labelled datasets, models are trained. After the completion of the training phase, the model is evaluated using test data, and it then makes output predictions.

Unsupervised Learning

Unsupervised learning is a type of machine learning in which models are trained using unlabelled datasets. It’s called unsupervised learning, that the machine does not mark the content in a certain direction during the entire learning. Unsupervised learning includes Clustering, k-means, NN and anomaly

Figure 1. Steps to build an AI model

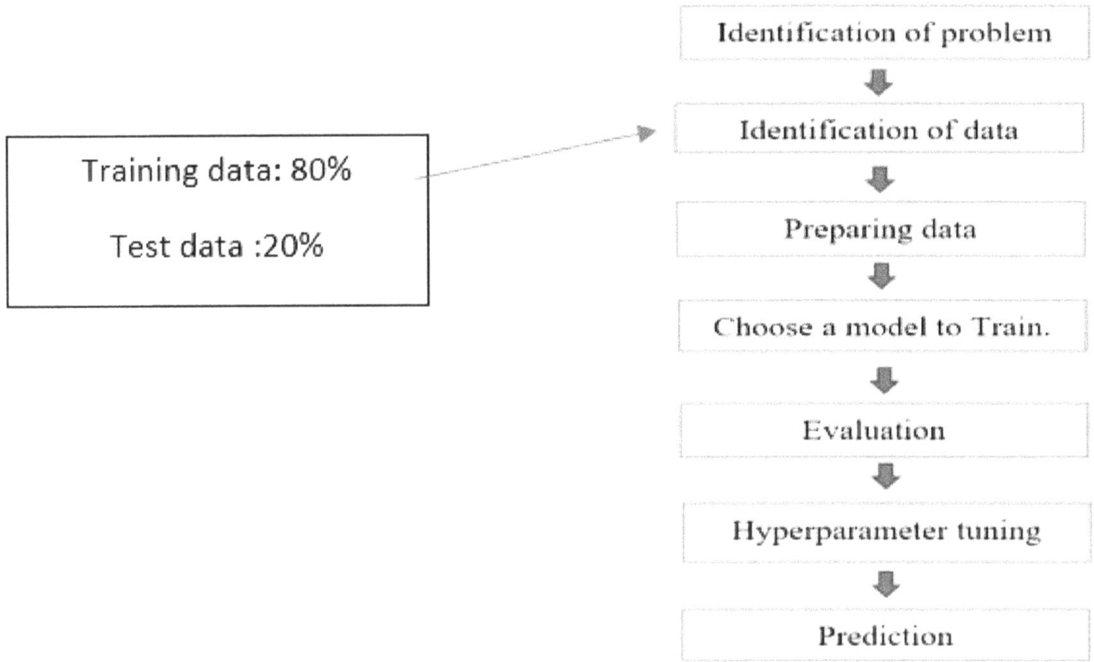

detection. Finding valuable insights from the data is made easier with the aid of unsupervised learning. Once the appropriate method has been applied, the algorithm splits the data objects into groups based on how similar and different the objects are.

Reinforcement Learning

Reinforcement learning in machine learning in addition to supervised learning and unsupervised learning is a type which deals with the systematic learning of a certain content. In reinforcement learning, the algorithms learn by hit and trial method, the machine interact with the environment to collect the experience from the learning system (Hammoudeh, 2018). Reinforcement learning is studied in many disciplines, such as game theory, control theory, operations research, information theory, simulation-based optimization, statistics etc. Algorithms of reinforcement learning includes Q-learning, SARSA, DQN, etc. Such methods are mainly used to solve the problems of robots.

Classification Algorithms

On the basis of training data, the Classification algorithm is a Supervised Learning technique that is used to categorise new observations. In classification, a programme makes use of the dataset or observations that are provided to learn how to categorise fresh observations into various classes or groups (Sarker, 2021). These algorithms classify the relevant variable into various groups before predicting the group for a specific input.

Naive Bayes: The Naive Bayes algorithm is based on the Bayes theorem and adopts a probabilistic approach. For each class, the algorithm has a set of prior probabilities. The computer changes these probabilities after receiving new data to create a concept known as posterior probability. This is helpful when determining whether or not the input corresponds to a list of classes.

Decision Tree: The decision tree algorithm is more like a flowchart, with nodes denoting the test on an input property and branches denoting the test result.

Random Forest: A random forest functions similarly to a bunch of trees. Different decision trees are used to process the segmented input data set. All decision tree outputs are averaged out and taken into account. When compared to the Decision tree technique, Random forests offers a classifier that is more accurate.

Support Vector Machines: SVM is an algorithm that classifies data using a hyperplane, making sure that the distance between the hyperplane and support vectors is maximum.

K Nearest Neighbours: To forecast the class of a new sample data point, the KNN algorithm employs a large collection of data points that have been divided into classes. Because it is very brief in comparison to other algorithms, it is known as the "lazy learning algorithm".

Regression Algorithms

Popular supervised machine learning methods include regression algorithms. Based on the input data points supplied into the learning system, regression algorithms can forecast the output values. Regression algorithms' primary uses include forecasting the weather and stock market prices. The most common algorithms under this section are:

Linear regression: By taking into account the reliable variables, it is used to gauge sincere qualities. Although it is the most straightforward regression procedure, it can only be used when there is a linear relationship or an issue that can be separated linearly. The best-fit line or regression line is a line drawn by the algorithm between data points and is used to forecast new values.

Lasso Regression: Finding the subset of predictors that minimises prediction error for a response variable is how the Lasso regression algorithm operates. This is accomplished by placing a restriction on the data points and allowing some of them to decrease in value to zero.

Logistic Regression: Binary classification is the main application of logistic regression. With the use of this technique, you can examine a group of variables and forecast a categorical result. Its main uses include estimating client lifetime value and home valuations, among other things.

Multivariate Regression: In situations where there are multiple predictor variables, this algorithm must be applied. This algorithm is widely used in product recommendation engines for the retail industry, where customers' preferred products depend on a variety of parameters, including brand, quality, price, and reviews.

Multiple Regression Algorithm: When using many explanatory variables as inputs, the Multiple Regression Algorithm combines linear regression and non-linear regression algorithms. The main uses include social science research, determining the validity of insurance claims, behavioural analysis, etc.

Clustering Algorithms

The process of grouping and arranging the data points according to commonalities among group members is known as clustering (Sarker, 2021). A component of unsupervised learning is this. Grouping like objects together is the main goal. For instance, based on certain attributes of the transaction, it can group all fraudulent transactions together. The most popular clustering algorithms are listed below:

K-Means Clustering: The unsupervised learning algorithm is the simplest here. The method groups together data points that are comparable and then clusters them. By determining the centroid of the group of data points and comparing each data point's distance to the centroid of the cluster, the clustering is completed. The examined data point is then assigned to the closest cluster based on distance. K-means refers to the number of clusters into which the data points are being divided.

Fuzzy C-means Algorithm: The FCM algorithm uses probabilities. Every piece of data is thought to have a chance of fitting into another cluster. Because data points lack an absolute membership over a certain cluster, the approach is known as fuzzy.

Expectation-Maximisation (EM) Algorithm: Its foundation is the Gaussian distribution that we studied in statistics. To address the issue, data are shown using a Gaussian distribution model. A point sample is calculated using expectation and maximisation equations after a probability has been assigned.

Hierarchical Clustering Algorithm: These algorithms learn the data points and identify resemblance patterns before sorting clusters in a hierarchical manner. It can be of two types

- Divisive clustering, for a top-down approach
- Agglomerative clustering, for a bottom-up approach

Smart Systems and Its Need

Smart systems offer technical innovations that have the potential to considerably increase the quality, dependability, and economic effectiveness of technical products. This is accomplished by locally integrating analogue or digital signal processing, greatly enhancing the functionality of a device or system (Sarker, 2022). The creation of a smart system typically requires knowledge from several different domains. The locally incorporated analogue or digital signal processing is the primary characteristic of a smart system that sets it apart from others. Applications such as robotics, auto industry, smart homes have severe limitations for dimension, necessitating increased miniaturisation of smart systems. In order to exchange information, the elements must be able to interact between the components of the system depends on communication.

Embedded Knowledge must be recorded on both information, environment and human experience. Depending on the requirements of the system, knowledge can be regarded in the context of smart systems in a variety of ways. An inference engine may employ a knowledge base created by human specialists, which is typical in Knowledge-based Systems (KBS).

Learning comprises the application of diverse methods and algorithms that allow to modify the knowledge of the system. Learning is accomplished by the system in an autonomous manner, updating its knowledge with minimal or any external intervention. Smart Systems must possess the ability to reason in order to make judgments using their information. a variety of methods, including expert systems, fuzzy logic, evolutionary algorithms, neural networks, etc. The system can predict how the environment will change in the future via reasoning, enabling it to take action before the changes become visible.

Smart Systems are designed to help people or to offer the required resources and procedures to enhance various facts of human life. Since they are frequently designed to collect data from users via sensors.

Different Domains

Smart City: The concept is of building an urban area with high economic sustainability and a good standard of living for its citizens. The idea behind smart cities is the interconnectedness of all possible city components, including residents and businesses, and their interface with daily usable devices such sensors, computer networks, and algorithmic decision-making. Smart Cities have a variety of application areas, including governance, energy, construction, and mobility. The concept is currently accompanied by a number of technological advancements, including 5G technologies, sensors, the Internet of Things, robotics, virtual and augmented reality, and artificial intelligence (Romero et al., 2020).

Smart Homes: These are based on the communication network that allows to connect and remotely control or monitor devices and services (Dayal et al., 2023). The devices used for Smart Homes includes air conditioning, ventilation, lightning, hardware, sound system and security systems. These devices are able to interact with tablets, mobile phones, etc. The objective behind the concept is to improve the lives of people through time assistance, guided by automated activities such as sensing, anticipating, and responding to events within a house.

Smart Industry: The concept underlines the smart factories, which are flexible, physical manufacturing facilities, security, that offer clients personalised services and operate on the basis of the usage of Big Data. It refers to a network of Smart factories that help in changing and encourages the involvement of researchers, developers, suppliers, distributors, and end users. The system is made up of people, information, organisations, and technologies that work together to achieve common goals are known as smart service systems.

Role of IoT and Cloud Computing in Smart Systems

Smart City Development

Outdoor Parking Management Services. The limited parking spaces that are available for a hypothetical smart city are managed by an outside parking management service. In order to accomplish this, ferromagnetic wireless sensors are deployed in every parking space at locations with high pedestrian traffic, such as airports, train stations, courthouses, etc. The city's integrated information center stores the data that these sensors collect, and the end user can access it via a smartphone application or through public displays that are located at each parking lot's entry gate and indicate the number of available parking spaces in various locations. Existing automated multilevel underground and above-ground parking systems are a good place to further integrate this technology.

Garden Irrigation Monitoring System

The garden irrigation monitoring system (Alouffi et al., 2021) gives park and garden administrators real-time information about the state of parks (Ramya, Palaniappan, George, 2012). To assess soil moisture and other pertinent characteristics, IoT devices should install subterranean sensors. When an abnormal condition is found, this service may notify other users in addition to submitting data to the appropriate authority. This service will lower the cost of energy, water, and labour to maintain the park.

Figure 2. Monitoring System

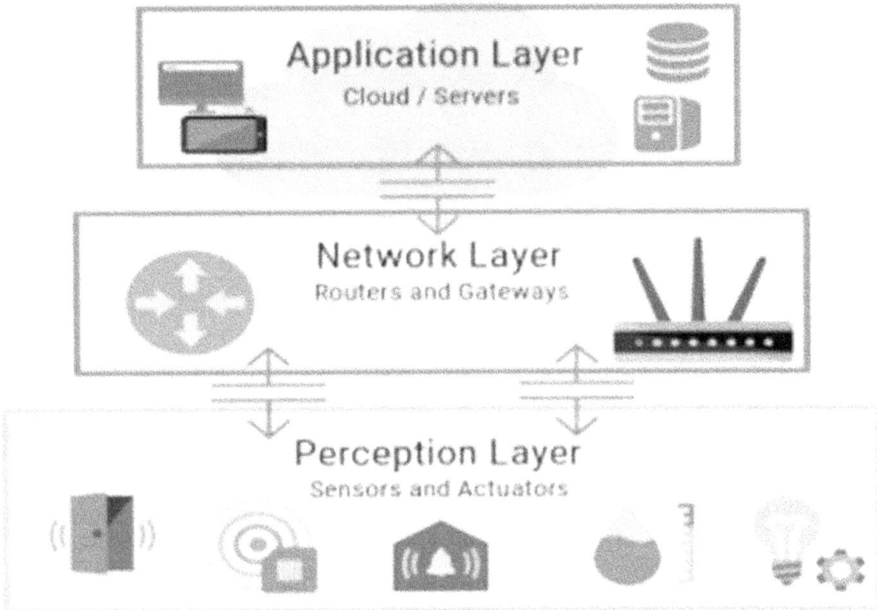

Traffic Control

Intelligent cities have become a reality in contemporary civilization thanks to the use of wireless networks, the internet of things, integrated communications, cloud computing, and other vital technologies (Kim, Evans, & Iversen, 2008). Because of the sharp rise in traffic, fuel prices, and pollution, research societies and business groups have begun to pay attention to improving urban traffic management systems in intelligent cities. Even while urban traffic management systems have benefited much from developing technologies, they are still excessively inefficient. In the era of cloud computing, analysis of many massive data types gathered from the internet of things, connected cars, and sensors/intelligent devices can open the door to offering beneficial services. Some well-known programs (like Google MAP) can facilitate loading and Providing conditions to observe the route traffic level by sharing images, videos, comments, and points. Intelligent transportation systems based on cloud computing can face the challenge of achieving an efficient intelligence that responds to motion and has high error tolerance and scalability. Intelligent transportation systems (ITS) enable vehicles to be intelligent. Research and production of intelligent cars (also known as driverless cars) began in the mid-1980s. Globally, research centers, vehicle suppliers, and universities are working together to design and manufacture intelligent vehicles (Kim, Evans, & Iversen, 2008). Intelligent cars have developed significantly over the past decade as a result of technological advancements in intelligent transportation systems. Traffic management and traffic safety can be improved with the help of intelligent transportation systems (ITS). Sensors and IoT will collect data in this system, which will be processed at a central location. A data analysis is performed in this center, and then the results are sent to an internet-connected device (Neto et al., 2018).

A remote sensor collects IoT data and sends it to a cloud environment, where it is stored or processed. In IoT services and products, information is generated from IoT data. When computing resources are

Figure 3. IoT in Transportation

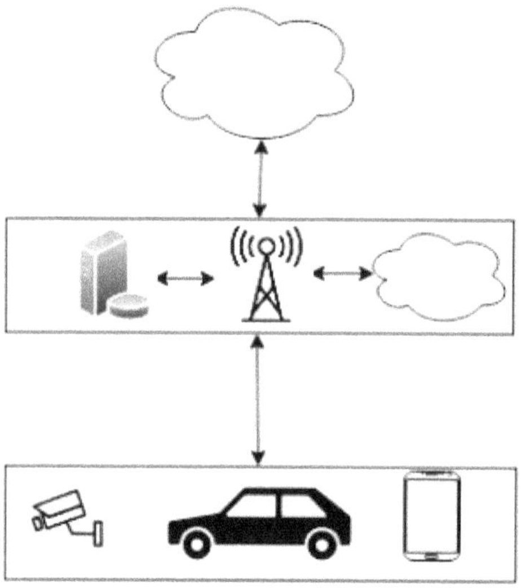

to be utilized, they must be provided at the location where the devices are located instead of at remote geographic locations in a cloud data center. As a result of edge computing, less data has to be transported back to the cloud. Furthermore, it reduces costs associated with data storage and analysis over metered networks. To avoid collisions, edge computing is needed for applications such as automatic breaking in smart cars, which cannot wait for cloud analysis and response. Smart cities must also optimize energy use and other resources, such as traffic patterns and parking spaces.

Participatory Sensing System

Data about the environment is gathered and stored by mobile phones, such as noise, temperature, and GPS coordinates. The information provides important information about an area underdeveloped and in need of attention. Applications focused on the ecosystem were built using IP multimedia subsystem (IMS), a platform based on Internet of Things. Short-range communications of smartphones like Bluetooth, RFID, and near-field communication (NFC) were the basic interest of IMS architecture. IoT was included in two layers, the application layer, and the perception layer (Ghosh, Charkaborty, & Law, 2018). Validation of the IoT-based IMS architecture was performed on the net field programmable gate array (Net FPGA) which is a low-cost open-source platform. By the use of same application of smartphone, user can subscribe special services from which he can get notification of specific kind of events like traffic jam, protest, car accident, etc., currently happening in city.

Environment Protection

Natural resources like water, air, lands and so on. Are getting either limited or very luxurious. It's miles crucial to combine answers to have better and extra controlled utilisation of these natural sources. Using technological structures which includes corporation useful resource planning (ERP), geographic records device (GIS) and IoT connectivity could be beneficial for green use of these sources. By way of the use of tracking structures, it will likely be simpler to mark waste factors and better distribute natural sources whilst controlling costs, and decreasing power and natural assets consumption. One of the essential factors of clever town programs is that they may be designed for interconnectivity between diverse IoT gadgets and facts collections which can also facilitate higher collaboration across packages and cloud services.

Air Quality Monitoring

Human illnesses and health disorders along with allergies are brought about in general via air pollution. IoT generation offers a one-of-a-type application for lowering the dangers posed by way of air pollutants. The smart air satisfactory monitoring answer collects vital statistics using sensor-based standards and transfers it to an interconnected gateway for green operation. It aids in processing facts for better evaluation and allows managers to screen air exceptional in actual-time. As a end result, IoT answers help firms in retaining required AQI degrees and monitoring air great in real-time. Any leakage or maybe the tiniest gas emission into the atmosphere can have catastrophic consequences. It may wreak havoc at the environment and create an surroundings imbalance as a way to cost a number of money to accurate. As a end result, using an IoT-powered fuel monitoring solution lowers the risk of catastrophic environmental consequences. Big-scale businesses use exceptional styles of gases to generate positive merchandise, and any carelessness may want to bring about devastating tragedies. This necessitates developing a complicated tracking system that detects gases in the environment and warns government to do so.

Environment Monitoring System

Poisonous gases normally go neglected by way of the human nose in enclosed areas like places of work, rooms, and lodges. As a result, adopting a sensor-based totally system might be quite beneficial in such instances. It monitors gases and offers notifications to the person's cellphone or smart device, letting them make knowledgeable selections. It aids in the right control of the surroundings and continues pleasant operating surroundings for the personnel. With its value-effective installation, IoT solutions for environmental tracking deliver a greater business capacity, and it is easy to improve profitability.

Citizen Security

IoT gadgets and service suppliers in smart houses, automobiles, health and entertainment are already go-connecting devices. Matters identification is needed to facilitate those services. Many straightforward IoT devices are hard or impossible to comfortable with, subsequently, things identification and authentication are crucial to IoT safety. Safety solutions want to identify and analysed matters linked to the community, and also proportion and observe wanted regulations.

Education

Smart education is an essential factor of clever metropolis offerings. In the long run, training may be the maximum important clever city carrier at all, for adults in addition to for children. As the sector unexpectedly globalizes, one of the handiest approaches to stay The position of big information, cloud computing, and IoT to make cities smarter 81 aggressive is to expand and hold to construct understanding-primarily based skills thru training. This consists of preliminary know-how, e.G., via school, vocational and university education in addition to lifelong getting to know. The role of colleges, schools, and universities is, consequently, a key element to bear in mind in the design of smart schooling answers.

Healthcare

Smarter healthcare control converts health-related information to e-health applications and offerings. Smarter towns are collaborating on their healthcare data to permit cozy communications and statistics sharing the usage of cloud-based technology. These e-records empower fitness professionals to enhance the productiveness of the service supplied to the point of contact of patients making use of IoT gadgets and services. Examples of clever healthcare consist of the availability of far-off opportunity diagnoses, faraway remedies or connectivity thru IoT services, online clinical services, asking for an appointment online or the possibility of having a virtual record via an electronic fitness management machine, far-off domestic offerings, alarm structures or maybe faraway affected person monitoring systems. Special interest and unique care are required to the persons with physical disabilities. Consequently, suitable support. Nature devices are required to those individuals. Nature of IoT devices needs to be very specific in keeping with the requirement. In a maximum of conditions bodily, disabled people are dependent on others. Therefore, SC ought to provide a special look after to men and women, and a probable IoT-primarily based architecture.

Water conservation

Water is a vital natural useful resource for agriculture, but it's far limited. A big part of water in a rustic like India is wanted for irrigation (Nawandar and Satpute, 2019). Crop irrigation is a extensive factor in influencing crop productiveness, as it is laid low with a selection of environmental elements together with air temperature, soil temperature, humidity, and soil moisture. For harvesting fields, farmers rely closely on human supervision and enjoy. The sector's water deliver need to be maintained. Water scarcity is a prime challenge in these days's world. Humans all throughout the world are already laid low with such shortage.

Records mining strategies can substantially benefit the choice-making method for many agricultural duties. The utility of affiliation rule guidelines to manage the amount of water in agricultural regions is this sort of operations (Ray and Goswami, 2020). Further, wi-fi sensor networks have developed as a brand-new precision agriculture method. In present day irrigation systems, smart sensor networks are used to accumulate discipline values for most advantageous plant irrigation. In many industries, including agriculture, healthcare, and logistics, gadget mastering is a vital method. This text additionally makes use of a gadget getting to know-based system to lessen freshwater waste. Smart water meter based on internet of factors (IoT) and Cloud computing ready with device gaining knowledge of algorithms, to differentiate among normal and immoderate water utilization at commercial, home and all different sectors having an abundance of water utilization, both for Indian and worldwide context. Recognizing that sharp metering of water has the capacity to modify purchaser engagement of water usage in urban and rural water resources, this paper fosters sustainable water management, a need of the present. With shrinking reserves of smooth water assets international, its miles becoming cumbersome to cater for this aid to masses in the coming years on a steady foundation. Using our clever water meter, water resources may be controlled efficaciously, and a top-of-the-line use may want to shop for water for future generations. Sensors will provide for actual time monitoring of hydraulic records, automated management, and alarming from the Cloud platform in case of activities together with water leakages, excessive utilization, and many others. Evaluation of the same will assist in taking significant movements. Consequently, we do advocate for a smart water metering era that may be utilized by Indian citizens, and worldwide, to diminish the wastage of water (Kanjo et al., 2009). With the ease of tracking and visualization of the information through the Cloud platform blended with system learning primarily based gear to hit upon extra water consumption, the server-much less structure we advise can be without difficulty followed and implemented on a large scale.

Retail

IoT improvements have found several applications in retailing. The enterprise evaluation indicates that 37% of meals and grocery organizations as of now explore distinct avenues regarding IoT innovation or have successfully begun IoT services or products and similarly fifty-eight% are intending to grow their use of the innovation. IoT assists with working on each internal responsibility and customer-centric problem. IoT presents diverse purposeful blessings including personalization, dynamic pricing, stock tracking, tracking, and customized tips. Energy-savvy indoor regulators and lighting are likewise utilized in various retail frameworks. Sensors deliver real-time inventory records that are then utilized to cater to purchaser needs and upgrade inventories. IoT in addition improves tracking and manipulation through coding and monitoring stock. That enables businesses to turn out to be greater efficient, speed up work techniques, decline mistakes, and avoid threats. The actual-time information given by means of sensors lets stores to make better purposeful as well as monetary decisions. IoT-enabled clever labels give possibilities to assist product identification in addition to offering vital records. This fact may be included with personalization and advice frameworks to enhance buying revel in with unavoidable immersive shows and smart items. IoT also offers ways to engage customers for the duration of the product lifecycle. For example, clever items can communicate with clever telephones to interpret biometric records. The lifestyles-cycle assist calls for a first-class IoT layout ensuring effective and at ease statistics dealing with technology. The eventual destiny of purchasing will incorporate these and other new thoughts. Consumer engagement as a factor of advanced exchange will hold on riding greater important personalization and simplification of retail endeavors. However, for this load of expectancies, in truth, this future is as of now here. Customers are all around associated and becoming tech-savvy. Therefore, the undertaking for outlets to conquer is making a powerful IoT atmosphere that adjusts and remains privy to the shoppers' needs and options (Hossain et al., 2021). For stores that embrace IoT by means of making use of superior, cell, sensor-empowered, and social media structures powered through an IoT environment, what is to come is these days.

Industry

It is far possible to monitor folks and manage utilities inside a corporation. Wi-fi sensors may be utilized by organizations for safety automation and weather management (Romero et al., 2020). Intelligence video surveillance will also generate masses of statistics. Cloud-based totally answers to fulfill the requirement of garage and processing become relevant in this instance. It will be possible to supply surveillance records over the cloud with the use of the elastic, on-demand benefits of cloud computing. Because the 'matters' growth in IoT, so additionally can be the call for cloud computing sources. It may every so often be difficult to know how great a deal resources are wanted for precise packages of IoT. Billions of items are predicted to be connected to the cloud thru IoT and this could in all likelihood make the garage on the cloud overstretched.

CHALLENGES AND OPPORTUNITIES

In current scenario AI is gaining popularity in manufacturing industries, healthcare and more. With respect to smart system AI presents some transparency challenges like Lack of explain ability of ML algorithms, imbalances in access to information, lack of knowledge of interacting with AI or human Agent and lack of transparency about the full range of intended and expected outcomes of AI development (Lee and Park, 2019). Apart from these challenges mentioned above AI is facing more challenges for smart system development. These challenges are

1. Finding the Data Set: AI system needs a good quality of data set. Implementation of AI algorithm should use the right set of data. It is necessary to determine the right data set for the AI algorithms. For determining the good data set, we may have to get in touch with the experts.
2. Data Security and Storage: AI application uses the data for learning and making decisions, so the data security is main challenge of smart systems. Data storage is also required for business operations and for automated application data storage is challenging task.
3. Infrastructure: Replacing the infrastructure with respect to smart systems infrastructure and their cost is challenging issues. These Smart Systems have high level of computational speed. This type of infrastructure is needed for high-speed smart systems.
4. Integration of AI into Existing Systems: Integration of AI into existing systems is again a challenging task because it includes various software and hardware components. To implement AI system into existing systems requires expertise and experience.
5. Algorithms and Training of AI models: The performance of smart system relies on the AI algorithms. After implementing the AI and ML models into the system it requires expert manpower to training ML or AI models which is again a challenging task for an organization.
6. Hardware installation and maintenance costs: For a smart system hardware and maintenance cost is high compare to the normal system, because smart system use cameras and sensors. These devices need regular maintenance. So for smart system we have to analyse the proper setup in term of future cost and how to optimize that cost. Natural damage and theft is also a major challenge for smart systems. It is necessary to make such kind of setup so that they can protect from theft and natural damage.

REFERENCES

Alouffi, B., Hasnain, M., Alharbi, A., Alosaimi, W., Alyami, H., & Ayaz, M. (2021). A systematic literature review on cloud computing security: Threats and mitigation strategies. *IEEE Access: Practical Innovations, Open Solutions*, *9*, 57792–57807. doi:10.1109/ACCESS.2021.3073203

Dayal, M., Chawla, A., Khari, M., & Mahajan, A. N. (2023). Artificial Intelligence-Based Smart Packet Filter. In *Proceedings of Third International Conference on Computing, Communications, and Cyber-Security* (pp. 791-801). Springer, Singapore.

Ghosh, A., Chakraborty, D., & Law, A. (2018). Artificial intelligence in Internet of things. *CAAI Transactions on Intelligence Technology*, *3*(4), 208–218. doi:10.1049/trit.2018.1008

Haddadin, S., & Knobbe, D. (2020). *Robotics and Artificial Intelligence—The Present and Future Visions. Algorithms and Law*. Cambridge University Press.

Hammoudeh, A. (2018). *A concise introduction to reinforcement learning*. Princess Suamaya University for Technology.

Hossain, M. S., Chisty, N. M. A., Hargrove, D. L., & Amin, R. (2021). Role of Internet of Things (IoT) in Retail Business and Enabling Smart Retailing Experiences. *Asian Business Review*, *11*(2), 75–80. doi:10.18034/abr.v11i2.579

Jin, W. (2020, May). Research on machine learning and its algorithms and development. *Journal of Physics: Conference Series*, *1544*(1), 012003. doi:10.1088/1742-6596/1544/1/012003

Kanjo, E., Bacon, J., Roberts, D., & Landshoff, P. (2009). MobSens: Making smart phones smarter. *IEEE Pervasive Computing*, *8*(4), 50–57. doi:10.1109/MPRV.2009.79

Kim, Y., Evans, R. G., & Iversen, W. M. (2008). Remote sensing and control of an irrigation system using a distributed wireless sensor network. *IEEE Transactions on Instrumentation and Measurement*, *57*(7), 1379–1387. doi:10.1109/TIM.2008.917198

Lee, D., & Park, J. H. (2019). Future trends of AI-based smart systems and services: Challenges, opportunities, and solutions. *Journal of Information Processing Systems*, *15*(4), 717–723.

Mahesh, B. (2020). Machine learning algorithms-a review. *International Journal of Science and Research (IJSR).[Internet]*, *9*, 381-386.

Nawandar, N. K., & Satpute, V. R. (2019). IoT based low cost and intelligent module for smart irrigation system. *Computers and Electronics in Agriculture*, *162*, 979–990. doi:10.1016/j.compag.2019.05.027

Neto, A. J., Zhao, Z., Rodrigues, J. J., Camboim, H. B., & Braun, T. (2018). Fog-based crime-assistance in smart iot transportation system. *IEEE Access: Practical Innovations, Open Solutions*, *6*, 11101–11111. doi:10.1109/ACCESS.2018.2803439

Ramya, V., Palaniappan, B., & George, B. (2012). Embedded system for automatic irrigation of cardamom field using Xbee-PRO technology. *International Journal of Computers and Applications*, *53*(14).

Ray, A., & Goswami, S. (2020, February). IoT and cloud computing based smart water metering system. In *2020 International Conference on Power Electronics & IoT Applications in Renewable Energy and its Control (PARC)* (pp. 308-313). IEEE. 10.1109/PARC49193.2020.236616

Romero, M., Guédria, W., Panetto, H., & Barafort, B. (2020). Towards a characterisation of smart systems: A systematic literature review. *Computers in Industry*, *120*, 103224. doi:10.1016/j.compind.2020.103224

Sarker, I. H. (2021). Machine learning: Algorithms, real-world applications and research directions. *SN Computer Science*, *2*(3), 1–21. doi:10.100742979-021-00592-x PMID:33778771

Sarker, I. H. (2022). Ai-based modeling: Techniques, applications and research issues towards automation, intelligent and smart systems. *SN Computer Science*, *3*(2), 1–20. doi:10.100742979-022-01043-x PMID:35194580

Chapter 16
5G Networks and Their Applications:
Ushering in a New Era

Mohit Dayal
Dr. Akhilesh Das Gupta Institute of Technology and Management, India

Kamal Upreti
Dr. Akhilesh Das Gupta Institute of Technology and Management, India

Dhananjay Singh Panwar
Dr. Akhilesh Das Gupta Institute of Technology and Management, India

Saijal Gupta
Dr. Akhilesh Das Gupta Institute of Technology and Management, India

Derick Phillips D'souza
Dr. Akhilesh Das Gupta Institute of Technology and Management, India

Rupender Duggal
Dr. Akhilesh Das Gupta Institute of Technology and Management, India

Gourav Sharma
Dr. Akhilesh Das Gupta Institute of Technology and Management, India

ABSTRACT

Developed by 3GPP, 5G is the technology standard for cellular broadband networks. Introduced in July of 2016 and commercialized almost 3 years later, it is seen as the successor to 4G networks. 5G uses device-to-device communication, dividing the service area into small geographical locations called cells. These cells are connected intensely through localized antennas. The connectivity of these antennas, despite being small, amounts to high-speed downloads ranging up to 10 Gbps. This chapter includes a brief introduction of 5G network; concise notes on the history of its predecessors such as 2G, 3G, and 4G; as well as a thorough look through the architecture of the 5G network model.

DOI: 10.4018/978-1-6684-6413-7.ch016

Copyright © 2023, IGI Global. Copying or distributing in print or electronic forms without written permission of IGI Global is prohibited.

BACKGROUND

By transmitting the letter "S"(. . .) over a distance of 3 km in the form of three dots of Morse code with the use of electromagnetic waves, Italian inventor G. Marconi opens the way for modern wireless communications. Following its introduction, wireless communications have grown to constitute a significant aspect of modern civilization (Wu et al., 2017). The way society functions has changed as a result of wireless communications, which have moved from satellite communication, television, and radio transmission to the ubiquitous mobile telephone. The 1G and 2G technologies uses circuit switching whereas 2.5G and 3G uses both packet and circuit switching and the next generation from 3.5G to 5G uses packet switching. Below is a summary of wireless technologies:

1. 1G
   This is basically the First Generation initiated in 1980's with a data rate up to 2.4 Kbps. It supports voice calls only. Voice calls were more vulnerable to unauthorized third-party eavesdropping because they were recorded and played in radio towers.
2. 2G
   This is the major upgrade in the cell phone industry. As introduced in late 1990's this took cell phone from analog to digital communications. With a data rate of up to 64kbps and was mostly utilized for voice transmission. It also offers e-mail and the Short Message Service (SMS) (Chochliouros et al., 2017). The three most important and prominent technologies were Global System for Mobile communications (GSM), Code Division Multiple Access (CDMA) and IS-95.
3. 2.5G and 2.75G
   The main and prominent technologies which firstly enabled data transfer with use of GPRS (General Packet Radio Service) and EDGE (Enhanced Data rate for GSM Evolution). Although it uses the 2G technology framework, packet switching and circuit switching are also used. With increase speed and slow data transmission possible with this. It can assist upto 144Kbps.
4. 3G
   The 3rd Generation systems combine Internet Protocol (IP) based services with high-speed mobile connectivity. With transmission rate up to 2 Mbps the era of Video calling and mobile internet access began. The inclusion off extra features like International Roaming and enhanced voice quality makes 3G a remarkable generation. With evolving technologies such as High-Speed Uplink/Downlink Packet Access (HSUPA/HSDPA) a bridge between 3G and 4G known as 3.5G emerge with improved data rate of 5-30Mbps.
5. 4G
   It is generally referred to as the offspring of 2G and 3G standards. With Long Term Evolution (LTE) Advanced and Mobile Worldwide Interoperability for Microwave Access (WIMAX) are being standardized by the 3GPP as the upcoming 4G standards. With inclusion of more amenities like Voice, Internet and multimedia available for the users everywhere and every time basis and at significantly higher data-rates than pervious.
6. 5G
   With user demand increasing exponentially, 4G will soon be superseded by 5G with an advanced access technology BDMA (Beam Division Multiple Access) and FBMC (Filter Bank Multi Carrier) multiple access. By examining the scenario of the base station interacting with the

mobile stations, the BDMA technique's theoretical underpinnings are clarified (Le et al., 2016). Each mobile station in this communication is given an orthogonal beam, and the BDMA technique divides that antenna beam based on the positions of the mobile stations to allow for numerous accesses to the mobile stations, which accordingly boosts the system's capacity. The recently introduced IEEE 802.11ac, 802.11ad and 802.11af standards are very beneficial and serve as a foundation for 5G.

Core Architecture

According to 3GPP specifications, the 5G core makes use of a cloud-aligned, service-based architecture (SBA) that covers all 5G interactions and functions, including session management, authentication, security, and traffic aggregation from end devices. The MEC infrastructure, which is crucial to 5G architectural principles, is used to deploy virtualized software functions, which is further highlighted in the 5G core (Khodashenas et al., 2016).

The core of 5G is a network of connected services and is shown in figure below:

Figure 1. Cisco Diagram for 5G core architecture

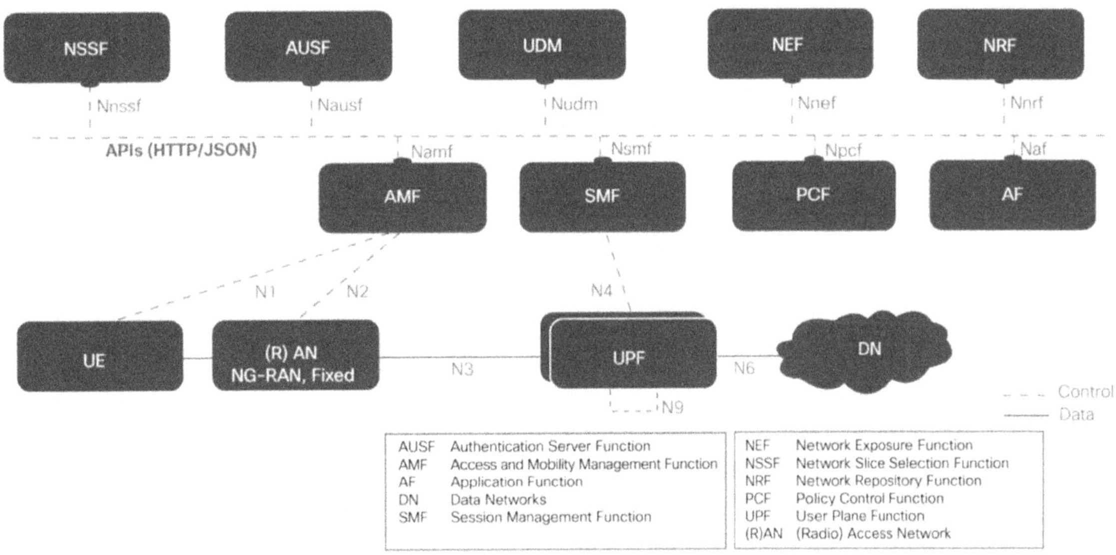

The Function of each node of the above network

1. AUSF

In a home network, the Authentication Server Function (AUSF) performs authentication with a UE. When 5G-AKA or EAP-AKA is used, it decides on UE authentication but depends on a backend service to compute the authentication data and keying materials. It implements the EAP (Extensible Authentication Protocol) Authentication Server and also stores the keys.

2. AMF

The AMF performs most of the tasks that the MME performs in a 4G network. One of the 5G core network's control plane network functions (NF) is the Access and Mobility Management Function (AMF) (Kumar et al., 2021). The 5G AMF is a development of the 4G MME, maintaining the separation of the User Plane and Control Plane while adding additional simplifications like relocating the Sessions Management functions to the SMF and supplying standard SBA interfaces.

- It Authenticates UE
- Connection Management
- Registration Management
- Mobility Management
- Reachability management
- NAS (Non-access stratum) signaling
- Terminates the RAN (Radio Access Network) CP interface (N2)

3. SMF

It performs the task of managing session functions which are handled by 4G MME, PGW-C and SGW-C. Its main duties include interacting with the decoupled data plane, initiating, updating, and terminating Protocol Data Unit (PDU) sessions, and managing session context with UPF (User Plane Function) (Lal et al., 2021). One SMF instance can support all or a portion of its functionalities. The SMF is capable of supporting the following functions, per 3GPP TS 23.501.

- Allocates the IP Address to UE
- Non access stratum (NAS) Signaling for SM (Session Management)
- Sets up the UPF's Traffic Steering to send traffic to the appropriate Data Network (DN).
- Serves as the Interface for communication offering to the user plane services.
- Downlink Data Notification

4. PCF

It performs the same function as the PCRF (Policy and Charging Rules Functions) in the 4G Network. PCF performs the following Functions:

- Provides policy guidelines for control plane operations. This covers network slicing, roaming, and mobility management.
- Supports the 5G QoS policy and charging control functions
- Accesses subscription data to inform UDR policy decisions

5. UPF

The data plane components of the SGW and PGW have basically been combined to create the UPF. Considering the CUPS (Control and User Plane Separation) architecture

EPC SGW-U + EPCC PGW-U ® 5G UPF

The UPF performs the following functions:

- Maintains and reports traffic
- Packet forwarding and routing
- For packet inspection and classification, the UPF may integrate a DPI (Deep Packet Inspection) Optionally
- Integrate Firewall and NAT (Network Address Translation) Functions

6. AF

   Application Function performs the same task as the EPC AF. It enables access to NEF (Network Exposure Function), interaction with a policy framework for policy control, and application influence over traffic routing.

7. UDM

   It performs the task as the 4G HSS function. The long-term security credentials required for AKA authentication are stored here to support the Authentication Credential Repository and Processing Function (ARPF). It also keeps track of subscription data.

8. NEF

   NEF offers a method for securely exposing 5G core functionalities and services (Gupta & Jha, 2015). It facilitates robust, secure, developer friendly access to the capabilities of 5G network, giving enterprises and third party developers the ability to design and tailor their own on-demand composite or specialized network services. The NEP performs the following functions:

   - Secure transmission of data to the 3GPP network from an external application
   - Translation of information
   - PFD (Packet Flow Description) Management. PFD consists of tuple of protocol, port number and server-side IP

9. NRF

   It supports Breakthrough mechanisms that let 5G elements find one another and obtain the most recent information about the elements they're interested in. The NRF carries out the following duties:

   - Keeps track of the profiles of the NF instances that are available and the services they support
   - Enable 5G core network's consumer NF occurrences to find other providers' NF instances
   - Network Functions can discover one another through the service registration and discovery functions.

10. NSSF

    AMF is set to serve the UE by the Network Slice Selection Function (NSSF), which also chooses the network slicing instance (NSI), establishes the permitted NSSAI (Network Slice Selection Assistance Information), and selects the NSSAI (Mahmood et al., 2022). As part of the UE initial registration and PDU establishment procedure, the AMF can retrieve the NRF, NSI ID, and target AMFs. The NSSF carries out the following duties:

    - Establishing the permitted NSSAI
    - Deciding on the AMF set to use in order to serve the UE
    - Selecting the instances of Network Slice to serve UE

Network slicing is Fundamental to 5G, and a number of new terms have been introduced during the standards development process. These include Slice/ Service Type (SST), Network Slice Template (NST), Network Slice Selection Assistance Information (NSSAI) and Network Slice Instance (NSI).

SECURITY

By heavily Utilization of cloud-based resources, virtualization, network slicing, and other cutting-edge technologies, 5G implementation will result in outstanding performance advantages and a variety of applications. These modifications expose more "attack surfaces" within the 5G security architecture and introduce new security risks. For vertical applications, communication encryption and integrity protection are essential. However, service dependability and availability are equally crucial for protecting device-controller sessions. In this way, the service-based architecture of 5G improves network resilience through redundancy, isolation, and duplication at RAN and network functions. Furthermore, conventional over the top security services cannot afford the delays that delay-sensitive URLLC traffic introduces in resource-constrained scenarios. Thus, it is crucial to take into account alternative strategies, such as physical layer security and intrusion detection systems for critical infrastructures. Network domain security, Network access security, User domain security, Service Based Architecture, Application domain security, Visibility and configurability of security and privacy are the main domains where security is required and more vulnerable.

Application Programming Interfaces (APIs) are used by the 5G services-based architecture to facilitate communication between service functions and to improve configuration, infrastructure deployment and management. Poorly coded or insecure APIs put the entire 5G network at risk by making core services vulnerable to attack.

With the advent of 5G, advanced enterprise services will replace conventional consumer smart phones. Greater workforce mobility, automation, and a plethora of new applications will be made possible by the widespread adoption of 5G in enterprise, industrial, and IoT use cases. The integration of 5G into these

Figure 2. 5G Security

environments necessitates a greater degree of end-user network and 5G service interface integration, posing new risks. In the end, an attack that successfully disrupts the network, steals or compromises the integrity of confidential data, or both, could have a much more significant economic and societal impact than in previous generations. Because they frequently rely on outdated technology and have inadequate cybersecurity protection, industrial and IoT devices, for instance, can be especially vulnerable. Attacks on exposed devices have a real-world impact on crucial services that rely on the 5G network. On the other hand, a sizable number of compromised IoT or industrial devices can be used to launch attacks against the 5G infrastructure and interfere with services for all users.

Edge computing is used in 5G to bring performance-sensitive applications close to end users in order to enable ultra-reliable, high-speed and low latency services. In practice, operators can split user-facing applications across virtualized data centers near mobile radio access nodes, inside enterprises customer data centers, and third-party cloud environments (Wijethilaka & Liyanage, 2021). This decentralized strategy has the potential to muddy operational lines and complicate security administration. Additionally, edge computing creates opportunities for attacks against 5G infrastructure-integrated application services and may open up new paths for attackers to direct their threats toward core 5G services.

Applications of 5G

The massive spike in cellular network that is caused due to the introduction of 5G networks is equally compensated by its merits. The objective of this paper is to not only assess the history and architecture of this impressive feat of technology, but also to extensively describe its applications (Al-Falahy & Alani, 2017). These applications prove that 5G technology has undeniably shattered expectations in not only date to date lives of its customers in the cellular network sector, but also several major industries as well. The magnitude of advancement and its uses are still being researched. 5G networks unfurl to us the endless possibilities as it introduces our latency to near real-time rendering several technologies that were supposedly state of the art, completely moot.

There is absolutely no doubt that the thousand-fold increase in the number of connected devices into this new architecture might face some challenges. Introduction of 5G to India will be the perfect test of the same fact. It is predicted to tackle these challenges by increasing the capacity, maintaining top-notch energy efficiency, cost and spectrum utilization as well. It also aims to improve scalability in order to help overcome the possible obstacles the immense traffic may cause.

The overall technical aim of this network system is to:

- Increase the data volume per area.
- Increase number of connected devices.
- Near real-time experience, reduced latency.
- Uniform data rate.
- A stable foundation for emerging technologies.

To progress into the use cases of the subject matter, we will now dive into some of the solutions that are proposed in the 5G network (Buzzi et al., 2016). These solutions are proposed so that the agenda, or to be more precise, the predicted goals of this new technology is reached as smoothly as possible.

Some of the solutions that are included into the 5G network are:

- Device-to-Device communication: Device-to-Device communication refers to the communication in cellular networks in which direct communication is established between two mobile users. The mobile users communicate without traversing through the base station or the core network. It was generally used for local networks, emergency communications and even the enhancement of Internet of Things (IoT) industry.
- Massive machine communication: Massive machine communication is the type of communication between machines over any wired or wireless network with minimal intervention from humans to facilitate this or promote data transfer. Generally classified under Machine type communication, massive machine communication deals particularly with wireless data transfers between a multitude of machines connected over a network. This type of communication is predicted to be the corner stone in the evolution of Internet of Things (IoT) to Internet of everything.
- Ultra-dense networks: Ultra-dense networks are networks which has cells that facilitate communication at a higher magnitude than the demand by its users. Ultra-dense networks are stable as they harder to overwhelm by the user traffic. They provide multiple access points marginally more than the density of active users.

After carefully deliberating through the technical aims and solutions that 5G network are predicted to provide, it's time we discuss the use cases of the subject matter. The real-world applications of this emerging technology, despite being at its younger stages, is already hypothesized and in certain instances proved to be the seed that certain industrial sectors needed to witness booming progress (Bonati et al., 2020). Several of these use cases include emerging sectors in not just the communication and IT sector, but as well as security, authentication, health, and safety as well. Some use cases of 5G include:

- Automobiles: Initially deployed with 4G, several automobile manufacturers have turned to 5G to improve latency rate. This technology developed will allow automobiles to communicate with other machines and infrastructure as well. Even leading automobile manufacturers like Tesla have implemented the 5G network technology into their self-driving cars.
- Public safety: Mission critical Push-to-Talk (MCPTT) is a product functionality enabled in most service operable devices in case of emergencies and public safety concerns. It allows reliable, low-latency calls that are one-to-one connected through device-to-device communication. Allowing nation-wide coverage, it makes sure that communication is thoroughly maintained during an emergency which can be crucial and at times the difference between life and death in some scenarios.
- Wireless video transmission for broadcast applications: Initiated in the 2020s, 5G broadcast have been tried out by several broadcasting companies including Sony (Azzaoui et al., 2020). This technology is being hypothesized to maintain a strong connection with television sets, but also mobile phones, computers, laptops and other devices that allow a connection. Moreover, this facilitates communication without consuming any data themselves and being easier to authenticate over any network.
- Internet of Things (IoT) Revolution: 5G even as an emergent technology predicted the advancement of IoT and its immense array of applications in the real-world. This includes IoT facilitated smart cities and traffic management, machine-to-machine communication in industries, or even drones (Liu et al., 2020). 5G has been the foundation that the IoT industry needed to imply on several levels. The ultra-reliable and top-notch network security architecture as well as high-speed data transfer rates that this technology provides is the solution to all of this industry's problems.

- Virtual Reality (VR) and Augmented Reality (AR): As not just a part of immersive gaming experiences, this sector has been able to facilitate our growth in digitising our realities at will. Augmented Reality is an up-and-coming field of the IT sector heavily researched for its implications by our tech giants. The use cases of this sector vary from navigation, communication, entertainment and so much more. 5G networks have been supporting the progress of this field providing minimal latency and the high-speed data it requires for its success.

These were just some of the examples of use cases the introduction of 5G networks has shown (Zhang et al., 2019). It has already impacted several industries positively such as healthcare, retail, manufacturing, agriculture, logistics, etc. It can be said with absolute faith that the use cases of this technology will only shine brighter in the coming years.

CONCLUSION

5G is an emerging technology which merits several industries by its introduction. Its merits not only help communication between humans, but also between machines and inanimate objects which can be made smarter. The possibility of self-driven cars, smart-homes, smarter traffic management systems, better navigation are just some of the vast possibilities this new technology has given rise to. The objective of this paper is to briefly educate the reader on 5G networks, keeping in mind the chronological sequence in which this major advancement in the field of communication has been made possible. It introduces to us the objectives of the 5G networks, and the goals this technology is hoping to achieve. It introduces us to the number of ideas used to bind this technology together, in order to overcome the hurdles, and to obtain its end goal. The endless use cases of this new network technology have been discussed extensively in this paper. This paper also dives into how the 5G networks is the reason why major industries can be automated. It has also been equated to the success of several IT industries like the artificial intelligence (AI), virtual and augmented reality (VR and AR), Internet of Things (IoT) etc. In turn these fields of IT will exponentially give rise to new technologies which made be made the new social standard for several activities like navigation, logistics, research etc. In the future 5G technologies might give rise to smart cities and so much more, only time can tell. The possibilities are endless.

REFERENCES

Al-Falahy, N., & Alani, O. (2017). Technologies for 5G networks: Challenges and opportunities. *IT Professional*, *19*(1), 12–20. doi:10.1109/MITP.2017.9

Azzaoui, A. E., Singh, S. K., Pan, Y., & Park, J. H. (2020). Block5gintell: Blockchain for AI-enabled 5G networks. *IEEE Access: Practical Innovations, Open Solutions*, *8*, 145918–145935. doi:10.1109/ACCESS.2020.3014356

Bonati, L., Polese, M., D'Oro, S., Basagni, S., & Melodia, T. (2020). Open, programmable, and virtualized 5G networks: State-of-the-art and The road ahead. *Computer Networks*, *182*, 107516. doi:10.1016/j.comnet.2020.107516

Buzzi, S., i, C.-L., Klein, T. E., Poor, H. V., Yang, C., & Zappone, A. (2016). A survey of energy-efficient techniques for 5G networks and challenges ahead. *IEEE Journal on Selected Areas in Communications*, *34*(4), 697–709. doi:10.1109/JSAC.2016.2550338

Chochliouros, I. P., Kostopoulos, A., Spiliopoulou, A. S., Dardamanis, A., Neokosmidis, I., Rokkas, T., & Goratti, L. (2017). *Business and market perspectives in 5G networks. In 2017 Internet of Things Business Models*. Users, and Networks. doi:10.1109/CTTE.2017.8260997

Gupta, A., & Jha, R. K. (2015). A survey of 5G network: Architecture and Emerging Technologies. *IEEE Access: Practical Innovations, Open Solutions*, *3*, 1206–1232. doi:10.1109/ACCESS.2015.2461602

Khodashenas, P. S., Aznar, J., Legarrea, A., Ruiz, C., Siddiqui, M. S., Escalona, E., & Figuerola, S. (2016). 5G network challenges and realization insights. *2016 18th International Conference on Transparent Optical Networks (ICTON)*. 10.1109/ICTON.2016.7550539

Kumar, P., Dayal, M., Khari, M., Fenza, G., & Gallo, M. (2021). NSL-BP: A meta classifier model based prediction of Amazon Product Reviews. *International Journal of Interactive Multimedia and Artificial Intelligence*, *6*(6), 95. doi:10.9781/ijimai.2020.10.001

Lal, N., Tiwari, S. M., Khare, D., & Saxena, M. (2021). Prospects for handling 5G network security: Challenges, recommendations and Future Directions. *Journal of Physics: Conference Series*, *1714*(1), 012052. doi:10.1088/1742-6596/1714/1/012052

Le, N. T., Hossain, M. A., Islam, A., Kim, D., Choi, Y.-J., & Jang, Y. M. (2016). Survey of promising technologies for 5G networks. *Mobile Information Systems*, *2016*, 1–25. doi:10.1155/2016/2676589

Liu, E., Effiok, E., & Hitchcock, J. (2020). Survey on health care applications in 5G networks. *IET Communications*, *14*(7), 1073–1080. doi:10.1049/iet-com.2019.0813

Mahmood, A., Beltramelli, L., Fakhrul Abedin, S., Zeb, S., Mowla, N. I., Hassan, S. A., Sisinni, E., & Gidlund, M. (2022). Industrial IOT in 5G-and-beyond networks: Vision, architecture, and design trends. *IEEE Transactions on Industrial Informatics*, *18*(6), 4122–4137. doi:10.1109/TII.2021.3115697

Wijethilaka, S., & Liyanage, M. (2021). Survey on network slicing for internet of things realization in 5G networks. *IEEE Communications Surveys and Tutorials*, *23*(2), 957–994. doi:10.1109/COMST.2021.3067807

Wu, Q., Li, G. Y., Chen, W., Ng, D. W., & Schober, R. (2017). An overview of Sustainable Green 5G Networks. *IEEE Wireless Communications*, *24*(4), 72–80. doi:10.1109/MWC.2017.1600343

Zhang, S., Wang, Y., & Zhou, W. (2019). Towards secure 5G networks: A survey. *Computer Networks*, *162*, 106871. doi:10.1016/j.comnet.2019.106871

Compilation of References

Alouffi, B., Hasnain, M., Alharbi, A., Alosaimi, W., Alyami, H., & Ayaz, M. (2021). A systematic literature review on cloud computing security: Threats and mitigation strategies. *IEEE Access: Practical Innovations, Open Solutions*, *9*, 57792–57807. doi:10.1109/ACCESS.2021.3073203

Zhang, S., Wang, Y., & Zhou, W. (2019). Towards secure 5G networks: A survey. *Computer Networks*, *162*, 106871. doi:10.1016/j.comnet.2019.106871

Bonati, L., Polese, M., D'Oro, S., Basagni, S., & Melodia, T. (2020). Open, programmable, and virtualized 5G networks: State-of-the-art and The road ahead. *Computer Networks*, *182*, 107516. doi:10.1016/j.comnet.2020.107516

Ramya, V., Palaniappan, B., & George, B. (2012). Embedded system for automatic irrigation of cardamom field using Xbee-PRO technology. *International Journal of Computers and Applications*, *53*(14).

Azzaoui, A. E., Singh, S. K., Pan, Y., & Park, J. H. (2020). Block5gintell: Blockchain for AI-enabled 5G networks. *IEEE Access: Practical Innovations, Open Solutions*, *8*, 145918–145935. doi:10.1109/ACCESS.2020.3014356

Liu, E., Effiok, E., & Hitchcock, J. (2020). Survey on health care applications in 5G networks. *IET Communications*, *14*(7), 1073–1080. doi:10.1049/iet-com.2019.0813

Neto, A. J., Zhao, Z., Rodrigues, J. J., Camboim, H. B., & Braun, T. (2018). Fog-based crime-assistance in smart iot transportation system. *IEEE Access: Practical Innovations, Open Solutions*, *6*, 11101–11111. doi:10.1109/ACCESS.2018.2803439

Kanjo, E., Bacon, J., Roberts, D., & Landshoff, P. (2009). MobSens: Making smart phones smarter. *IEEE Pervasive Computing*, *8*(4), 50–57. doi:10.1109/MPRV.2009.79

Wijethilaka, S., & Liyanage, M. (2021). Survey on network slicing for internet of things realization in 5G networks. *IEEE Communications Surveys and Tutorials*, *23*(2), 957–994. doi:10.1109/COMST.2021.3067807

Chochliouros, I. P., Kostopoulos, A., Spiliopoulou, A. S., Dardamanis, A., Neokosmidis, I., Rokkas, T., & Goratti, L. (2017). *Business and market perspectives in 5G networks. In 2017 Internet of Things Business Models.* Users, and Networks. doi:10.1109/CTTE.2017.8260997

Ray, A., & Goswami, S. (2020, February). IoT and cloud computing based smart water metering system. In *2020 International Conference on Power Electronics & IoT Applications in Renewable Energy and its Control (PARC)* (pp. 308-313). IEEE. 10.1109/PARC49193.2020.236616

Hossain, M. S., Chisty, N. M. A., Hargrove, D. L., & Amin, R. (2021). Role of Internet of Things (IoT) in Retail Business and Enabling Smart Retailing Experiences. *Asian Business Review*, *11*(2), 75–80. doi:10.18034/abr.v11i2.579

Jin, W. (2020, May). Research on machine learning and its algorithms and development. *Journal of Physics: Conference Series*, *1544*(1), 012003. doi:10.1088/1742-6596/1544/1/012003

Khodashenas, P. S., Aznar, J., Legarrea, A., Ruiz, C., Siddiqui, M. S., Escalona, E., & Figuerola, S. (2016). 5G network challenges and realization insights. *2016 18th International Conference on Transparent Optical Networks (ICTON)*. 10.1109/ICTON.2016.7550539

Hammoudeh, A. (2018). *A concise introduction to reinforcement learning*. Princess Suamaya University for Technology.

Lal, N., Tiwari, S. M., Khare, D., & Saxena, M. (2021). Prospects for handling 5G network security: Challenges, recommendations and Future Directions. *Journal of Physics: Conference Series*, *1714*(1), 012052. doi:10.1088/1742-6596/1714/1/012052

Mahmood, A., Beltramelli, L., Fakhrul Abedin, S., Zeb, S., Mowla, N. I., Hassan, S. A., Sisinni, E., & Gidlund, M. (2022). Industrial IOT in 5G-and-beyond networks: Vision, architecture, and design trends. *IEEE Transactions on Industrial Informatics*, *18*(6), 4122–4137. doi:10.1109/TII.2021.3115697

Sarker, I. H. (2022). Ai-based modeling: Techniques, applications and research issues towards automation, intelligent and smart systems. *SN Computer Science*, *3*(2), 1–20. doi:10.100742979-022-01043-x PMID:35194580

Ghosh, A., Chakraborty, D., & Law, A. (2018). Artificial intelligence in Internet of things. *CAAI Transactions on Intelligence Technology*, *3*(4), 208–218. doi:10.1049/trit.2018.1008

Kumar, P., Dayal, M., Khari, M., Fenza, G., & Gallo, M. (2021). NSL-BP: A meta classifier model based prediction of Amazon Product Reviews. *International Journal of Interactive Multimedia and Artificial Intelligence*, *6*(6), 95. doi:10.9781/ijimai.2020.10.001

Al-Falahy, N., & Alani, O. (2017). Technologies for 5G networks: Challenges and opportunities. *IT Professional*, *19*(1), 12–20. doi:10.1109/MITP.2017.9

Mahesh, B. (2020). Machine learning algorithms-a review. *International Journal of Science and Research (IJSR).[Internet]*, *9*, 381-386.

Haddadin, S., & Knobbe, D. (2020). *Robotics and Artificial Intelligence—The Present and Future Visions. Algorithms and Law*. Cambridge University Press.

Wu, Q., Li, G. Y., Chen, W., Ng, D. W., & Schober, R. (2017). An overview of Sustainable Green 5G Networks. *IEEE Wireless Communications*, *24*(4), 72–80. doi:10.1109/MWC.2017.1600343

Buzzi, S., i, C.-L., Klein, T. E., Poor, H. V., Yang, C., & Zappone, A. (2016). A survey of energy-efficient techniques for 5G networks and challenges ahead. *IEEE Journal on Selected Areas in Communications*, *34*(4), 697–709. doi:10.1109/JSAC.2016.2550338

Romero, M., Guédria, W., Panetto, H., & Barafort, B. (2020). Towards a characterisation of smart systems: A systematic literature review. *Computers in Industry*, *120*, 103224. doi:10.1016/j.compind.2020.103224

Le, N. T., Hossain, M. A., Islam, A., Kim, D., Choi, Y.-J., & Jang, Y. M. (2016). Survey of promising technologies for 5G networks. *Mobile Information Systems*, *2016*, 1–25. doi:10.1155/2016/2676589

Sarker, I. H. (2021). Machine learning: Algorithms, real-world applications and research directions. *SN Computer Science*, *2*(3), 1–21. doi:10.100742979-021-00592-x PMID:33778771

Abawi, G. S., & Widmer, T. L. (2000). Impact of soil health management practices on soilborne pathogens, nematodes and root diseases of vegetable crops. *Applied Soil Ecology*, *15*(1), 37–47. doi:10.1016/S0929-1393(00)00070-6

Abbasi, R., Martinez, P., & Ahmad, R. (2022). The digitization of agricultural industry–a systematic literature review on agriculture 4.0. *Smart Agricultural Technology*, 100042.

Abdullahi, H. S., Sheriff, R., & Mahieddine, F. (2017). Convolution neural network in precision agriculture for plant image recognition and classification. In *2017 Seventh International Conference on Innovative Computing Technology (INTECH)* (Vol. 10, pp. 256-272). Ieee

Accorsi, R., Cholette, S., Manzini, R., & Tufano, A. (2018). A hierarchical data architecture for sustainable food supply chain management and planning. *Journal of Cleaner Production*, *203*, 1039–1054. doi:10.1016/j.jclepro.2018.08.275

Adamchuk, V. I., Hummel, J. W., Morgan, M. T., & Upadhyaya, S. K. (2004). On-the-go soil sensors for precision agriculture. *Computers and Electronics in Agriculture*, *44*(1), 71–91. doi:10.1016/j.compag.2004.03.002

Adam, L. (2007). *Information and Communication Technologies*. Knowledge Management and Indigenous Knowledge.

Adão, T., Hruška, J., Pádua, L., Bessa, J., Peres, E., Morais, R., & Sousa, J. J. (2017). Hyperspectral imaging: A review on UAV-based sensors, data processing and applications for agriculture and forestry. *Remote Sensing*, *9*(11), 1110. doi:10.3390/rs9111110

Agatonovic-Kustrin, S., & Beresford, R. (2000). Basic concepts of artificial neural network (ANN) modeling and its application in pharmaceutical research. *Journal of Pharmaceutical and Biomedical Analysis*, *22*(5), 717–727. doi:10.1016/S0731-7085(99)00272-1 PMID:10815714

Agriinfo. (2018). Soil –Plant – Water – Relationships. *Agriinfo*. https://agriinfo.in/soil-plant-water-relationships-2206/

Aguilar-Manjarrez, J., & Ross, L. G. (1995). Geographical information system (GIS) environmental models for aquaculture development in Sinaloa State, Mexico. *Aquaculture International*, *3*(2), 103–115.

Aha, D. W., Kibler, D., & Albert, M. K. (1991). Instance-based learning algorithms. *Machine Learning*, *6*(1), 37–66. doi:10.1007/BF00153759

Ahirwar, S., Swarnkar, R., Bhukya, S., & Namwade, G. (2019). Application of drones in agriculture. *International Journal of Current Microbiology and Applied Sciences*, *8*(01), 2500–2505. doi:10.20546/ijcmas.2019.801.264

Ahmad, I., Shahabuddin, S., Sauter, T., Harjula, E., Kumar, T., Meisel, M., Juntti, M., & Ylianttila, M. (2021). The Challenges of Artificial Intelligence in Wireless Networks for the Internet of Things: Exploring Opportunities for Growth. *IEEE Industrial Electronics Magazine*, *15*(1), 16–29. doi:10.1109/MIE.2020.2979272

Ahmad, L., & Mahdi, S. S. (2018). Components of Precision Agriculture. In *Satellite Farming* (pp. 19–30). Springer.

Ahmad, L., & Nabi, F. (2021). *Agriculture 5.0: Artificial Intelligence, IoT and Machine Learning*. CRC Press.

Ahn, H.S., Dayoub, F., Popovic, M., MacDonald, B.A., Siegwart, R.Y., & Sa, I. (2018). An Overview of Perception Methods for Horticultural Robots: From Pollination to Harvest. ArXiv, abs/1807.03124.

Aiello, G., Catania, P., Vallone, M., & Venticinque, M. (2022). Worker safety in agriculture 4.0: A new approach for mapping operator's vibration risk through Machine Learning activity recognition. *Computers and Electronics in Agriculture*, *193*, 106637. doi:10.1016/j.compag.2021.106637

Aitkenhead, M. J., Mcdonald, A. J. S., Dawson, J. J., Couper, G., Smart, R. P., Billett, M., Hope, D., & Palmer, S. (2003). A novel method for training neural networks for time-series prediction in environmental systems. *Elsevier*, *162*, 87–95. doi:10.1016/S0304-3800(02)00401-5

Akyildiz, I.F., Gutierrez-Estevez, D.M., Reyes, E.C. (2010). The evolution to 4G cellular systems: LTE-Advanced. *Phys. Commun.* *3*(4), 217–244. doi:.phycom.2010.08.001. doi:10.1016/j

Alam, E. N. (2020). IoT in Agriculture Industry. *Jurnal Sistem Cerdas*, *3*(1), 36–42. doi:10.37396/jsc.v3i1.57

Albawi, S., Mohammed, T. A., & Al-Zawi, S. (2017). *Understanding of a convolutional neural network.*

Alderfasi, A. A., & Nielsen, D. C. (2001). Use of crop water stress index for monitoring water status and scheduling irrigation in wheat. *Agricultural Water Management*, *47*(1), 69–75. doi:10.1016/S0378-3774(00)00096-2

Al-Omary, A. Y., & Jamil, M. S. (2006). A new approach of clustering based machine-learning algorithm. *Knowledge-Based Systems*, *19*(4), 248–258. doi:10.1016/j.knosys.2005.10.011

Alston, J. M., & Pardey, P. G. (2014). Agriculture in the Global Economy %J. *The Journal of Economic Perspectives*, *28*(1), 121–146. doi:10.1257/jep.28.1.121

Amatya, S., Karkee, M., Gongal, A., Zhang, Q., & Whiting, M. D. (2016). Detection of cherry tree branches with full foliage in planar architecture for automated sweet-cherry harvesting. *Biosystems Engineering*, *146*, 3–15. doi:10.1016/j.biosystemseng.2015.10.003

Ambika, N. (2019). Energy-Perceptive Authentication in Virtual Private Networks Using GPS Data. In M. Z. (eds), Security, privacy and trust in the IoT environment (pp. 25-38). Cham: Springer. doi:10.1007/978-3-030-18075-1_2

Ambika, N. (2020). SYSLOC: Hybrid Key Generation in Sensor Network. In S. P., B. B., P. M., K. N., & H. W. (eds), Handbook of Wireless Sensor Networks: Issues and Challenges in Current Scenario's. Advances in Intelligent Systems and Computing (Vol. 1132, pp. 325-347). Cham: Springer.

Ampatzidis, Y., Partel, V., & Costa, L. (2020). Agroview: Cloud-based application to process, analyze and visualize UAV-collected data for precision agriculture applications utilizing artificial intelligence. *Computers and Electronics in Agriculture*, *174*, 105457. doi:10.1016/j.compag.2020.105457

Anand, A., de Veciana, G., & Shakkottai, S. (2020). Joint Scheduling of URLLC and eMBB Traffic in 5G Wireless Networks. *IEEE/ACM Transactions on Networking*, *28*(2), 477–490. doi:10.1109/TNET.2020.2968373

Andrews, J. G., Buzzi, S., Choi, W., Hanly, S. V., Lozano, A., Soong, A. C. K., & Zhang, J. C. (2014). What Will 5G Be? *IEEE Journal on Selected Areas in Communications*, *32*(6), 1065–1082. doi:10.1109/JSAC.2014.2328098

Andrio, A. (2019, August). Development of UAV technology in seed dropping for aerial revegetation practices in Indonesia. *IOP Conference Series. Earth and Environmental Science*, *308*(1), 012051. doi:10.1088/1755-1315/308/1/012051

Appleton, J. (2021, April 6). The Use Of Ai For Smart Urban Services In Smart Cities. *Bee Smart City*. https://www.beesmart.city/en/solutions/the-use-of-ai-for-smart-urban-services-in-smart-cities

Arif, C., Indra Setiawan, B., Mizoguchi, M., & Doi, R. (2012a). Estimation of soil moisture in paddy field using Artificial Neural Networks. *International Journal of Advanced Research in Artificial Intelligence*, *1*(1). doi:10.14569/IJARAI.2012.010104

Arokiaraj, D. (2015). A Study on Environmental Responsibility of the Stakeholders of Auto Industry in Chennai. International *Journal of Innovative Technology and Exploring Engineering, 9*(1).

Arokiaraj, D., Ramyar, R. A., Ganeshkumar, C., & Gomathi Sankar, J. (2020b). An empirical analysis of consumer behaviour towards organic food products purchase in India. *Calitatea Qual Access Success, 21.*

Arokiaraj, D., Ganeshkumar, C., & Paul, P. V. (2020). Innovative management system for environmental sustainability practices among Indian auto-component manufacturers. *International Journal of Business Innovation and Research*, *23*(2), 168–182. doi:10.1504/IJBIR.2020.110095

Arras, L., Osman, A., Müller, K.-R., & Samek, W. (2019). Evaluating recurrent neural network explanations. doi:10.18653/v1/W19-4813

Åstrand, B., & Baerveldt, A. J. (2002). An agricultural mobile robot with vision-based perception for mechanical weed control. *Autonomous Robots*, *13*(1), 21–35. doi:10.1023/A:1015674004201

Atzori, L., Iera, A., & Morabito, G. (2010). The Internet of Things: A survey. *Computer Networks*, *54*(15), 2787–2805. doi:10.1016/j.comnet.2010.05.010

Ayamga, M., Tekinerdogan, B., & Kassahun, A. (2021). Exploring the challenges posed by regulations for the use of drones in agriculture in the African context. *Land (Basel)*, *10*(2), 164. doi:10.3390/land10020164

Ayamga, M., Tekinerdogan, B., Kassahun, A., & Rambaldi, G. (2021b). Developing a policy framework for the adoption and management of drones for agriculture in Africa. *Technology Analysis and Strategic Management*, *33*(8), 970–987. doi:10.1080/09537325.2020.1858047

Ayaz, M., Ammad-Uddin, M., Sharif, Z., Mansour, A., & Aggoune, E. H. M. (2019). Internet-of-Things (IoT)-based smart agriculture: Toward making the fields talk. *IEEE Access: Practical Innovations, Open Solutions*, *7*, 129551–129583.

Ayim, C., Kassahun, A., Tekinerdogan, B., & Addison, C. (2020). Adoption of ICT innovations in the agriculture sector in Africa: A Systematic Literature Review. *Agriculture and Food Security, 11*(22).

Ayodele, T. O. (2010). Types of machine learning algorithms. *New advances in machine learning, 3*, 19-48.

Babbar, N., Kumar, A., & Verma, V. K. (2022). *Crop management: Wheat yield prediction and disease detection using an intelligent predictive algorithms and metrological parameters. Deep Learning for Sustainable Agriculture*. Academic Press.

Babu, S. (2013). A software model for precision agriculture for small and marginal farmers. *Global Humanitarian Technology Conference: South Asia Satellite,* 352–355. IEEE. 10.1109/GHTC-SAS.2013.6629944

Bacco, M., Berton, A., Ferro, E., Gennaro, C., Gotta, A., Matteoli, S., & Zanella, A. (2018). Smart farming: Opportunities, challenges and technology enablers. *IoT Vertical and Topical Summit on Agriculture-Tuscany (IOT Tuscany)*, 1-6.

Bacco, M., Barsocchi, P., Ferro, E., Gotta, A., & Ruggeri, M. (2019). The Digitisation of Agriculture: A Survey of Research Activities on Smart Farming. *Array*, *3–4*, 100009. doi:10.1016/j.array.2019.100009

Bakar, Z. A., Tahir, N. M., & Yassin, I. M. (2010, May). Classification of parkinson's disease based on multilayer perceptrons neural network. In *6th International Colloquium on Signal Processing & its Applications* (pp. 1-4). IEEE.

Bakker, T., van Asselt, K., Bontsema, J., Müller, J., & van Straten, G. (2006). An autonomous weeding robot for organic farming. *Springer Tracts in Advanced Robotics*, *25*, 579–590. doi:10.1007/978-3-540-33453-8_48

Bak, T., & Jakobsen, H. (2004). Agricultural robotic platform with four wheel steering for weed detection. *Biosystems Engineering*, *87*(2), 125–136. doi:10.1016/j.biosystemseng.2003.10.009

Banumathi, M., & Arokiaraj, D. (2011). Eco-labeling–The Need for Sustainable Marketing. In *National Conference in the era of Global Recovery,*(pp. 511-515).

Barasa, P. M., Botai, C. M., Botai, J. O., & Mabhaudhi, T. (2021). A Review of Climate-Smart Agriculture Research and Applications in Africa. *Agronomy (Basel)*, *11*(6), 1255. doi:10.3390/agronomy11061255

Barbedo, J. G. A. (2016). Expert systems applied to plant disease diagnosis: Survey and critical view. *IEEE Latin America Transactions*, *14*(4), 1910–1922. doi:10.1109/TLA.2016.7483534

Barlow, H. B. (1989). Unsupervised learning. *Neural Computation*, *1*(3), 295–311. doi:10.1162/neco.1989.1.3.295

Batani, J., Musungwini, S., & Rebanowako, T. G. (2019). An Assessment of the Use of mobile phones as sources of Agricultural information by tobacco Smallholder farmers in Zimbabwe. *International Journal of Systems Integration*, *3*(1), 1–21. doi:10.20470/jsi.v10i3.375

Beaman, J. A., & Johnson, A. J. (2006). Food distribution channel overview: a guide for new manufacturers. *OSU*. https://ir.library.oregonstate.edu/concern/open_educational_resources/8623hz116

Beer, S. (1981). *Brain of the Firm*. Wiley.

Bekele, W., & Drake, L. (2003). Soil and water conservation decision behavior of subsistence farmers in the Eastern Highlands of Ethiopia: A case study of the Hunde-Lafto area. *Ecological Economics*, *46*(3), 437–451. doi:10.1016/S0921-8009(03)00166-6

Belal, A. A. EL-Ramady, H., Jalhoum, M., Gad, A., & Mohamed, E. S. (2021). Precision Farming Technologies to Increase Soil and Crop Productivity. In Agro-Environmental Sustainability in MENA Regions (pp. 117-154). Springer, Cham.

Bellon-Maurel, V., Brossard, L., Garcia, F., Mitton, N., Termier, A., & Barbier, S. (2022). Agriculture and Digital Technology.

Ben Ayed, R., & Hanana, M. (2021). Artificial Intelligence to Improve the Food and Agriculture Sector. *Journal of Food Quality, 2021*(Ml). doi:10.1155/2021/5584754

Beuchelt, T. D., & Badstue, L. (2013). Gender, nutrition- and climate-smart food production: Opportunities and trade-offs. *Food Security*, *5*(5), 709–721. doi:10.100712571-013-0290-8

Bhanu, K. N., Jasmine, H. J., & Mahadevaswamy, H. S. (2020, June). Machine learning implementation in IoT based intelligent system for agriculture. In *2020 International Conference for Emerging Technology (INCET)* (pp. 1-5). IEEE. 10.1109/INCET49848.2020.9153978

Bhardwaj, H., Tomar, P., Sakalle, A., & Sharma, U. (2021). Artificial Intelligence and Its Applications in Agriculture with the Future of Smart Agriculture Techniques. In *Artificial Intelligence and IoT-Based Technologies for Sustainable Farming and Smart Agriculture* (pp. 25–39). IGI Global. doi:10.4018/978-1-7998-1722-2.ch002

Bhattacharyya, S. S., Maitra, D., & Deb, S. (2021). Study of adoption and absorption of emerging technologies for smart supply chain management: A dynamic capabilities perspective. [IJAL]. *International Journal of Applied Logistics*, *11*(2), 14–54.

Bhbosale, S., Pujari, V., & Multani, Z. J. A. I. I. R. J. (2020). Advantages And Disadvantages Of Artificial Intellegence. 227-230.

Biau, G., & Scornet, E. (2016). A random forest guided tour. *Test*, *25*(2), 197–227. doi:10.100711749-016-0481-7

Bilgili, M. (2011). The use of artificial neural networks for forecasting the monthly mean soil temperatures in Adana, Turkey. *Turkish Journal of Agriculture and Forestry*, *35*(1), 83–93. https://dergipark.org.tr/en/pub/tbtkagriculture/issue/11593/138197

Bioucas-Dias, J. M., Plaza, A., Camps-Valls, G., Scheunders, P., Nasrabadi, N. M., & Chanussot, J. (2013). Hyperspectral remote sensing data analysis and future challenges. *Geoscience and Remote Sensing Magazine, 1*(2), 6–36. IEEE . doi:10.1109/MGRS.2013.2244672

Birk, A., & Kenn, H. (2002). Roboguard, a teleoperated mobile security robot. *Control Engineering Practice*, *10*(11), 1259–1264. doi:10.1016/S0967-0661(02)00088-6

Blasco, J., Aleixos, N., Roger, J. M., Rabatel, G., & Moltó, E. (2002). AE—Automation and emerging technologies: Robotic weed control using machine vision. *Biosystems Engineering*, *83*(2), 149–157.

Bloice, M. D., Stocker, C., & Holzinger, A. (2017). Augmentor: an image augmentation library for machine learning. *arXiv preprint arXiv:1708.04680.*

Bogue, R. (2021). Robots are poised to transform agriculture. *Industrial Robot: the international journal of robotics research and application.*

Böhning, D. (1992). Multinomial logistic regression algorithm. *Annals of the Institute of Statistical Mathematics*, *44*(1), 197–200. doi:10.1007/BF00048682

Bonaccorso, G. (2017). *Machine learning algorithms.* Packt Publishing Ltd.

Booth, S., Tompkin, J., Pfister, H., Waldo, J., Gajos, K., & Nagpal, R. (2017). Piggybacking robots: Human-robot overtrust in university dormitory security. *ACM/IEEE International Conference on Human-Robot Interaction* (pp. 426-434). ACM. 10.1145/2909824.3020211

Borgelt, C. (2005). *An Implementation of the FP-growth Algorithm.*

Bouras, C., Kollia, A., & Papazois, A. (2017). Dense Deployments and DAS in 5G: A TechnoEconomic Comparison. *Wireless Personal Communications*, *94*(3), 1777–1797. doi:10.100711277-016-3711-0

Bralts, V. F., Driscoll, M. A., Shayya, W. H., & Cao, L. (1993). An expert system for the hydraulic analysis of microirrigation systems. *Computers and Electronics in Agriculture*, *9*(4), 275–287.

Brighton, H., & Mellish, C. (2002). Advances in instance selection for instance-based learning algorithms. *Data Mining and Knowledge Discovery*, *6*(2), 153–172. doi:10.1023/A:1014043630878

Brown, G. (2017). Service-based architecture for 5G core networks. *Huawei White Paper, 1.*

Budiharto, W., Chowanda, A., Gunawan, A. A. S., Irwansyah, E., & Suroso, J. S. (2019, December). A review and progress of research on autonomous drone in agriculture, delivering items and geographical information systems (GIS). In *2nd world symposium on communication engineering (WSCE)* (pp. 205-209). IEEE.

Bu, F., & Wang, X. (2019). A smart agriculture IoT system based on deep reinforcement learning. [doi:https://doi.org/10.1016/j.future.2019.04.041]. *Future Generation Computer Systems*, *99*, 500–507.

Bünger, L. (2021). *Robotic waste sorting.* Worcester Polytechnic Institute.

Busch, G. (2012). GIS-based Tools for Regional Assessments and Planning Processes Regarding Potential Environmental Effects of Poplar SRC. *BioEnergy Research* *5*(3), 584–605. doi:10.1007/S12155-012-9224-0

Cakir, Y., Kirci, M., & Gunes, E. O. (2014). Yield prediction of wheat in south-east region of Turkey by using artificial neural networks. *The 3rd International Conference on Agro-Geoinformatics.* Agro-Geoinformatics . doi:10.1109/AGRO-GEOINFORMATICS.2014.6910609

Călina, J., Călina, A., Miluţ, M., Croitoru, A., Stan, I., & Buzatu, C. (2020). Use of drones in cadastral works and precision works in silviculture and agriculture. *Publisher NARDI Fundulea, România*, *37*, 273–284.

Campello, R. J. G. B., Kröger, P., Sander, J., & Zimek, A. (2020). Density-based clustering. *Wiley Interdisciplinary Reviews. Data Mining and Knowledge Discovery*, *10*(2), e1343. doi:10.1002/widm.1343

Carbonell, J. G., Michalski, R. S., & Mitchell, T. M. (1983). 1 - AN OVERVIEW OF MACHINE LEARNING. In R. S. Michalski, J. G. Carbonell, & T. M. Mitchell (Eds.), *Machine Learning* (pp. 3–23). Morgan Kaufmann.

Castaldi, F., Pelosi, F., Pascucci, S., & Casa, R. (2017). Assessing the potential of images from unmanned aerial vehicles (UAV) to support herbicide patch spraying in maize. *Precision Agriculture*, *18*(1), 76–94. doi:10.100711119-016-9468-3

Castelli, M. (2018). Supervised Learning: Classification, Reference Module in Life Sciences. *Informatica, 31.*

Chang, C. & Robotics, K. L. (2018). Smart agricultural machine with a computer vision-based weeding and variable-rate irrigation scheme. *Mdpi.Com.* doi:10.3390/robotics7030038

Chang, D. H., & Islam, S. (2000). Estimation of soil physical properties using remote sensing and artificial neural network. *Remote Sensing of Environment*, *74*(3), 534–544.

Charoen-Ung, P., & Mittrapiyanuruk, P. (2019). Sugarcane Yield Grade Prediction Using Random Forest with Forward Feature Selection and Hyper-parameter Tuning. *Advances in Intelligent Systems and Computing, 769*, 33–42. doi:10.1007/978-3-319-93692-5_4

Chattopadhyay, P. B., & Rangarajan, R. (2014). Application of ANN in sketching spatial nonlinearity of unconfined aquifer in agricultural basin. [doi:https://doi.org/10.1016/j.agwat.2013.11.007]. *Agricultural Water Management*, *133*, 81–91.

Cheng, H., Peng, H., & Liu, S. (2013, March). An improved K-means clustering algorithm in agricultural image segmentation. In PIAGENG: Image Processing and Photonics for Agricultural Engineering (Vol. 8761, pp. 86-90). SPIE.

Cheng, X., Zhang, Y., Chen, Y., Wu, Y., & Yue, Y. (2017). Pest identification via deep residual learning in complex background. [doi:https://doi.org/10.1016/j.compag.2017.08.005]. *Computers and Electronics in Agriculture*, *141*, 351–356.

Chen, J., Chen, J., Zhang, D., Sun, Y., & Nanehkaran, Y. A. (2020). Using deep transfer learning for image-based plant disease identification. *Computers and Electronics in Agriculture*, *173*, 105393. https://doi.org/10.1016/J.COMPAG.2020.105393

Chen, J., & Yang, A. (2019). Intelligent Agriculture and Its Key Technologies Based on Internet of Things Architecture. *IEEE Access: Practical Innovations, Open Solutions*, *7*, 77134–77141. doi:10.1109/ACCESS.2019.2921391

Chen, L., Zhang, H., Xiao, J., Nie, L., Shao, J., Liu, W., & Chua, T. S. (2017). Sca-cnn: Spatial and channel-wise attention in convolutional networks for image captioning. In *Proceedings of the conference on computer vision and pattern recognition* (pp. 5659-5667). IEEE.

Chen, P. H., Lin, C. J., & Schölkopf, B. (2005). A tutorial on ν-support vector machines. *Applied Stochastic Models in Business and Industry*, *21*(2), 111–136. doi:10.1002/asmb.537

Chlingaryan, A., Sukkarieh, S., & Whelan, B. (2018). Machine learning approaches for crop yield prediction and nitrogen status estimation in precision agriculture: A review. *Computers and Electronics in Agriculture*, *151*, 61–69. doi:10.1016/j.compag.2018.05.012

Chu, X., Yang, W., Ouyang, W., Ma, C., Yuille, A., & Wang, X. (2017). Multicontext attention for human pose estimation. In *Conference on Computer Vision and Pattern Recognition,* 5669–5678. IEEE .

Coble, K. H., Mishra, A. K., Ferrell, S., & Griffin, T. (2018). Big data in agriculture: A challenge for the future. *Applied Economic Perspectives and Policy*, *40*(1), 79–96. doi:10.1093/aepp/ppx056

Cui, S., Ling, P., Zhu, H., & Keener, H. M. (2018). Plant Pest Detection Using an Artificial Nose System: A Review. *Sensors 18*(2), 378. doi:10.3390/S18020378

Cunha, R. L., Silva, B., & Netto, M. A. (2018, October). A scalable machine learning system for pre-season agriculture yield forecast. In *2018 IEEE 14th International Conference on e-Science (e-Science)* (pp. 423-430). IEEE. 10.1109/eScience.2018.00131

Cunningham, P., Cord, M., & Delany, S. J. (2008). Supervised learning. In *Machine learning techniques for multimedia* (pp. 21–49). Springer. doi:10.1007/978-3-540-75171-7_2

Daheim, C., Poppe, K., & Schrijver, R. (2019). *Precision agriculture and the future of farming in Europe: scientific foresight study.* European Parliament. https://data.europa.eu/doi/10.2861/175493

Dahlman, E., Parkvall, S., & Skold, J. (2013). *4G: LTE/LTE-advanced for mobile broadband.* Academic press.

Daponte, P., De Vito, L., Glielmo, L., Iannelli, L., Liuzza, D., Picariello, F., & Silano, G. (2019, May). A review on the use of drones for precision agriculture. *IOP Conference Series. Earth and Environmental Science*, *275*(1), 012022. doi:10.1088/1755-1315/275/1/012022

Daugherty, P. R., & Wilson, H. J. (2018). *Human+ machine: Reimagining work in the age of AI.* Harvard Business Press.

Davenport, T. H. (1993). *Process innovation: reengineering work through information technology.* Harvard Business Press.

David, A. (2020$_a$). Air Pollution Control Monitoring & Delivery Rate Escalated by Efficient use of Markov Process in Manet Networks: to Measure Quality of Service Parameters. *Test Engineering & Management, The Mattingley Publishing Co., Inc.*

David, A., & Ravi, S. (2017). The direness of cultivable land spotted on agricultural: A special reference to rice production in South India. *Abhinav National Monthly Refereed Journal of Research in Commerce & Management, 6*(09), 55-59.

David, A., Muthukumar, A., Ayyappan, D., & Ravi, S. (2020). Accident Avoidance System by Using Sensors Module. *Test Engineering & Management, The Mattingley Publishing Co., Inc.*

David, A. (2020b). The consumer purchasing process of organic food product: An empirical analysis. [QAS]. *Journal of Management System-Quality Access to Success*, *21*(177), 128–132.

David, A. (2020c). *Corporate and individual environmental responsibility towards the automobile*. Book Rivers.

David, A., Kumar, C. G., & Paul, P. V. (2022). Blockchain technology in the food supply chain: Empirical analysis. [IJISSCM]. *International Journal of Information Systems and Supply Chain Management*, *15*(3), 1–12. doi:10.4018/IJISSCM.290014

Dayal, M., Chawla, A., & Khari, M. (2022). Coalescence of Neural Networks and Blockchain. In *Handbook of Green Computing and Blockchain Technologies* (pp. 31–44). CRC Press.

Dayal, M., Chawla, A., Khari, M., & Mahajan, A. N. (2023). Artificial Intelligence-Based Smart Packet Filter. In *Proceedings of Third International Conference on Computing, Communications, and Cyber-Security* (pp. 791-801). Springer, Singapore.

Dayal, M., Chawla, A., Singh, D., & Garg, R. K. (2021). Combination of Convolutional Neural Networks and Blockchain in Handwritten Digit Recognition. In *Handbook of Green Computing and Blockchain Technologies* (pp. 55–67). CRC Press. doi:10.1201/9781003107507-5

Dayal, M., Garg, R. K., Chaudhary, S., Chaubey, C., & Chawla, A. (2022). Blockchain for Medicine and Healthcare. In *Handbook of Green Computing and Blockchain Technologies* (pp. 171–181). CRC Press.

Dayan, P., & Niv, Y. (2008). Reinforcement learning: The good, the bad and the ugly. *Current Opinion in Neurobiology*, *18*(2), 185–196. doi:10.1016/j.conb.2008.08.003 PMID:18708140

Dell'Uomo, L., & Scarrone, E. (2002, October). An all-IP solution for QoS mobility management and AAA in the 4G mobile networks. In The 5th International Symposium on Wireless Personal Multimedia Communications (Vol. 2, pp. 591-595). IEEE.

Devapriya, P., Ferrell, W., & Geismar, N. (2017). Integrated production and distribution scheduling with a perishable product. *European Journal of Operational Research*, *259*(3), 906–916. https://doi.org/10.1016/J.EJOR.2016.09.019

Devaraj, A., Rathan, K., Jaahnavi, S., & Indira, K. (2019). Identification of plant disease using image processing technique. *Proceedings of the 2019 International Conference on Communication and Signal Processing*, 749–753. IEEE. doi:10.1109/ICCSP.2019.8698056

Dewi, T., Risma, P., & Oktarina, Y. (2020). Fruit sorting robot based on color and size for an agricultural product packaging system. *Bulletin of Electrical Engineering and Informatics*, *9*(4), 1438–1445.

Dickson, M. A., Ni, B., & Han, S. (2004). *U.S. Patent No. 6,686,951*. Washington, DC: U.S. Patent and Trademark Office.

Dieber, B., Breiling, B., Taurer, S., Kacianka, S., Rass, S., & Schartner, P. (2017). Security for the robot operating system. *Robotics and Autonomous Systems*, *98*, 192–203. doi:10.1016/j.robot.2017.09.017

Dileep, M. R., Navaneeth, A. V., Ullagaddi, S., & Danti, A. (2020, November). A study and analysis of various types of agricultural drones and their applications. In *Fifth International Conference on Research in Computational Intelligence and Communication Networks (ICRCICN)* (pp. 181-185). IEEE.

Dimitriadis, S., & Goumopoulos, C. (2008, August). Applying machine learning to extract new knowledge in precision agriculture applications. In *2008 Panhellenic Conference on Informatics* (pp. 100-104). IEEE. 10.1109/PCI.2008.30

Diwate, S. K., Nitnaware, V. N., & Argulwar, K. (2018). Design and development of application specific drone machine for seed sowing. *International Research Journal of Engineering and Technology*, *5*(05), 4003–4007.

Donzia, S. K. Y., Kim, H. K., & Hwang, H. J. (2018, July). A software model for precision agriculture framework based on smart farming system and application of iot gateway. In *International Conference on Computational Science/Intelligence & Applied Informatics* (pp. 49-58). Springer, Cham.

dos Santos Ferreira, A., Freitas, D. M., da Silva, G. G., Pistori, H., & Folhes, M. T. (2019). Unsupervised deep learning and semi-automatic data labeling in weed discrimination. [doi:https://doi.org/10.1016/j.compag.2019.104963]. *Computers and Electronics in Agriculture*, *165*, 104963.

Doshi, J., & Patel, T., & kumar Bharti, S. (2019). Smart Farming using IoT, a solution for optimally monitoring farming conditions. *Procedia Computer Science*, *160*, 746–751.

DotEcon Ltd & Axon Partners. (2018). Study on Implications of 5G Deployment on Future Business Models. Axon Partners Group, p. 116.

Dove, G., Halskov, K., Forlizzi, J., & Zimmerman, J. (2017). UX design innovation: Challenges for working with machine learning as a design material. *Conference on Human Factors in Computer Systems.*

Dukes, M. D., & Perry, C. (2006). Uniformity testing of variable-rate center pivot irrigation control systems. *Precision Agriculture*, *7*(3), 205–218. doi:10.100711119-006-9020-y

Durmus¸, H., Gunes¸, E. O., & Kirci, M. (2017). Disease detection on the ¨ leaves of the tomato plants by using deep learning. In *6th International Conference on Agro-Geoinformatics*, 1–5.

Dursun, M., & Ozden, S. (2011). A wireless application of drip irrigation automation supported by soil moisture sensors. *Scientific Research and Essays*, *6*(7), 1573–1582.

Dutta, A., & Hammad, E. (2020, September). 5G security challenges and opportunities: a system approach. In *2020 IEEE 3rd 5G World Forum (5GWF)* (pp. 109-114). IEEE.

Dzino-Silajdzic, V. (2018). *Practical Guide; Focus group discussions*. 40. https://www.crs.org/sites/default/files/tools-research/fgds_april_24_final_lo_res_.pdf

Eberly, L. E. (2007). Multiple linear regression. *Topics in Biostatistics*, 165-187.

Ebihara, Y., Uematsu, S., & Nomiya, S. (2010). Control of verticillium dahlia at a strawberry nursery by paddy-upland rotation. *Journal of General Plant Pathology*, *76*(1), 7–20. doi:10.100710327-009-0205-x

Eli-Chukwu, N. C. (2019). Applications of artificial intelligence in agriculture: A review. *Engineering, Technology &. Applied Scientific Research*, *9*(4), 4377–4383.

Elijah, O., Leow, C. Y., Rahman, T. A., Nunoo, S., & Iliya, S. Z. (2016). A Comprehensive Survey of Pilot Contamination in Massive MIMO—5G System. *IEEE Communications Surveys and Tutorials*, *18*(2), 905–923. doi:10.1109/COMST.2015.2504379

Elshorbagy, A., & Parasuraman, K. (2008). On the relevance of using artificial neural networks for estimating soil moisture content. *Journal of Hydrology (Amsterdam)*, *362*(1-2), 1–18.

Esposito, M., Crimaldi, M., Cirillo, V., Sarghini, F., & Maggio, A. (2021). Drone and sensor technology for sustainable weed management: A review. *Chemical and Biological Technologies in Agriculture*, *8*(1), 1–11. doi:10.118640538-021-00217-8

Eusuff, H., Lansey, M., & Pasha, F. (2006). Shuffled frog-leaping algorithm: A memetic meta-heuristic for discrete optimization. *Engineering Optimization*, *38*(2), 129–154.

Evans, R. G., LaRue, J., Stone, K. C., & King, B. A. (2013). Adoption of site-specific variable rate sprinkler irrigation systems. *Irrigation Science*, *31*(4), 871–887. doi:10.100700271-012-0365-x

FAO. (2019, April 3). New standards to curb the global spread of plant pests and diseases. FAO. https://www.fao.org/news/story/en/item/1187738/icode/

FAO. (2020). Agriculture 4.0: Start Agricultural robotics and automated equipment for sustainable crop production. *Integrated Crop Management*, 24.

Faraci, G., Raciti, A., Rizzo, S., & Schembra, G. (2018). A 5G platform for Unmanned Aerial Monitoring in Rural Areas: Design and Performance Issues. *Conf. Netw. Softwarization Work. NetSoft 2018* (pp. 356–361). IEEE.

Fedorova, E., Darbasov, V., & Okhlopkov, M. (2020). The role of agricultural economists in study on problems related to regional food safety. In *E3S Web of Conferences* (Vol. 176, p. 05011). EDP Sciences.

Feleen, F., & David, A. (2021). A Comparative Study of Work from Home Vs Work from Office: Preference of Women Employees in IT Industry. *Design Engineering (London)*, *7*(1), 5763–5775.

Fennimore, S. A., Slaughter, D. C., Siemens, M. C., Leon, R. G., & Saber, M. N. (2016). Technology for automation of weed control in specialty crops. *Cambridge.Org, 30*, 823–837. doi:10.1614/WT-D-16-00070.1

Ferentinos, K. P. (2018). Deep learning models for plant disease detection and diagnosis. *Computers and Electronics in Agriculture*, *145*, 311–318. https://doi.org/10.1016/J.COMPAG.2018.01.009

Filimonau, V., Todorova, E., Mzembe, A., Sauer, L., & Yankholmes, A. (2020). A comparative study of food waste management in full service restaurants of the United Kingdom and the Netherlands. *Journal of Cleaner Production*, *258*, 120775.

Fortes, E. P. (2017). Seed plant drone for reforestation. *The Graduate Review*, *2*(1), 13–26.

Fox, G., Mooney, J., Rosati, P., & Lynn, T. (2021). AgriTech Innovators: A Study of Initial Adoption and Continued Use of a Mobile Digital Platform by Family-Operated Farming Enterprises. *Agriculture, 11*(12), 1283. doi:10.3390/agriculture11121283

Frasconi, C., Raffaelli, M., Emmi, L., Fontanelli, M., Martelloni, L., & Peruzzi, A. (2017). An automatic machine able to perform variable rate application of flame weeding: design and assembly. *Chemical Engineering Transactions, 58.*

Friha, O., Ferrag, M. A., Shu, L., Maglaras, L., & Wang, X. (2021). Internet of Things for the Future of Smart Agriculture: A Comprehensive Survey of Emerging Technologies. *IEEE/CAA Journal of Automatica Sinica, 8*(4), 718–752. doi:10.1109/JAS.2021.1003925

Frost, A. R., Schofield, C. P., Beaulah, S. A., Mottram, T. T., Lines, J. A., & Wathes, C. M. (1997). A review of livestock monitoring and the need for integrated systems. *Computers and Electronics in Agriculture, 17*(2), 139–159.

Fuentes, A., Yoon, S., Kim, S. C., & Park, D. S. (2017). A Robust Deep-Learning-Based Detector for Real-Time Tomato Plant Diseases and Pests Recognition. *Sensors, 17*(9), 2022. doi:10.3390/S17092022

Fujita, E., Kawasaki, Y., Uga, H., Kagiwada, S., & Iyatomi, H. (2016). Basic investigation on a robust and practical plant diagnostic system. In *15th International Conference on Machine Learning and Applications (ICMLA)*, 989–992. IEEE. 10.1109/ICMLA.2016.0178

Ganesh Kumar, C., Murugaiyan, P., & Madanmohan, G. (2017). Agri-food supply chain management: A literature review. *Intelligent Information Management, 9*(02), 68–96. doi:10.4236/iim.2017.92004

Ganeshkumar, C., David, A., & Jebasingh, D. R. (2022). Digital Transformation: Artificial Intelligence Based Product Benefits and Problems of Agritech Industry. In *Agri-Food 4.0.* Emerald Publishing Limited. doi:10.1108/S1877-636120220000027010

Ganeshkumar, C., Jena, S. K., Sivakumar, A., & Nambirajan, T. (2021). Artificial intelligence in the agricultural value chain: Review and future directions. *Journal of Agribusiness in Developing and Emerging Economies.* doi:10.1108/JADEE-07-2020-0140

Ganeshkumar, C., & Khan, A. (2021). Mapping of Agritech Companies in Indian Agricultural Value Chain. In *Proceedings of the Second International Conference on Information Management and Machine Intelligence* (pp. 155-161). Springer, Singapore. 10.1007/978-981-15-9689-6_18

Ganeshkumar, C., Prabhu, M., & Abdullah, N. N. (2019). Business analytics and supply chain performance: Partial least squares-structural equation modeling (PLS-SEM) Approach. *International Journal of Management and Business Research, 9*(1), 91–96.

Ganeshkumar, C., Prabhu, M., Reddy, P. S., & David, A. (2020). Value Chain Analysis of Indian Edible Mushrooms. *International Journal of Technology, 11*(3), 599–607. doi:10.14716/ijtech.v11i3.3979

Garg, D., Khan, S., & Alam, M. (2020). Integrative use of IoT and deep learning for agricultural applications. *Proceedings of ICETIT*, 521–531.

Gašparović, M., Zrinjski, M., Barković, Đ., & Radočaj, D.Gaˇsparovi'c. (2020). An automatic method for weed mapping in oat fields based on UAV imagery. *Computers and Electronics in Agriculture, 173*, 105385. doi:10.1016/j.compag.2020.105385

Gavai, P., Musungwini, S., & Mugoniwa, B. (2018). A Model for the Adoption and Effective Utilization of ICTs in Commercial Agriculture in Zimbabwe. *Journal of Systems Integration, 9*(4), 40–58. doi:10.20470/jsi.v9i4.356

Gebeyehu, M. N. (2019). Remote sensing and GIS application in agriculture and natural resource management. *International Journal of Environmental Sciences & Natural Resources, 19*(2), 45–49. doi:10.19080/IJESNR.2019.19.556009

Ghahramani, Z. (2003). Unsupervised learning. *Advanced Lectures on Machine Learning.*

Gilbert, N. (2011, March). Climate-smart agriculture is needed. *Nature.* doi:10.1038/news.2011.131

Given, L. M. (2008). Naturalistic Inquiry. In *Encyclopedia of Research Design.* SAGE Publications, Inc., doi:10.4135/9781412961288.n262

Goap, A., Sharma, D., Shukla, A. K., & Rama Krishna, C. (2018). An IoT based smart irrigation management system using Machine learning and open source technologies. *Computers and Electronics in Agriculture, 155*, 41–49. https://doi.org/10.1016/J.COMPAG.2018.09.040

Greasley, A., & Edwards, J. S. (2021). Enhancing discrete-event simulation with big data analytics: A review. *The Journal of the Operational Research Society, 72*(2), 247–267.

Grogan, A. (2012). Smart farming. *Engineering & Technology, 7*(6), 38–40.

Grogan, A. (2012, July). Smart farming. *Eng. Technol., 7*(6), 38–40. doi:10.1049/et.2012.0601

GSMA. (2018). Road to 5G: Introduction and Migration. *Gsma*, p. 54.: https://www.gsma.com/futurenetworks/wp-content/uploads/ 2018 /04/Road-to-5G-Introduction-and-Migration_FINAL.pdf

Gupta, A., & Jha, R. K. (2015). A Survey of 5G Network: Architecture and Emerging Technologies. *IEEE Access: Practical Innovations, Open Solutions, 3*, 1206–1232. doi:10.1109/ACCESS.2015.2461602

Gupta, M., Abdelsalam, M., Khorsandroo, S., & Mittal, S. (2020). Security and privacy in smart farming: Challenges and opportunities. *IEEE Access: Practical Innovations, Open Solutions, 8*, 34564–34584.

Gu, Z., Qi, Z., Ma, L., Gui, D., Xu, J., Fang, Q., Yuan, S., & Feng, G. (2017). Development of an irrigation scheduling software based on model predicted crop water stress. *Computers and Electronics in Agriculture, 143*, 208–221. https://doi.org/10.1016/J.COMPAG.2017.10.023

Hady, M. F. A., & Schwenker, F. (2013). Semi-supervised Learning. In M. Bianchini, M. Maggini, & L. C. Jain (Eds.), *Handbook on Neural Information Processing* (pp. 215–239). Springer Berlin Heidelberg.

Haghverdi, A., & Washington-Allen, R. (2018). Prediction of cotton lint yield from phenology of crop indices using artificial neural networks. *Elsevier.* doi:10.1016/j.compag.2018.07.021

Hakkim, V. A., Joseph, E. A., Gokul, A. A., & Mufeedha, K. (2016). Precision farming: the future of Indian agriculture. *Journal of Applied Biology and Biotechnology, 4*(6), 0-7.

Harishankar, S., Kumar, R. S., Sudharsan, K. P., Vignesh, U., & Viveknath, T. (2014). Solar powered smart irrigation system. *Advance in electronic and electric engineering, 4*(4), 341-346.

Harrell, F. E. (2015). Ordinal logistic regression. In *Regression modeling strategies* (pp. 311–325). Springer. doi:10.1007/978-3-319-19425-7_13

Harris, D. C. (1990). Control of verticillium wilt and other soil-borne diseases of strawberry in britain by chemical soil disinfestation. *Journal of Horticultural Science, 65*(4), 401–408. doi:10.1080/00221589.1990.11516072

Harris, D., Pathan, A. K., Gothkar, P., Joshi, A., Chivasa, W., & Nyamudeza, P. (2001). On-farm seed priming: Using participatory methods to revive and refine a key technology. [doi:https://doi.org/10.1016/S0308-521X(01)00023-3]. *Agricultural Systems*, *69*(1), 151–164.

Harun, A. N., Kassim, M. R. M., Mat, I., & Ramli, S. S. (2015, May). Precision irrigation using wireless sensor network. In *International Conference on Smart Sensors and Application (ICSSA)* (pp. 71-75). IEEE. 10.1109/ICSSA.2015.7322513

Hassan, W. H. (2019). Current research on Internet of Things (IoT) security: A survey. *Computer Networks*, *148*, 283–294. doi:10.1016/j.comnet.2018.11.025

Hassebo, A., Obaidat, M., & Ali, M. A. (2018). Commercial 4G LTE cellular networks for supporting emerging IoT applications. In *Advances in Science and Engineering Technology International Conferences (ASET)*, pp. 1–6. 10.1109/ICASET.2018.8376832

Hastie, T., Tibshirani, R., & Friedman, J. (2009). Unsupervised learning. In *The elements of statistical learning* (pp. 485–585). Springer. doi:10.1007/978-0-387-84858-7_14

Hedley, C. B., & Yule, I. J. (2009). Soil water status mapping and two variable-rate irrigation scenarios. *Precision Agriculture*, *10*(4), 342–355. doi:10.100711119-009-9119-z

Heisel, T., Schou, J., Christensen, S., & Andreasen, C. (2001). Cutting weeds with a CO 2 laser. *Weed Research*, *41*(1), 19–29. https://doi.org/10.1046/J.1365-3180.2001.00212.X

Hinnell, A., Lazarovitch, N., & Furman, A. (2010). Neuro-Drip: estimation of subsurface wetting patterns for drip irrigation using neural networks. *Springer, 28*(6), 535–544. doi:10.1007/s00271-010-0214-8

Honrubia, M. (2021). *4 Agriculture technology projects to transform the industry.*

Hristoski, I., Kostoska, O., Dimovski, T., & Kotevski, Z. (2017). Farm management software for increased productivity and competitiveness.

Huang, J., Qian, F., Gerber, A., Mao, Z. M., Sen, S., & Spatscheck, O. (2012, June). A close examination of performance and power characteristics of 4G LTE networks. In *Proceedings of the 10th international conference on Mobile systems, applications, and services* (pp. 225-238). 10.1145/2307636.2307658

Hughes, D., & Salathé, M. (2015). An open access repository of images on plant health to enable the development of mobile disease diagnostics. *arXiv preprint arXiv:1511.08060.*

Hui, Y. Y., Choy, K. L., Ho, G. T. S., Leung, K. H., & Lam, H. Y. (2016). A cloud-based location assignment system for packaged food allocation in e-fulfillment warehouse. *Sage Journals, 8*, 1–15. doi:10.1177/1847979016684832

Hung, M. C., Wu, J., Chang, J. H., & Yang, D. L. (2005). An Efficient k-Means Clustering Algorithm Using Simple Partitioning. *Journal of Information Science and Engineering*, *21*(6), 1157–1177.

Ibáñez, M. D., & Blázquez, M. A. (2019). Ginger and Turmeric Essential Oils for Weed Control and Food Crop Protection. *Plants*, *8*(3). Advance online publication. doi:10.3390/plants8030059

IBEF. (2022). *Indian drone industry reaching the skies. India Brand Equity Foundation. Ministry of Commerce and Industry*. Government of India.

ICRISAT. (2021). Digital Agriculture: Pathway to Prosperity. In: International Crops Research Institute For The Semi-Arid Tropics. *ISCRISAT.* https://www.icrisat.org/digital-agriculture/

Iera, A., Molinaro, A., Polito, S., & Ruggeri, G. (2005). End-to-end QoS provisioning in 4G with mobile hotspots. *IEEE Network*, *19*(5), 26–34. doi:10.1109/MNET.2005.1509949

Infosys. (2018). AI for Healthcare: Balancing Efficiency and Ethics. *Infosys.* https://www.infosys.com/smart-automation/docpdf/ai-healthcare.pdf

Iost Filho, F. H., Heldens, W. B., Kong, Z., & de Lange, E. S. (2020). Drones: Innovative technology for use in precision pest management. *Journal of Economic Entomology, 113*(1), 1–25. doi:10.1093/jee/toz268 PMID:31811713

Iqbal, M. A., & Talukder, K. H. (2020). Detection of Potato Disease Using Image Segmentation and Machine Learning. *International Conference on Wireless Communications, Signal Processing and Networking,* 43–47. doi:10.1109/WISPNET48689.2020.9198563

Ivancic, W., Murawski, R., Matheou, K., & Downey, A. (2019). Flying Drones beyond Visual Line of Sight Using 4G LTE: Issues and Concerns. *Integr. Commun. Navig. Surveill. Conf.* (pp. 1-13). 10.1109/ICNSURV.2019.8735246

Jagtap, S. T., Phasinam, K., Kassanuk, T., Jha, S. S., Ghosh, T., & Thakar, C. M. (2022). Towards application of various machine learning techniques in agriculture. *Materials Today: Proceedings, 51,* 793-797. doi:10.1016/j.matpr.2021.06.236

Jain, D., Krishna, P. V., & Saritha, V. (2012). A study on Internet of Things based applications.

Jan, B., Farman, H., Javed, H., Montrucchio, B., Khan, M., & Ali, S. (2017). Energy efficient hierarchical clustering approaches in wireless sensor networks: A survey. *Wireless Communications and Mobile Computing*, 2017.

Jawahar, R., & Umarani, R. (2019). Development of Seed Cube Technology with Enhanced Seeds of Albizia lebbeck for Rapid Propagation in Fallow Lands. *International Journal of Current Microbiology and Applied Sciences*, *8*(6), 1603–1613.

Jayaraman, P. P., Yavari, A., Georgakopoulos, D., Morshed, A., & Zaslavsky, A. (2016). Internet of things platform for smart farming: Experiences and lessons learnt. *Sensors (Basel), 16*(11), 1884. doi:10.339016111884 PMID:27834862

Jayne, T. S. (2012). Managing food price instability in East and Southern Africa. *Global Food Security, 1*(2), 143–149. doi:10.1016/j.gfs.2012.10.002

Jha, K., Doshi, A., Patel, P., & Shah, M. (2019). A comprehensive review on automation in agriculture using artificial intelligence. *Artificial Intelligence in Agriculture, 2,* 1–12.

Johnston, A. E., Poulton, P. R., & Coleman, K. (2009). *Soil Organic Matter.*

Joshi, E., Sasode, D. S., Singh, N., & Chouhan, N. (2020). Revolution of Indian Agriculture through Drone Technology. *Biotica Research Today, 2*(5 Spl.), 174-176.

Jungnickel, V., Manolakis, K., Zirwas, W., Panzner, B., Braun, V., Lossow, M., Sternad, M., Apelfrojd, R., & Svensson, T. (2014). The role of small cells, coordinated multipoint, and massive MIMO in 5G. *IEEE Communications Magazine, 52*(5), 44–51. doi:10.1109/MCOM.2014.6815892

Kaelbling, L. P., Littman, M. L., & Moore, A. W. (1996). Reinforcement learning: A survey. *Journal of Artificial Intelligence Research, 4,* 237–285. doi:10.1613/jair.301

Kakani, V., Nguyen, V. H., Kumar, B. P., Kim, H., & Pasupuleti, V. R. (2020). A critical review on computer vision and artificial intelligence in food industry. *Journal of Agriculture and Food Research, 2,* 100033. https://doi.org/10.1016/J.JAFR.2020.100033

Kameshwaran, K., & Malarvizhi, K. (2014). Survey on clustering techniques in data mining. *International Journal of Computer Science and Information Technologies*, *5*(2), 2272–2276.

Kamilaris, A., Gao, F., Prenafeta-Boldu, F. X., & Ali, M. I. (2016, December). Agri-IoT: A semantic framework for Internet of Things-enabled smart farming applications. In *2016 IEEE 3rd world forum on internet of things (WF-IoT)* (pp. 442-447). IEEE. 10.1109/WF-IoT.2016.7845467

Kamilaris, A., Gao, F., Prenafeta-Boldu, F. X., & Ali, M. I. (2016, December). Agri-IoT: A semantic framework for Internet of Things-enabled smart farming applications. In *3rd world forum on internet of things (WF-IoT)* (pp. 442-447). IEEE.

Kamilaris, A., & Prenafeta-Boldú, F. X. (2018). Deep learning in agriculture: A survey. *Computers and Electronics in Agriculture*, *147*, 70–90. https://doi.org/10.1016/J.COMPAG.2018.02.016

Kamutuezu, E. U., Winschiers-Theophilus, H., & Peters, A. (2021). An Exploration of Factors Influencing the Adoption of ICT Enabled Entrepreneurship Applications in Namibian Rural Communities. *Adoption of Rural Entrepreneurship Digital Services in Namibia.*

Kanchana, S. (2018). IoT in Agriculture : Smart Farming. *International Journal of Scientific Research in Computer Science. Engineering and Information Technology*, (May), 181–184. doi:10.32628/CSEIT183856

Kante, M., Oboko, R., & Chepken, C. (2019). An ICT model for increased adoption of farm input information in developing countries: A case in Sikasso, Mali. *Information Processing in Agriculture*, *6*(1), 26–46. doi:10.1016/j.inpa.2018.09.002

Kanungo, T., Mount, D. M., Netanyahu, N. S., Piatko, C. D., Silverman, R., & Wu, A. Y. (2002). An efficient k-means clustering algorithm: Analysis and implementation. *IEEE Transactions on Pattern Analysis and Machine Intelligence*, *24*(7), 881–892. doi:10.1109/TPAMI.2002.1017616

Karar, M. E., Alsunaydi, F., Albusaymi, S., & Alotaibi, S. (2021). A new mobile application of agricultural pests recognition using deep learning in cloud computing system. *Alexandria Engineering Journal, 60*(5), 4423-4432. doi:10.1016/j.aej.2021.03.009

Kasinathan, T., Singaraju, D., & Uyyala, S. R. (2021). Insect classification and detection in field crops using modern machine learning techniques. [doi:https://doi.org/10.1016/j.inpa.2020.09.006]. *Information Processing in Agriculture*, *8*(3), 446–457.

Katarya, R., Raturi, A., Mehndiratta, A., & Thapper, A. (2020, February). Impact of machine learning techniques in precision agriculture. In *2020 3rd International Conference on Emerging Technologies in Computer Engineering: Machine Learning and Internet of Things (ICETCE)* (pp. 1-6). IEEE. 10.1109/ICETCE48199.2020.9091741

Kecman, V. (2005). Support vector machines–an introduction. In *Support vector machines: theory and applications* (pp. 1–47). Springer. doi:10.1007/10984697_1

Kehui, X., Deqin, X., & Xiwen, L. (2010). Smart water-saving irrigation system in precision agriculture based on wireless sensor network. *Nongye Gongcheng Xuebao (Beijing)*, *26*(11), 170–175. https://doi.org/10.3969/j.issn.1002-6819.2010.11.030

Kemiklioglu, E., & Ozen, O. (2018). Design of a sensor to detect fruit freshness. *International Journal of Scientific and Technological Research*, *4*(01), 1–6.

Khan, A., Sohail, A., Zahoora, U., & Qureshi, A. S. (2020). A survey of the recent architectures of deep convolutional neural networks. *Artificial Intelligence Review*, *53*(8), 5455–5516. doi:10.100710462-020-09825-6

Khanal, S., Fulton, J., & Shearer, S. (2017). An overview of current and potential applications of thermal remote sensing in precision agriculture. *Computers and Electronics in Agriculture*, *139*, 22–32. doi:10.1016/j.compag.2017.05.001

Khanal, S., Kc, K., Fulton, J. P., Shearer, S., & Ozkan, E. (2020). Remote sensing in agriculture—Accomplishments, limitations, and opportunities. *Remote Sensing*, *12*(22), 3783. doi:10.3390/rs12223783

Khare, V., & Seth, A. (2020). The case for AI in smallholder farmer water management. In AI holds water: A case for AI-enabled water management in agriculture. [White Paper, Wadhwani Institute for Artificial Intelligence, India].

Khattab, A., Abdelgawad, A., & Yelmarthi, K. (2016, December). Design and implementation of a cloud-based IoT scheme for precision agriculture. In *2016 28th International Conference on Microelectronics (ICM)* (pp. 201-204). IEEE.

Khedr, A. E., Kadry, M., & Walid, G. (2015). Proposed Framework for Implementing Data Mining Techniques to Enhance Decisions in Agriculture Sector Applied Case on Food Security Information Center Ministry of Agriculture, Egypt. *Procedia Computer Science*, *65*, 633–642. https://doi.org/10.1016/J.PROCS.2015.09.007

Khorsandi, A., Alimardani, A., Vahidi, B., & Hosseinian, S. H. (2011). Hybrid shuffled frog leaping algorithm and nelder–mead simplex search for optimal reactive power dispatch. *IET Generation, Transmission & Distribution*, *5*(2), 249–256.

Khosla, R. (2008). Myths of Precision Farming. *From the ground up, agronomy news.*

Kiani, F., Seyyedabbasi, A., Nematzadeh, S., Candan, F., Çevik, T., Anka, F., Randazzo, G., Lanza, S., & Muzirafuti, A. (2022). Adaptive metaheuristic-based methods for autonomous robot path planning: Sustainable agricultural applications. *Applied Sciences (Basel, Switzerland)*, *12*(3), 943. doi:10.3390/app12030943

Kim, M. agriculture, J. G.-C. and electronics in, & 2008, undefined. (2008). Artificial Neural Network estimation of soil erosion and nutrient concentrations in runoff from land application areas. *Elsevier*. doi:10.1016/j.compag.2008.05.021

Kim, K., Bae, S., & Huh, K. (2010). Intelligent surveillance and security robot systems. *IEEE Workshop on Advanced Robotics and its Social Impacts* (pp. 70-73). IEEE.

Kim, Y. G., Kim, H. K., Lee, S. G., & Lee, K. D. (2006). Ubiquitous home security robot based on sensor network. *IEEE/WIC/ACM International Conference on Intelligent Agent Technology* (pp. 700-704). IEEE. 10.1109/IAT.2006.128

Kim, Y. G., Kim, H. K., Yoon, S. H., Lee, S. G., & Lee, K. D. (2006). Home security robot based on Sensor Network. *SICE-ICASE International Joint Conference* (pp. 5977-5982). IEEE.

Kim, Y., Evans, R. G., & Iversen, W. M. (2008). Remote sensing and control of an irrigation system using a distributed wireless sensor network. *IEEE Transactions on Instrumentation and Measurement*, *57*(7), 1379–1387. doi:10.1109/TIM.2008.917198

King, J. E. (2008). Binary logistic regression. *Best practices in quantitative methods*, 358-384.

Kingsford, C., & Salzberg, S. L. (2008). What are decision trees? *Nature Biotechnology*, *26*(9), 1011–1013. doi:10.1038/nbt0908-1011 PMID:18779814

Kirkpatrick, K. (2019). Technologizing agriculture. *Communications of the ACM*, *62*(2), 14–16. doi:10.1145/3297805

Klauser, F., & Pauschinger, D. (2021). Entrepreneurs of the air: Sprayer drones as mediators of volumetric agriculture. *Journal of Rural Studies*, *84*, 55–62. doi:10.1016/j.jrurstud.2021.02.016

Komarek, A. M., De Pinto, A., & Smith, V. H. J. A. S. (2020). A review of types of risks in agriculture: What we know and what we need to know. *Agricultural Systems, 178*, 102738.

Korlahalli, K. B., Hangal, M.A., Jituri, N., Rego, P.F., Raykar, S.M. (n.d.). An Automatically Controlled Drone Based Aerial Pesticide Sprayer. *Project Reference No.39S_BE_0564*

Kotsiantis, S. B., Zaharakis, I., & Pintelas, P. (2007). Supervised machine learning: A review of classification techniques. *Emerging artificial intelligence applications in computer engineering, 160*(1), 3-24.

Kouadio, L., Deo, R. C., Byrareddy, V., Adamowski, J. F., & Mushtaq, S. (2018). Artificial intelligence approach for the prediction of Robusta coffee yield using soil fertility properties. *Computers and Electronics in Agriculture, 155*, 324–338.

Krisztin, T. (2018). Semi-parametric spatial autoregressive models in freight generation modeling. *Transportation Research Part E, Logistics and Transportation Review, 114*, 121–143. https://doi.org/10.1016/J.TRE.2018.03.003

Kukreja, H., Bharath, N., Siddesh, C., & Kuldeep, S. J. I. J. A. R. I. I. E. (2016). An introduction to artificial neural network. 1, 27-30.

Kumar, I., Rawat, J., Mohd, N., & Husain, S. (2021). Opportunities of Artificial Intelligence and Machine Learning in the Food Industry. *Journal of Food Quality*. doi:10.1155/2021/4535567

Kumar, P., Dayal, M., Khari, M., Fenza, G., & Gallo, M. (2021). Nsl-bp: A meta classifier model based prediction of amazon product reviews.

Kumar, R., Singh, M. P., Kumar, P., & Singh, J. P. (2015, May). Crop Selection Method to maximize crop yield rate using machine learning technique. In 2015 international conference on smart technologies and management for computing, communication, controls, energy and materials (ICSTM) (pp. 138-145). IEEE. doi:10.1109/ICSTM.2015.7225403

Kumar, A., Rani, M., & Kumar, P. (2022). Drone Technology in Sustainable Agriculture: The Future of Farming Is Precision Agriculture and Mapping. In *Agriculture, Livestock Production and Aquaculture* (pp. 3–12). Springer. doi:10.1007/978-3-030-93262-6_1

Kumari, R., & Saini, K. (2021, March). Advanced Automobile Manufacturing: An Industry 4.0. In *2021 8th International Conference on Computing for Sustainable Global Development (INDIACom)* (pp. 899-904). IEEE.

Kumari, R., & Saini, K. (2021, March). Advanced Automobile Manufacturing: An Industry 4.0. In *International Conference on Computing for Sustainable Global Development (INDIACom)* (pp. 899-904). IEEE.

Kwak, C., & Clayton-Matthews, A. (2002). Multinomial logistic regression. *Nursing Research, 51*(6), 404–410. doi:10.1097/00006199-200211000-00009 PMID:12464761

Labbé, S., Lebourgeois, V., Jolivot, A., & Marti, R. (2012). Thermal infra-red remote sensing for water stress estimation in agriculture. *Options Méditerranéennes. Série B: Études et Recherches, 67*, 175–184.

Lamm, R. D., Slaughter, D. C., & Giles, D. K. (2002). Precision weed control system for cotton. *Transactions of the ASAE. American Society of Agricultural Engineers, 45*(1), 231.

Larsen, R. A., Schaalje, G. B., & Lawson, J. S. (2009). Food shelf life: estimation and optimal design. *Journal of Statistical Computation and Simulation, 80*(2), 143–157. doi:10.1080/00949650802549135

Latha, M., Poojith, A., Reddy, B. A., & Kumar, G. V. (2014). Image processing in agriculture. *International journal of innovative research in electrical, electronics, instrumentation and control engineering, 2*(6).

Laumond, J. P. (1998). *Robot motion planning and control* (Vol. 229). Springer. doi:10.1007/BFb0036070

Lauzon, F. Q. (2012). An introduction to deep learning. *International Conference on Information Science, Signal Processing and their Applications.*

Lauzon, F. Q. (2012, 2-5 July 2012). An introduction to deep learning. Paper presented at the *International Conference on Information Science, Signal Processing and their Applications* (ISSPA).

LaValley, M. P. (2008). Logistic regression. *Circulation, 117*(18), 2395–2399. doi:10.1161/CIRCULATIONAHA.106.682658 PMID:18458181

Learning, D. (2020). *Deep learning*. High-Dimensional Fuzzy Clustering.

LeCun, Y., Bengio, Y., & Hinton, G. (2015). Deep learning. *Nature*, *521*(7553), 436–444. doi:10.1038/nature14539

Lee, M., Hwang, J., & Yoe, H. (2013). Agricultural Production System Based on IoT. In *16th International Conference on Computational Science and Engineering,* pp. 833–837. doi:10.1109/CSE.2013.126

Lee, D., & Park, J. H. (2019). Future trends of AI-based smart systems and services: Challenges, opportunities, and solutions. *Journal of Information Processing Systems*, *15*(4), 717–723.

Lee, S. K., Bae, M., & Kim, H. (2017). Future of IoT networks: A survey. *Applied Sciences (Basel, Switzerland)*, *7*(10), 1072. doi:10.3390/app7101072

Lezoche, M., Hernandez, J. E., Díaz, M. D. M. E. A., Panetto, H., & Kacprzyk, J. (2020). Agri-food 4.0: A survey of the supply chains and technologies for the future agriculture. *Computers in Industry*, *117*, 103187.

Liakos, K. G., Busato, P., Moshou, D., Pearson, S., & Bochtis, D. (2018). Machine Learning in Agriculture: A Review. *Sensors (Basel)*, *18*(8), 2674. https://doi.org/10.3390/s18082674

Li, B., Fei, Z., & Zhang, Y. (2019). UAV Communications for 5G and Beyond: Recent Advances and Future Trends. *IEEE Internet Things J.*, *6*(2), 2241–2263.

Likas, A., Vlassis, N., & Verbeek, J. J. (2003). The global k-means clustering algorithm. *Pattern Recognition*, *36*(2), 451–461. doi:10.1016/S0031-3203(02)00060-2

Li, M., & Yost, R. S. (2000). Management-oriented modeling: Optimizing nitrogen management with artificial intelligence. *Agricultural Systems*, *65*(1), 1–27.

Li, N., Shepperd, M., & Guo, Y. (2020). A systematic review of unsupervised learning techniques for software defect prediction. [doi:https://doi.org/10.1016/j.infsof.2020.106287]. *Information and Software Technology*, *122*, 106287.

Linaza, M. T., Posada, J., Bund, J., Eisert, P., Quartulli, M., Döllner, J., Pagani, A., Olaizola, I. G., Barriguinha, A., Moysiadis, T., & Lucat, L. (2021). Data-driven artificial intelligence applications for sustainable precision agriculture. *Agronomy (Basel)*, *11*(6). doi:10.3390/agronomy11061227

Lin, X., Yajnanarayana, V., Muruganathan, S., Gao, S., Asplund, H., Maattanen, H.-L., & Wang, Y.-P. (2018). The Sky Is Not the Limit: LTE for Unmanned Aerial Vehicles. *IEEE Communications Magazine*, *56*(4), 204–210.

Lipper, L., Thornton, P., Campbell, B. M., Baedeker, T., Braimoh, A., Bwalya, M., & Torquebiau, E. F. (2014). Climate-smart agriculture for food security. *Nature Climate Change*, *4*(12), 1068–1072.

Li, T., Johansen, K., & McCabe, M. F. (2022). A machine learning approach for identifying and delineating agricultural fields and their multi-temporal dynamics using three decades of Landsat data. [doi:https://doi.org/10.1016/j.isprsjprs.2022.02.002]. *ISPRS Journal of Photogrammetry and Remote Sensing*, *186*, 83–101.

Liu, J. N., Wang, M., & Feng, B. (2005). iBotGuard: An Internet-based intelligent robot security system using invariant face recognition against intruder. *IEEE Transactions on Systems, Man and Cybernetics. Part C, Applications and Reviews*, *35*(1), 97–105. doi:10.1109/TSMCC.2004.840051

Livingstone, D. J. (Ed.). (2008). *Artificial neural networks: methods and applications* (pp. 185–202). Humana Press.

Li, Y., & Chao, X. (2020). ANN-Based Continual Classification in Agriculture. *Agriculture*, *10*(5). doi:10.3390/agriculture10050178

Lokhorst, C., De Mol, R. M., & Kamphuis, C. (2019). Invited review: Big Data in precision dairy farming. *Animal*, *13*(7), 1519–1528. doi:10.1017/S1751731118003439 PMID:30630546

López, E. M., García, M., Schuhmacher, M., & Domingo, J. L. (2008). A fuzzy expert system for soil characterization. *Environment International*, *34*(7), 950–958.

López-Granados, F. (2011). Weed detection for site-specific weed management: Mapping and real-time approaches. *Weed Research*, *51*(1), 1–11. doi:10.1111/j.1365-3180.2010.00829.x

López-Granados, F., Torres-Sánchez, J., Serrano-Pérez, A., de Castro, A. I., Mesas-Carrascosa, F., & Peña, J. M. (2016). Early season weed mapping in sunflower using UAV technology: Variability of herbicide treatment maps against weed thresholds. *Precision Agriculture*, *17*(2), 183–199. doi:10.100711119-015-9415-8

López-Morales, J. A., Martínez, J. A., & Skarmeta, A. F. (2020). Digital transformation of agriculture through the use of an interoperable platform. *Sensors (Basel)*, *20*(4), 1153. doi:10.339020041153 PMID:32093147

Loures, L., Chamizo, A., Ferreira, P., Loures, A., Castanho, R., & Panagopoulos, T. (2020). Assessing the effectiveness of precision agriculture management systems in mediterranean small farms. *Sustainability (Switzerland)*, *12*(9), 1–15. doi:10.3390u12093765

Lu, Y.-C., Daughtry, C., Hart, G., & Watkins, B. (1997). The current state of precision farming. *Food Rev. Int. 13*(2), 141–162. 87559129709541104. doi:10.1080/

Luangkesorn, K. L., Klein, G., & Bidanda, B. (2016). Analysis of production systems with potential for severe disruptions. *International Journal of Production Economics*, *171*, 478–486. https://doi.org/10.1016/J.IJPE.2015.09.014

Luo, R. C., & Chou, Y. T. L. C., Lai, C. C., & Tsai, A. C. (2007). NCCU security warrior: An intelligent security robot system. *33rd Annual Conference of the IEEE Industrial Electronics Society* (pp. 2960-2965). IEEE.

Luo, R. C., Hsu, T. Y., Lin, T. Y., & Su, K. L. (2005). The development of intelligent home security robot. *International Conference on Mechatronics* (pp. 422-427). Taipei, Taiwan: IEEE. 10.1109/ICMECH.2005.1529294

Luo, R. C., Liao, C. T., Su, K. L., & Lin, K. C. (2005). Automatic docking and recharging system for autonomous security robot. *RSJ International Conference on Intelligent Robots and Systems* (pp. 2953-2958). IEEE. 10.1109/IROS.2005.1545197

Luo, R. C., Lin, T. Y., & Su, K. L. (2009). Multisensor based security robot system for intelligent building. *Robotics and Autonomous Systems*, *57*(3), 330–338. doi:10.1016/j.robot.2008.10.025

Lu, Y. C., Daughtry, C., Hart, G., & Watkins, B. (1997). The current state of precision farming. *Food Reviews International*, *13*(2), 141–162.

Madaan, G., Swapna, H. R., Kumar, A., Singh, A., & David, A. (2021). Enactment of sustainable technovations on healthcare sectors. *Asia Pacific Journal of Health Management*, *16*(3), 184–192. doi:10.24083/apjhm.v16i3.989

Magomadov, V. S. (2019). Deep learning and its role in smart agriculture. *Journal of Physics: Conference Series*, *1399*, 044109. https://doi.org/10.1088/1742-6596/1399/4/044109

Mahdavinejad, M. S., Rezvan, M., Barekatain, M., Adibi, P., Barnaghi, P., & Sheth, A. P. (2018). Machine learning for Internet of Things data analysis: A survey. *Digital Communications and Networks*, *4*(3), 161–175. doi:10.1016/j.dcan.2017.10.002

Ma, J., Du, K., Zheng, F., Zhang, L., Gong, Z., & Sun, Z. (2018). A recognition method for cucumber diseases using leaf symptom images based on deep convolutional neural network. *Computers and Electronics in Agriculture*, *154*, 18–24.

Malavade, V. N., & Akulwar, P. K. (2016). Role of IoT in agriculture. *IOSR Journal of Computer Engineering, 2016*, 2278–0661.

Mananze, S., Pôças, I., & Cunha, M. (2018). Retrieval of Maize Leaf Area Index Using Hyperspectral and Multispectral Data. *Remote Sensing, 10*(12), 1942. doi:10.3390/RS10121942

Mannila, H. (1996). Data mining: machine learning, statistics, and databases. *Proceedings of 8th International Conference on Scientific and Statistical Data Base Management.*

Martin, J., Amin, R., Eltawil, A., & Hussien, A. (2011, June). Limitations of 4G wireless systems. In *Proceedings of Virginia Tech Wireless Symposium* (Blacksburg, VA).

Maulud, D. H., & Abdulazeez, A. M. (2020). *A Review on Linear Regression Comprehensive in Machine Learning.*

Mays, N., & Pope, C. (1995). Qualitative Research: Rigour and qualitative research. *BMJ (Clinical Research Ed.), 311*(6997), 109–112. doi:10.1136/bmj.311.6997.109 PMID:7613363

Mazoyer, M., & Roudart, L. (2006). *A history of world agriculture: from the neolithic age to the current crisis.* NYU Press.

McCalla, A. F. J. U. A., & Economics, R. (2001). Challenges to world agriculture in the 21st century. *Agricultural and Resource Economics, 4*(3), 1-2.

Mckinley, R., & Talboys, P. W. (1979). Effects of Pratylenchus penetrans on development of strawberry wilt caused by Verticillium dahliae. *Annals of Applied Biology, 92*(3), 347–357. doi:10.1111/j.1744-7348.1979.tb03884.x

McQueen, R. J., Garner, S. R., Nevill-Manning, C. G., & Witten, I. H. (1995). Applying machine learning to agricultural data. *Computers and Electronics in Agriculture, 12*(4), 275–293. doi:10.1016/0168-1699(95)98601-9

Medela, A., Cendón, B., Gonzalez, L., Crespo, R., & Nevares, I. (2013, July). IoT multiplatform networking to monitor and control wineries and vineyards. In Future Network & Mobile Summit (pp. 1-10). IEEE.

Mehra, M., Saxena, S., Sankaranarayanan, S., Tom, R. J., & Veeramanikandan, M. (2018). IoT based hydroponics system using Deep Neural Networks. *Computers and Electronics in Agriculture, 155*, 473–486.

Meivel, S., Maguteeswaran, D. R., Gandhiraj, N., & Srinivasan, G. (2016). Quadcopter UAV Based Fertilizer and Pesticide Spraying System *International Academic Research Journal of Engineering Sciences.*

Menard, S. (2002). Applied logistic regression analysis. *Sage (Atlanta, Ga.).*

Meng'Ao, L., Dongxue, M., Songyuan, G., & Shufen, L. (2015). *Research and improvement of DBSCAN cluster algorithm.*

Meyer, D., & Wien, F. T. (2015). Support vector machines. *The Interface to libsvm in package e1071, 28*, 20.

Miller, H. G., & Mork, P. (2013). From data to decisions: A value chain for big data. *IT Professional, 15*(1), 57–59. doi:10.1109/MITP.2013.11

Milner, G. (2016). What is GPS? *Journal of Technology in Human Services, 34*(1), 9–12.

Mishra, M., & Srivastava, M. (2014). A view of artificial neural network. Paper presented at the *International Conference on Advances in Engineering & Technology Research (ICAETR).*

Misra, N. N., Dixit, Y., Al-Mallahi, A., Bhullar, M. S., Upadhyay, R., & Martynenko, A. (2020). *IoT, big data and artificial intelligence in agriculture and food industry.* IEEE Internet of Things Journal.

Mitton, N., Brossard, L., Bouadi, T., Garcia, F., Gautron, R., Hilgert, N., & Bellon Maurel, V. (2022). Foundations and state of play. In *Agriculture and Digital Technology: getting the most out of digital technology to contribute to the transition to sustainable agriculture and food systems*. INRAE, p. 30-75.

Mogili, U. R., & Deepak, B. B. V. L. (2018). Review on application of drone systems in precision agriculture. *Procedia Computer Science, 133*, 502–509. doi:10.1016/j.procs.2018.07.063

Mogili, U. R., & Deepak, B. B. V. L. (2020). An intelligent drone for agriculture applications with the aid of the mavlink protocol. In *Innovative Product Design and Intelligent Manufacturing Systems* (pp. 195–205). Springer. doi:10.1007/978-981-15-2696-1_19

Mohamed, E. S., Belal, A. A., Abd-Elmabod, S. K., El-Shirbeny, M. A., Gad, A., & Zahran, M. B. (2021). Smart farming for improving agricultural management. *The Egyptian Journal of Remote Sensing and Space Sciences.*

Mohanty, S. P., Hughes, D. P., & Salathé, M. (2016). Using deep learning for image-based plant disease detection. *Frontiers in Plant Science, 7*(September), 1419. https://doi.org/10.3389/FPLS.2016.01419/BIBTEX

Moisture, S. (2017). Importance of Soil Water. *Agriinfo.* https://agriinfo.in/soil-moisture-importance-of-soil-water-263/

Mojjada, R. K., Kiran Kumar, K., Yadav, A., & Satya Vara Prasad, B. V. V. (2020). Detection of plant leaf disease using digital image processing. *Materials Today: Proceedings.* doi:10.1016/J.MATPR.2020.11.115

Molden, D. (2013). *Water for food water for life: A comprehensive assessment of water management in agriculture.* Routledge.

Monga, T. (2018). Estimating Vineyard Grape Yield from Images. Lecture Notes in Computer Science (Including Subseries Lecture Notes in Artificial Intelligence and Lecture Notes in Bioinformatics). LNAI, 339–343. doi:10.1007/978-3-319-89656-4_37

Montas, H., & Madramootoo, C. A. (1992). A decision support system for soil conservation planning. *Computers and Electronics in Agriculture, 7*(3), 187–202.

Monteiro, A. L., Freitas Souza, M. d., Lins, H. A., Teófilo, T. M. S., Barros Júnior, A. P., Silva, D. V., & Mendonça, V. (2021). A new alternative to determine weed control in agricultural systems based on artificial neural networks (ANNs). [doi:https://doi.org/10.1016/j.fcr.2021.108075]. *Field Crops Research, 263*, 108075.

Montgomery, D. C., Peck, E. A., & Vining, G. G. (2021). *Introduction to linear regression analysis.* John Wiley & Sons.

Moore, M. M., Slonimsky, E., Long, A. D., Sze, R. W., & Iyer, R. S. (2019). Machine learning concepts, concerns and opportunities for a pediatric radiologist. *Pediatric Radiology, 49*(4), 509–516. doi:10.100700247-018-4277-7 PMID:30923883

Msendema, M. B., & Nyirenda, G. B. (2019). Unravelling Development Capabilities Enhanced by ICT in Informal Micro-Finance Settings. Case of Women Village Banks in Blantyre, Malawi. *2019 IST-Africa Week Conference. IST-Africa, 2019*, 1–8. doi:10.23919/ISTAFRICA.2019.8764871

Mucherino, A., & Papajorgji, P. Research, P. P.-O., & 2009, undefined. (2009). A survey of data mining techniques applied to agriculture. *Springer, 9*(2), 121–140. doi:10.1007/s12351-009-0054-6

Mugandani, R., Wuta, M., Makarau, A., & Chipindu, B. (2012). Re-Classification of the Agro-Ecological Regions of Zimbabwe in Conformity With Climate Variability and Change. *African Crop Science Journal, 20*(S2), 361–369. doi:10.4314/acsj.v20i2

Munyaradzi Gocha, S. M. and L. C. (2014). The Use Of Information Communication Technology By A-Level Teachers In Selected Secondary Schools In Gweru, Zimbabwe. *Journal of Technology Management.*, *1*(1), 1–21.

Munz, J. M., Gindele, N., & Doluschitz, R. (2020). Exploring the characteristics and utilisation of Farm Management Information Systems (FMIS) in Germany. *Computers and Electronics in Agriculture*, *170*, 105246.

Muraru, S. L., Cardei, P., Muraru, V., Sfîru, R., & Condruz, P. (2019). Researches regarding the use of drones in agriculture. *International Multidisciplinary Scientific GeoConference: SGEM*, *19*(6.2), 683-690.

Musungwini, S. (2021). *Inclusive Digital Transformation for the Marginalized Communities in a Developing Context.* IGI Global, 95–122. doi:10.4018/978-1-6684-3901-2.ch005

Musungwini, S., Zyl, I. van, & Kroeze, J. H. (2022). *The Perceptions of Smallholder Farmers on the Use of Mobile Technology: A Naturalistic Inquiry in Zimbabwe*. 530–544. doi:10.1007/978-3-030-98015-3_37

Musungwini, S. (2016). A model for harnessing the power of the Mobile Phone Technology to improve Smallholder Agriculture in Zimbabwe. *Proceedings of the 5th International Conference on M4D Mobile Communication Technology for Development: M4D 2016, General Tracks, December*, 237–252.

Musungwini, S. (2018). Mobile Phone Use by Zimbabwean Smallholder Farmers : A Baseline Study. *The African Journal of Information and Communication*, *22*(22), 29–52. doi:10.23962/10539/26171

Musungwini, S., & Van Zyl, I. (2017). ` Mobile Technology for Development ' Experiences from Zimbabwe Vending Markets a Naturalistic Enquiry. *International Journal of Business and Management Studies.*, *06*(01), 101–111.

Mutale, M. (2018). Challenges and opportunities of information and communication technologies (ICTs) in Binga District, Zimbabwe. *African Journal of Social Work*, *8*(2), 55–62.

Nabi, F., Jamwal, S., & Padmanbh, K. (2020). Wireless sensor network in precision farming for forecasting and monitoring of apple disease: a survey. *International Journal of Information Technology*, 1-12.

Nagaraj, A. (2021). Introduction to Sensors in IoT and Cloud Computing Applications. UAE: Bentham Science Publishers. doi:10.2174/97898114793591210101

Nagasubramanian, G., Sakthivel, R. K., Patan, R., Sankayya, M., Daneshmand, M., & Gandomi, A. H. (2021). Ensemble classification and IoT-based pattern recognition for crop disease monitoring system. *IEEE Internet of Things Journal*, *8*(16), 12847–12854.

Nair, K. P. (2019). Soil Fertility and Nutrient Management. In K. P. Nair (Ed.), *Intelligent Soil Management for Sustainable Agriculture: The Nutrient Buffer Power Concept* (pp. 165–189). Springer International Publishing.

Nakai, S., & Yamada, Y. (2014). Development of a weed suppression robot for rice cultivation: Weed suppression and posture control. *Int J Electr Comput Electron Commun Eng*, *8*, 1736–1740.

Nasiakou, A., Vavalis, M., & Zimeris, D. (2016). Smart energy for smart irrigation. *Computers and Electronics in Agriculture*, *129*, 74–83. https://doi.org/10.1016/J.COMPAG.2016.09.008

Nasteski, V. (2017). An overview of the supervised machine learning methods. *Horizons. b*, *4*, 51-62.

Navarro, E., Costa, N., & Pereira, A. (2020). A systematic review of IoT solutions for smart farming. *Sensors (Basel)*, *20*(15), 4231. doi:10.339020154231 PMID:32751366

Navarro-Hellín, H., Martínez-del-Rincon, J., Domingo-Miguel, R., Soto-Valles, F., & Torres-Sánchez, R. (2016). A decision support system for managing irrigation in agriculture. *Computers and Electronics in Agriculture*, *124*, 121–131. https://doi.org/10.1016/J.COMPAG.2016.04.003

Nawandar, N. K., & Satpute, V. R. (2019). IoT based low cost and intelligent module for smart irrigation system. *Computers and Electronics in Agriculture*, *162*, 979–990. https://doi.org/10.1016/J.COMPAG.2019.05.027

Nayyar, A., Nguyen, B. L., & Nguyen, N. G. (2020). The internet of drone things (IoDT): future envision of smart drones. In *First international conference on sustainable technologies for computational intelligence* (pp. 563-580). Springer.

Nayyar, A., Nguyen, B. L., & Nguyen, N. G. (2020). The internet of drone things (IoDT): future envisions of smart drones. In *First international conference on sustainable technologies for computational intelligence* (pp. 563-580). Springer, Singapore.

Ndiaye, M., Salhi, S., & Madani, B. (2020). When Green Technology Meets Optimization Modeling: The Case of Routing Drones in Logistics, Agriculture, and Healthcare. In *Modeling and Optimization in Green Logistics* (pp. 127–145). Springer. doi:10.1007/978-3-030-45308-4_7

Neethirajan, S., & Kemp, B. (2021). Digital livestock farming. *Sensing and Bio-Sensing Research*, *32*, 100408.

Negash, L., Kim, H. Y., & Choi, H. L. (2019, November). Emerging UAV applications in agriculture. In *2019 7th International Conference on Robot Intelligence Technology and Applications (RiTA)* (pp. 254-257). IEEE. 10.1109/RITAPP.2019.8932853

Neményi, M., Mesterházi, P. Á., Pecze, Z., & Stépán, Z. (2003). The role of GIS and GPS in precision farming. *Computers and Electronics in Agriculture*, *40*(1-3), 45–55.

Nettle, C. E. M. A. R., & Dela Rue, B. (2019). Making sense in the cloud: Farm advisory services in a smart farming future NJAS-Wageningen J. *Life Sciences*.

Neufeldt, H., Jahn, M., Campbell, B. M., Beddington, J. R., DeClerck, F., De Pinto, A., & Zougmoré, R. (2013). Beyond climate-smart agriculture: Toward safe operating spaces for global food systems. *Agriculture & Food Security*, *2*(1), 1–6.

Nevavuori, P., Narra, N., & Lipping, T. (2019). Crop yield prediction with deep convolutional neural networks. *Computers and Electronics in Agriculture*, *163*, 104859. https://doi.org/10.1016/J.COMPAG.2019.104859

Next Generation Mobile Networks Alliance. (2015). 5G Initiative. *5G white paper.* Next generation mobile networks.

Ngo, H. C., Hashim, U. R., Sek, Y. W., Kumar, Y. J., & Ke, W. S. (2019). Weeds detection in agricultural fields using convolutional neural network. *International Journal of Innovative Technology and Exploring Engineering*, *8*(11), 292–296. https://doi.org/10.35940/IJITEE.K1327.0981119

Nigam, A., Garg, S., Agrawal, A., & Agrawal, P. (2019). Crop Yield Prediction Using Machine Learning Algorithms. *Fifth International Conference on Image Information Processing (ICIIP)*, pp. 125-130, 2019. 10.1109/ICIIP47207.2019.8985951

Nikolenko, S., Kadurin, A., & Arkhangelskaya, E. J. S. P. (2018). Deep learning. *Journal of the American Heart Association.*

Nishant, P. S., Venkat, P. S., Avinash, B. L., & Jabber, B. (2020, June). Crop yield prediction based on indian agriculture using machine learning. In *2020 International Conference for Emerging Technology (INCET)* (pp. 1-4). IEEE. 10.1109/INCET49848.2020.9154036

Noon, S. K., Amjad, M., Qureshi, M. A., & Mannan, A. (2020). Use of deep learning techniques for identification of plant leaf stresses: A review. *Sustain. Comput. Informatics Syst.*, *28*, 100443.

Nørremark, M., & Griepentrog, H. (2004). Analysis and definition of the close-to-crop area in relation to robotic weeding. *Orgprints.* https://orgprints.org/4834/

Norton, G. W., & Alwang, J. (2020). Changes in Agricultural Extension and Implications for Farmer Adoption of New Practices. *Applied Economic Perspectives and Policy, 42*(1), 8–20. doi:10.1002/aepp.13008

O'Shea, K., & Nash, R. (2015). An introduction to convolutional neural networks.

O'Sullivan, C. A., Bonnett, G. D., McIntyre, C. L., Hochman, Z., & Wasson, A. P. (2019). Strategies to improve the productivity, product diversity and profitability of urban agriculture. [doi:https://doi.org/10.1016/j.agsy.2019.05.007]. *Agricultural Systems*, *174*, 133–144.

OADA. (2020, Jul 28). OADA/oada-docs: OADA API Related Documentation and References. *Github*. https://github.com/oada/oada-docs#readme

OECD/FAO. (2016). Agriculture in Sub-Saharan Africa. Prospects and challenges for the next decade. *Agricultural Outlook 2016-2025, 181*(November 1947), 59–93. doi:10.1787/agr_outlook-2016-en

Ojha, T., Misra, S., & Raghuwanshi, N. S. (2015). Wireless sensor networks for agriculture: The state-of-the-art in practice and future challenges. *Computers and Electronics in Agriculture*, *118*, 66–84. https://doi.org/10.1016/j.compag.2015.08.011

Onishchuk, M. O. (2020). Opto-mechanical sorting of municipal solid waste [Doctoral dissertation, BHTY, USA].

Oshiro, T. M., Perez, P. S., & Baranauskas, J. A. (2012). *How many trees in a random forest?* Quinlan, J. R. (1986). Induction of decision trees. *Machine Learning*, *1*(1), 81–106.

Osman, T., Psyche, S. S., Kamal, M. R., Tamanna, F., Haque, F., & Rahman, R. M. (2017). Predicting Early Crop Production by Analysing Prior Environment Factors. *Advances in Intelligent Systems and Computing*, 470–479. doi:10.1007/978-3-319-49073-1_51

Ouafiq, E. M., Elrharras, A., Mehdary, A., Chehri, A., Saadane, R., & Wahbi, M. (2021). IoT in smart farming analytics, big data based architecture. In *Human Centred Intelligent Systems* (pp. 269–279). Springer. doi:10.1007/978-981-15-5784-2_22

Pachayappan, M., Ganeshkumar, C., & Sugundan, N. (2020). Technological implication and its impact in the agricultural sector: An IoT Based Collaboration framework. *Procedia Computer Science*, *171*, 1166–1173. doi:10.1016/j.procs.2020.04.125

Padilla, N. E., Payne, J. A. G., Simbulan, V. S., Lapastura, R. J. S., & Cadeliña, E. J. A. (2020). Capacity building for dairy farmers towards commercialization of green corn silage, haylage and urea molasses mineral block. *Universal Journal of Agricultural Research*, *8*(6), 233–240. doi:10.13189/ujar.2020.080604

Pallathadka, H., Mustafa, M., Sanchez, D. T., Sekhar Sajja, G., Gour, S., & Naved, M. (2021). IMPACT OF MACHINE learning ON Management, healthcare AND AGRICULTURE. *Materials Today: Proceedings*. doi:10.1016/j.matpr.2021.07.042

Palm, C., Blanco-Canqui, H., DeClerck, F., Gatere, L., & Grace, P. (2014). Conservation agriculture and ecosystem services: An overview. [doi:https://doi.org/10.1016/j.agee.2013.10.010]. *Agriculture, Ecosystems & Environment*, *187*, 87–105.

Pandey, P. C., Tripathi, A. K., & Sharma, J. K. (2021). An evaluation of GPS opportunity in market for precision agriculture. In *GPS and GNSS Technology in Geosciences* (pp. 337–349). Elsevier.

Panpatte, S., & Ganeshkumar, C. (2021). Artificial intelligence in agriculture sector: case study of blue river technology. In *Proceedings of the Second International Conference on Information Management and Machine Intelligence* (pp. 147-153). Springer, Singapore. 10.1007/978-981-15-9689-6_17

Pant, J., Pant, R. P., Kumar Singh, M., Pratap Singh, D., & Pant, H. (2021). Analysis of agricultural crop yield prediction using statistical techniques of machine learning. *Materials Today: Proceedings, 46*, 10922-10926. doi:10.1016/j.matpr.2021.01.948

Pantazi, X. E., Moshou, D., Alexandridis, T., Whetton, R. L., & Mouazen, A. M. (2016). Wheat yield prediction using machine learning and advanced sensing techniques. *Computers and Electronics in Agriculture*, *121*, 57–65. https://doi.org/10.1016/J.COMPAG.2015.11.018

Partel, V., Kakarla, S. C., & Ampatzidis, Y. (2019). Development and evaluation of a low-cost and smart technology for precision weed management utilizing artificial intelligence. *Computers and Electronics in Agriculture*, *157*, 339–350.

Pathak, H., Kumar, G. A. K., Mohapatra, S. D., Gaikwad, B. B., & Rane, J. (2020). *Use of drones in agriculture: Potentials, Problems and Policy Needs*. ICAR-National Institute of Abiotic Stress Management.

Patil, A., Beldar, M., Naik, A., & Deshpande, S. (2016, March). Smart farming using Arduino and data mining. In *3rd International Conference on Computing for Sustainable Global Development (INDIACom)* (pp. 1913-1917). IEEE.

Patil, V. C., Al-Gaadi, K. A., Biradar, D. P., & Rangaswamy, M. (2012). Internet of things (Iot) and cloud computing for agriculture: An overview. *Proceedings of agro-informatics and precision agriculture (AIPA 2012)*, 292-296. .

Paul, V., Ganeshkumar, C., & Jayakumar, L. (2019). Performance evaluation of population seeding techniques of permutation-coded GA traveling salesman problems based assessment: Performance evaluation of population seeding techniques of permutation-coded GA. [IJAMC]. *International Journal of Applied Metaheuristic Computing*, *10*(2), 55–92. doi:10.4018/IJAMC.2019040103

Paunova-Hubenova, E., & Trichkova-Kashamova, E. (2021, October). Smart Solutions for Control and Management in Livestock Farms. In Big Data, Knowledge and Control Systems Engineering (BdKCSE) (pp. 1-7). IEEE.

Payaswini, P., & Manjaiah, D. H. (2014). Challenges and issues. In *4G Networks Mobility Management.* https://arxiv.org/abs/1402.3985

Payero, J. O., Mirzakhani-Nafchi, A., Khalilian, A., Qiao, X., & Davis, R. (2017). Development of a Low-Cost Internet-of-Things (IoT) System for Monitoring Soil Water Potential Using Watermark 200SS Sensors. *AIT 07*(03), 71–86. doi:10.4236/ait.2017.73005

Perez, L., & Wang, J. (2017). The effectiveness of data augmentation in image classification using deep learning. *arXiv preprint arXiv:1712.04621.*

Peteinatos, G. G., Weis, M., Andújar, D., Rueda Ayala, V., & Gerhards, R. (2014). Potential use of ground-based sensor technologies for weed detection. *Pest Management Science*, *70*(2), 190–199. doi:10.1002/ps.3677 PMID:24203911

Petkovic, S., Petkovic, D., & Petkovic, A. (2017). IoT devices VS. drones for data collection in agriculture. *DAAAM International Scientific Book*, *16*, 63–80. doi:10.2507/daaam.scibook.2017.06

Placidi, P., Gasperini, L., Grassi, A., Cecconi, M., & Scorzoni, A. (2020, June). Characterization of Low-Cost Capacitive Soil Moisture Sensors for IoT Networks. *Sensors (Basel)*, *20*(12), 3585. https://doi.org/10.3390/s20123585

Pobkrut, T., Eamsa-Ard, T., & Kerdcharoen, T. (2016, June). Sensor drone for aerial odour mapping for agriculture and security services. In *13th International Conference on Electrical Engineering/Electronics, Computer, Telecommunications and Information Technology (ECTI-CON)* (pp. 1-5). IEEE.

Popa, M., Prostean, O., & Popa, A.S. (2019). Machine Learning Approach for Agricultural IoT In Proc. *International Journal of Recent Technology and Engineering (IJRTE)*, 22-29.

Popovski, P., Trillingsgaard, K. F., Simeone, O., & Durisi, G. (2018). 5G Wireless Network Slicing for eMBB, URLLC, and mMTC: A Communication-Theoretic View. *IEEE Access: Practical Innovations, Open Solutions*, *6*, 55765–55779. https://doi.org/10.1109/ACCESS.2018.2872781

Prasanna, S., & Jebapriya, J. (2020). IoT based agriculture monitoring and smart farming using drones. *Mukt Shabd, 9.*

Pratheepkumar, P., Sharmila, J. J., & Arokiaraj, D. (2017). Towards mobile opportunistic in cloud computing. *Indian Journal of Scientific Research, 17*(02), 2250–0138.

Prathibha, S. R., Hongal, A., & Jyothi, M. P. (2017, March). IoT based monitoring system in smart agriculture. In 2017 international conference on recent advances in electronics and communication technology (ICRAECT) (pp. 81-84). IEEE.

Prathibha, S. R., Hongal, A., & Jyothi, M. P. (2017). IOT Based Monitoring System in Smart Agriculture. *International Conference on Recent Advances in Electronics and Communication Technology, ICRAECT.*

Psirofonia, P., Samaritakis, V., Eliopoulos, P., & Potamitis, I. (2017). Use of Unmanned Aerial Vehicles for Agricultural Applications with Emphasis on Crop Protection: Three Novel Case - studies. *International Journal of Agricultural Science and Technology*, *5*(1), 30–39. doi:10.12783/ijast.2017.0501.03

Puri, V., Nayyar, A., & Raja, L. (2017). Agriculture drones: A modern breakthrough in precision agriculture. *Journal of Statistics and Management Systems, 20*(4), 507–518. doi:10.1080/09720510.2017.1395171

Qiang, L., & Jiuping, X. (2008). A Study on Vehicle Routing Problem in the Delivery of Fresh Agricultural Products under Random Fuzzy Environment. *International Journal of Information and Management Sciences, 19*(4), 673–690.

Quarta, D., Pogliani, M., Polino, M., Maggi, F., Zanchettin, A. M., & Zanero, S. (2017). An experimental security analysis of an industrial robot controller. *IEEE Symposium on Security and Privacy (SP)* (pp. 268-286). IEEE. 10.1109/SP.2017.20

Queiroz, D. M. D., Coelho, A. L. D. F., Valente, D. S. M., & Schueller, J. K. (2021). Sensors applied to Digital Agriculture: A review. *Revista Ciência Agronômica*, 51.

Rachmawati, S., Putra, A. S., Priyatama, A., Parulian, D., Katarina, D., Habibie, M. T., & Valentino, V. H. (2021, August). Application of Drone Technology for Mapping and Monitoring of Corn Agricultural Land. In *International Conference on ICT for Smart Society (ICISS)* (pp. 1-5). IEEE. 10.1109/ICISS53185.2021.9533231

Radhakrishnan, V. (n.d.). Seed Ball Technique. *Bangalore conference.*

Rahmann, G., Azim, K., Brányiková, I., Chander, M., David, W., Erisman, J. W., Grimm, D., Hammermeister, A., Ji, L., Kuenz, A., Løes, A. K., Wan-Mohtar, W. A. A. Q. I., Neuhoff, D., Niassy, S., Olowe, V., Schoeber, M., Shade, J., Ullmann, J., & van Huis, A. (2021). Innovative, sustainable, and circular agricultural systems for the future. *Organic Agriculture, 11*(2), 179–185. doi:10.100713165-021-00356-0

Raj, P., Saini, K., & Surianarayanan, C. (2022). *Edge/Fog Computing Paradigm: The Concept, Platforms and Applications.* Academic Press. .

Rajasekaran, T., & Anandamurugan, S. (2019). Challenges and applications of wireless sensor networks in smart farming—a survey. *Advances in big data and cloud computing*, 353-361.

Raj, P., Saini, K., & Surianarayanan, C. (2022). *Edge/Fog Computing Paradigm: The Concept, Platforms and Applications.* Academic Press.

Ram Kumar, R. P., Sanjeeva, P., & Vijay Kumar, B. (2018). Transforming the traditional farming into smart farming using drones. In *Proceedings of the Second International Conference on Computational Intelligence and Informatics* (pp. 589-598). Springer, Singapore.

Ramcharan, A., Baranowski, K., McCloskey, P., Ahmed, B., Legg, J., & Hughes, D. P. (2017). Deep learning for image-based cassava disease detection. *Frontiers in Plant Science, 8*, 1852. https://doi.org/10.3389/FPLS.2017.01852/BIBTEX

Ramos, P. J., Prieto, F. A., Montoya, E. C., & Oliveros, C. E. (2017). Automatic fruit count on coffee branches using computer vision. *Computers and Electronics in Agriculture*, *137*, 9–22. https://doi.org/10.1016/J.COMPAG.2017.03.010

Rani, A., Chaudhary, A., Sinha, N., Mohanty, M., & Chaudhary, R. (2019). Drone: The green technology for future agriculture. *Harit Dhara*, *2*(1), 3–6.

Ratnaparkhi, S., Khan, S., Arya, C., Khapre, S., Singh, P., Diwakar, M., & Shankar, A. (2020). Smart agriculture sensors in IOT: A review. *Materials Today: Proceedings*.

Ravi, S., David, A., & Imaduddin, M. (2018). Controlling & calibrating vehicle-related issues using RFID technology. *International Journal of Mechanical and Production Engineering Research and Development*, *8*(2), 1125–1132. doi:10.24247/ijmperdapr2018130

Ray, P. P. (2017). Internet of things for smart agriculture: Technologies, practices and future direction. AIS 9 (4), 395–420. doi:10.3233/AIS-170440

Ray, S. (2015). Six Easy Steps to Learn Naive Bayes Algorithm (with code in Python). *Analytics Vidhya*. https://www.analyticsvidhya. com/blog/2015/09/naive-bayes

Razaak, M. (2019). An integrated precision farming application based on 5G, UAV and deep learning technologies. *Communications in Computer and Information Science*, *1089*, 109–119.

Regan, Á. (2019). 'Smart farming'in Ireland: A risk perception study with key governance actors. *NJAS Wageningen Journal of Life Sciences*, *90*, 100292.

Reger, M., Bauerdick, J., & Bernhardt, H. (2018). Drones in Agriculture: Current and future legal status in Germany, the EU, the USA and Japan. *Landtechnik*, *73*(3), 62–79.

Reich, Y., & Barai, S. V. (1999). Evaluating machine learning models for engineering problems. *Artificial Intelligence in Engineering*, *13*(3), 257–272. doi:10.1016/S0954-1810(98)00021-1

Rejeb, A., Abdollahi, A., Rejeb, K., & Treiblmaier, H. (2022). Drones in agriculture: A review and bibliometric analysis. *Computers and Electronics in Agriculture*, *198*, 107017. doi:10.1016/j.compag.2022.107017

Rendle, K. A., Garrett, S. B., & Halley, M. C. (2017). *Beyond Exploratory : A Tailored Framework for Assessing Rigor in Qualitative Health Services Research*. PMC.

Ren, Q., Zhang, R., Cai, W., Sun, X., & Cao, L. (2020, February). Application and development of new drones in agriculture. In *IOP Conference Series: Earth and Environmental Science* (Vol. 440, No. 5, p. 052041). IOP Publishing.

Ren, Q., Zhang, R., Cai, W., Sun, X., & Cao, L. (2020, February). Application and development of new drones in agriculture. *IOP Conference Series. Earth and Environmental Science*, *440*(5), 052041. doi:10.1088/1755-1315/440/5/052041

Research, R. Z.-P. J. of W. S., & 2010, undefined. (2014). Ethics for weed science. *Researchgate.Net*. doi:10.1017/S0043174500089633

Reuters. (2019, April 11). 5G-connected cows test milking parlor of the future. *Reuters*. https://www.reuters.com/article/us-telecoms-5g-cows/5g-connected-cows-test-milking-parlor-of-the-future-idUSKCN1RN1IY (accessed Jun. 30, 2020).

Ribarics, P. (2016). Big Data and its impact on agriculture. *Ecocycles*, *2*(1), 33–34. doi:10.19040/ecocycles.v2i1.54

Rico-Fernández, M. P., Rios-Cabrera, R., Castelán, M., Guerrero-Reyes, H. I., & Juarez-Maldonado, A. (2019). A contextualized approach for segmentation of foliage in different crop species. *Computers and Electronics in Agriculture*, *156*, 378–386.

Rigatti, S. J. (2017). Random forest. *Journal of Insurance Medicine (New York, N.Y.)*, *47*(1), 31–39. doi:10.17849/insm-47-01-31-39.1 PMID:28836909

Rijswijk, K., Klerkx, L., & Turner, J. A. (2019). Digitalisation in the New Zealand Agricultural Knowledge and Innovation System: Initial understandings and emerging organisational responses to digital agriculture. *NJAS Wageningen Journal of Life Sciences*, *90–91*(1), 100313. doi:10.1016/j.njas.2019.100313

Roshanianfard, A., Noguchi, N., Okamoto, H., & Ishii, K. (2020). A review of autonomous agricultural vehicles (The experience of Hokkaido University). *Journal of Terramechanics*, *91*, 155–183.

Roy, S. (2019). A peek into China's smart IoT and 5G-enabled tomato farms in Wuzhen. *Techwire Asia*. https://techwireasia.com/2019/10/a-peek-int o-chinas-smart-iot-and-5g-enabled-tomato-farms-in-wuzhen/

Ruan, J., Wang, Y., Chan, F. T. S., Hu, X., Zhao, M., Zhu, F., Shi, B., Shi, Y., & Lin, F. (2019). A life cycle framework of green IoT-based agriculture and its finance, operation, and management issues. *IEEE Communications Magazine*, *57*(3), 90–96. doi:10.1109/MCOM.2019.1800332

Rumpf, T., Mahlein, A. K., Steiner, U., Oerke, E. C., Dehne, H. W., & Plümer, L. (2010). Early detection and classification of plant diseases with Support Vector Machines based on hyperspectral reflectance. *Computers and Electronics in Agriculture*, *74*(1), 91–99. https://doi.org/10.1016/J.COMPAG.2010.06.009

Sachitra, V., & Padmini, C. (2021). COVID-19 Challenging Period and Agriculture Sector in Sri Lanka: Way to Lead. *Asian Journal of Advances in Agricultural Research*, (July), 21–34. doi:10.9734/ajaar/2021/v15i430160

Sadgrove, E. J., Falzon, G., Miron, D., & Lamb, D. W. (2018). Real-time object detection in agricultural/remote environments using the multiple-expert colour feature extreme learning machine (MEC-ELM). *Computers in Industry*, *98*, 183–191.

Sadiku, M. N. O., Ashaolu, T. J., & Musa, S. M. (2020). Emerging Technologies in Agriculture. *International Journal Of Scientific Advances*, *1*(1), 1–5. doi:10.51542/ijscia.v1i1.6

Sadowski, S., & Spachos, P. (2020). Wireless technologies for smart agricultural monitoring using internet of things devices with energy harvesting capabilities. *Computers and Electronics in Agriculture*, *172*, 105338. https://doi.org/10.1016/j.compag.2020.105338

Saha, A. K., Saha, J., Ray, R., Sircar, S., Dutta, S., Chattopadhyay, S. P., & Saha, H. N. (2018, January). IOT-based drone for improvement of crop quality in agricultural field. In *IEEE 8th Annual Computing and Communication Workshop and Conference (CCWC)* (pp. 612-615). IEEE. 10.1109/CCWC.2018.8301662

Sai Sravani, E., Sreehitha, A. V., Konda Babu, A., & Nandan, D. (2020). Evaluation and Study of IoT Entrances. In *Micro-Electronics and Telecommunication Engineering* (pp. 431–438). Springer.

Saini, K., & Raj, P. (2021, December). *Advancing smarter and more secure industrial applications using AI, IoT, and blockchain technology*. IGI Global. doi:10.4018/978-1-7998-8367-8

Saini, K., & Raj, P. (Eds.). (2021). *Advancing Smarter and More Secure Industrial Applications Using AI, IoT, and Blockchain Technology*. IGI Global.

Saini, K., Mishra, I., & Srivastava, S. (2021). Farmer's E-mart: An e-commerce store for crops. *3rd International Conference on Advances in Computing, Communication Control. And Networking* (ICAC3N), (pp. 346–350). doi:10.1109/ICAC3N53548.2021.9725783

Saini, K., Mishra, I., & Srivastava, S. (2021, December). Farmer's E-mart: An E-Commerce Store For Crops. In *3rd International Conference on Advances in Computing, Communication Control and Networking (ICAC3N)* (pp. 346-350). IEEE.

Saini, K. (2019). Evolution of blockchain technology in business applications. [JETIR]. *Journal of Emerging Technologies and Innovative Research, 6*(9), 240–244.

Saiz-Rubio, V., & Rovira-Más, F. (2020). From smart farming towards agriculture 5.0: A review on crop data management. *Agronomy (Basel), 10*(2), 207. doi:10.3390/agronomy10020207

Saldana, J. (2009). An Introduction to Codes and Coding. *The Coding Manual for Qualitative Researchers., 2006*, 1–31. doi:10.1519/JSC.0b013e3181ddfd0a

Saldaña-Robles, A. L., Bustos-Gaytán, A., Diosdado-De la Peña, J. A., Saldaña-Robles, A., Alcántar-Camarena, V., Balvantín-García, A., & Saldaña-Robles, N. (2020). Structural design of an agricultural backhoe using TA, FEA, RSM and ANN. *Computers and Electronics in Agriculture, 172*, 105278.

Sanchez-Iborra, R., Sanchez-Gomez, J., Ballesta-Viñas, J., Cano, M. D., & Skarmeta, A. F. (2018). Performance evaluation of LoRa considering scenario conditions. *Sensors (Basel), 18*(3), 772.

Sangeetha, B. Parvathi, N. K., Ambresh P., Ambalgi, S. L., Abdul, H., Thilagam, K., & Vijayakumar. P. (2022). IOT based smart irrigation management system for environmental sustainability in India. Sustainable Energy Technologies and Assessments, 52, 101973.

Sanjeevi, P., Prasanna, S., Siva Kumar, B., Gunasekaran, G., Alagiri, I., & Vijay Anand, R. (2020). Precision agriculture and farming using Internet of Things based on wireless sensor network. *Transactions on Emerging Telecommunications Technologies, 31*(12), e3978. doi:10.1002/ett.3978

Sarker, M. N. I., Islam, M. S., Murmu, H., & Rozario, E. (2020). Role of big data on digital farming. *Int J Sci Technol Res, 9*(4), 1222–1225.

Sawant, S. S., & Prabukumar, M. (2020). A review on graph-based semi-supervised learning methods for hyperspectral image classification. [doi:https://doi.org/10.1016/j.ejrs.2018.11.001]. *The Egyptian Journal of Remote Sensing and Space Sciences, 23*(2), 243–248.

Saxena, A., Suna, T., & Saha, D. (2020). Application of Artificial Intelligence in Indian Agriculture. In Souvenir: 19 national convention–artificial intelligence in agriculture: Indian perspective. RCA Alumni Association, xvi.

Schwenker, F., & Trentin, E. (2014). Pattern classification and clustering: A review of partially supervised learning approaches. [doi:https://doi.org/10.1016/j.patrec.2013.10.017]. *Pattern Recognition Letters, 37*, 4–14.

Sciforce. (2020, June 22). Smart Farming: The Future of Agriculture. *IOT Fort All.* https://www.iotforall.com/smart-farming-future-of-agricultur e

Seeman, M., Broxvall, M., Saffiotti, A., & Wide, P. (2006). An autonomous spherical robot for security tasks. *IEEE International Conference on Computational Intelligence for Homeland Security and Personal Safety* (pp. 51-55). IEEE. 10.1109/CIHSPS.2006.313312

Sengupta, S., & Lee, W. S. (2014). Identification and determination of the number of immature green citrus fruit in a canopy under different ambient light conditions. *Biosystems Engineering, 117*(C), 51–61. https://doi.org/10.1016/J.BIOSYSTEMSENG.2013.07.007

Sen, P. C., Hajra, M., & Ghosh, M. (2020). Supervised classification algorithms in machine learning: A survey and review. In *Emerging technology in modelling and graphics* (pp. 99–111). Springer. doi:10.1007/978-981-13-7403-6_11

Shah, R. (2020). Pesticides and Human Health. In *Emerging Contaminants.* IntechOpen.

Shanmugapriya, P., Rathika, S., Ramesh, T., & Janaki, P. (2019). Applications of remote sensing in agriculture-A Review. *International Journal of Current Microbiology and Applied Sciences*, *8*(1), 2270–2283. doi:10.20546/ijcmas.2019.801.238

Sharma, R. (2021). Review on Application of Drone Systems in Precision Agriculture. *Journal of Advanced Research in Electronics Engineering and Technology*, *7*(2), 5–7.

Shinde, P. P., & Shah, S. (2018, 16-18 Aug. 2018). A Review of Machine Learning and Deep Learning Applications. Paper presented at the *Fourth International Conference on Computing Communication Control and Automation (ICCUBEA)*.

Shivaji, C. P., Tanaji, J. K., Satish, N. A., & Mone, P. P. (2017). Agriculture drone for spraying fertilizer and pesticides. *Int. J. Res. Trends Innov*, *2*, 34–36.

Shouman, M., Turner, T., & Stocker, R. (2012). Applying k-nearest neighbour in diagnosing heart disease patients. *International Journal of Information and Education Technology (IJIET)*, *2*(3), 220–223. doi:10.7763/IJIET.2012.V2.114

Shukla, M., & Jharkharia, S. (2014). An inventory model for continuously deteriorating agri-fresh produce: An artificial immune system-based solution approach. *International Journal of Integrated Supply Management*, *9*(1–2), 110–135. https://doi.org/10.1504/IJISM.2014.064362

Sicat, R. S., Carranza, E. J., & Nidumolu, U. B. (2005). Fuzzy modeling of farmers' knowledge for land suitability classification. *Agricultural Systems*, *83*, 49–75.

Siddhartha, T., Nambirajan, T., & Ganeshkumar, C. (2021). Self-help group (SHG) production methods: insights from the union territory of Puducherry community. *Journal of Enterprising Communities: People and Places in the Global Economy*.

Siddhartha, T., Nambirajan, T., & Ganeshkumar, C. (2019). Production and retailing of self-help group products. *Global Business and Economics Review*, *21*(6), 814–835. doi:10.1504/GBER.2019.102590

Singh, H. (2019). *Practical Machine Learning and Image Processing: For Facial Recognition, Object Detection, and Pattern Recognition Using Python*. Apress. doi:10.1007/978-1-4842-4149-3

Sinha, R., Wei, Y., & Hwang, S.-H. (2017). A survey on LPWA technology: LoRa and NBIoT. *ICT Express*, *3*(1), 14–21.

Sinwar, D., Dhaka, V. S., Sharma, M. K., & Rani, G. (2020). AI-Based Yield Prediction and Smart Irrigation, 155–180. doi:10.1007/978-981-15-0663-5_8

Sishodia, R. P., Ray, R. L., & Singh, S. K. (2020). Applications of remote sensing in precision agriculture: A review. *Remote Sensing*, *12*(19), 3136.

Song, G., Yin, K., Zhou, Y., & Cheng, X. (2009). A surveillance robot with hopping capabilities for home security. *IEEE Transactions on Consumer Electronics*, *55*(4), 2034–2039. doi:10.1109/TCE.2009.5373766

Sonka, S. T. (2020). Digital Technologies, Big Data, and Agricultural Innovation. *The innovation revolution in Agriculture*, 207.

Sood, K., Singh, S., Rana, R. S., Rana, A., Kalia, V., & Kaushal, A. (2015, October). Application of GIS in precision agriculture. In Paper presented as lead lecture in national seminar on "Precision farming technologies for high Himalayas (pp. 04-05).

Srivastava, A. R., & Venkatesan, M. (2020). Tea leaf disease prediction using texture-based image processing. *Advances in Intelligent Systems and Computing, 1054,* 17–25. doi:10.1007/978-981-15-0135-7_3

Srivastava, K., Pandey, P. C., & Sharma, J. K. (2020). An approach for route optimization in applications of precision agriculture using UAVs. *Drones (Basel)*, *4*(3), 58. doi:10.3390/drones4030058

Srivastava, V., Singh, A. K., David, A., & Rai, N. (2022). Modelling Student Employability on an Academic Basis: A Supervised Machine Learning Approach with R. In *Handbook of Research on Innovative Management Using AI in Industry 5.0* (pp. 179–191). IGI Global. doi:10.4018/978-1-7998-8497-2.ch012

Steenwerth, K. L., Hodson, A. K., Bloom, A. J., Carter, M. R., Cattaneo, A., Chartres, C. J., & Jackson, L. E. (2014). Climate-smart agriculture global research agenda: Scientific basis for action. *Agriculture & Food Security*, *3*(1), 1–39.

Stehr, N. J. (2015). Drones: The newest technology for precision agriculture. *Natural Sciences Education*, *44*(1), 89–91. doi:10.4195/nse2015.04.0772

Steinwart, I., & Christmann, A. (2008). *Support vector machines*. Springer Science & Business Media.

Stočes, M., Vaněk, J., Masner, J., & Pavlík, J. (2016). Internet of things (iot) in agriculture-selected aspects. *Agris on-line Papers in Economics and Informatics, 8*(665-2016-45107), 83-88.

Subramanian, B., Chen, S. S., & Reddy, K. R. (2019). *Emerging Technologies for Agriculture and Environment.* Springer.

Suciu, G., Marcu, I., Balaceanu, C., Dobrea, M., & Botezat, E. (2019, June). Efficient IoT system for precision agriculture. In *15th International Conference on Engineering of Modern Electric Systems (EMES)* (pp. 173-176). IEEE.

Sureephong, P., Wiangnak, P., & Wicha, S. (2017, March). The comparison of soil sensors for integrated creation of IOT-based Wetting front detector (WFD) with an efficient irrigation system to support precision farming. In *International Conference on Digital Arts, Media and Technology (ICDAMT)* (pp. 132-135). IEEE.

Sutton, A. (2018). 5G network architecture. *J. Inst. Telecommun. Prof.*, *12*, 8–15. https://doi.org/10.1016/b978-0-08-102267-2.00001-4

Su, X., Yan, X., & Tsai, C. L. (2012). Linear regression. *Wiley Interdisciplinary Reviews: Computational Statistics*, *4*(3), 275–294. doi:10.1002/wics.1198

Syers, J. K. (1997). Managing soils for long-term productivity. *Philosophical Transactions of the Royal Society of London. Series B, Biological Sciences*, *352*(1356), 1011–1021. https://doi.org/10.1098/RSTB.1997.0079

Tajik, S., Ayoubi, S., & Nourbakhsh, F. (2012). Prediction of soil enzymes activity by digital terrain analysis: Comparing artificial neural network and multiple linear regression models. *Environmental Engineering Science*, *29*(8), 798–806. https://doi.org/10.1089/EES.2011.0313

Talaviya, T., Shah, D., Patel, N., Yagnik, H., & Shah, M. (2020). Implementation of artificial intelligence in agriculture for optimisation of irrigation and application of pesticides and herbicides. *Artificial Intelligence in Agriculture*, *4*, 58–73. doi:10.1016/j.aiia.2020.04.002

Tang, L., & Tian, L. ASAE, B. S.-T. of the, & 2000, undefined. (2000). Color image segmentation with genetic algorithm for in-field weed sensing. *Elibrary.Asabe.Org.* https://elibrary.asabe.org/abstract.asp?aid=2970

Tang, Y., Dananjayan, S., Hou, C., Guo, Q., Luo, S., & He, Y. (2021). A survey on the 5G network and its impact on agriculture: Challenges and opportunities. *Computers and Electronics in Agriculture*, *180*, 105895. https://doi.org/10.1016/j.compag.2020.105895

Teddlie, C., & Tashakkori, A. (2009). *Foundations of mixed methods research.* SAGE Publications.

Tetila, E. C., Machado, B. B., Menezes, G. K., Da Silva Oliveira, A., Alvarez, M., Amorim, W. P., De Souza Belete, N. A., Da Silva, G. G., & Pistori, H. (2020). Automatic recognition of soybean leaf diseases using uav images and deep convolutional neural networks. *IEEE Geoscience and Remote Sensing Letters*, *17*(5), 903–907.

Thalluri, L. N., Adapa, S. D., Priyanka, D., Sarma, A. V. N., & Venkat, S. N. (2021, October). Drone Technology Enabled Leaf Disease Detection and Analysis system for Agriculture Applications. In *2nd International Conference on Smart Electronics and Communication (ICOSEC)* (pp. 1079-1085). IEEE. 10.1109/ICOSEC51865.2021.9591837

Tian, Y., Yang, G., Wang, Z., Wang, H., Li, E., & Liang, Z. (2019). Apple detection during different growth stages in orchards using the improved YOLO-V3 model. *Computers and Electronics in Agriculture*, *157*, 417–426. https://doi.org/10.1016/J.COMPAG.2019.01.012

Tomar, P., Kaur, G., & Singh, P. (2018). A prototype of IoT-based real time smart street parking system for smart cities. In *Internet of Things and Big Data Analytics Toward Next-Generation Intelligence* (pp. 243–263). Springer.

Tong, W., Feng, X., & Chen, X. J. (2019). Research on Remote Control and Management Based on '4G Network' In Modern and High Efficiency Agriculture, pp. 233–241

Too, E. C., Yujian, L., Njuki, S., & Yingchun, L. (2019). A comparative study of fine-tuning deep learning models for plant disease identification. *Computers and Electronics in Agriculture*, *161*, 272–279. https://doi.org/10.1016/J.COMPAG.2018.03.032

Torres Vega, M., Liaskos, C., Abadal, S., Papapetrou, E., Jain, A., Mouhouche, B., Kalem, G., Ergüt, S., Mach, M., Sabol, T., Cabellos-Aparicio, A., Grimm, C., De Turck, F., & Famaey, J. (2020). Immersive interconnected virtual and augmented reality: A 5G and IoT perspective. *Journal of Network and Systems Management*, *28*(4), 796–826. doi:10.100710922-020-09545-w

Tracey, B., Duraisamy, K., & Alonso, J. (2013). *Application of supervised learning to quantify uncertainties in turbulence and combustion modeling.*

Tranmer, M., & Elliot, M. (2008). Multiple linear regression. [CCSR]. *The Cathie Marsh Centre for Census and Survey Research*, *5*(5), 1–5.

Trendov, M., Varas, S., & Zeng, M. (2019). Digital technologies in agriculture and rural areas: status report. *Digital technologies in agriculture and rural areas: status report.*

Trendov, N. M., Varas, S., & Zeng, M. (2019). *Digital Technologies In Agriculture And Rural Areas.* FAO.

Tripathi, S., Shukla, S., Attrey, S., Agrawal, A., & Bhadoria, V. S. (2020). Smart Industrial Packaging and Sorting System. 245–254. doi:10.1007/978-981-15-3647-2_18

Tripathi, P. S., & Prasad, R. (2018). Spectrum for 5G services. *Wireless Personal Communications*, *100*(2), 539–555. doi:10.100711277-017-5217-9

Tripicchio, P., Satler, M., Dabisias, G., Ruffaldi, E., & Avizzano, C. A. (2015, July). *Towards smart farming and sustainable agriculture with drones. In international conference on intelligent environments.* IEEE.

Tsan, M., Totapally, D. S., Hailu, D. M., Addom, B. K., & CTA. (2019). *Digitalisation for Agriculture Report* [Data set].

Tudi, M., Daniel Ruan, H., Wang, L., Lyu, J., Sadler, R., Connell, D., & Phung, D. T. (2021). Agriculture Development, Pesticide Application and Its Impact on the Environment. *International Journal of Environmental Research and Public Health*, *18*(3). doi:10.3390/ijerph18031112

Twinomurinzi, H., Schofield, A., Hagen, L., Ditsoane-Molefe, S., & Tshidzumba, N. A. (2017). Towards a shared worldview on e-skills: A discourse between government, industry and academia on the ICT skills paradox. *South African Computer Journal*, *29*(3), 215–237. doi:10.18489acj.v29i3.408

Ugochukwu, N. C., Carlu, W., & Abiodun, O. (2020). *Rural Smallholder Farmers' Awareness and Use of ICT-based market information Sources in South Africa. May*. doi:10.20944/preprints202005.0159.v1

Usery, E. L., Pocknee, S., & Boydell, B. (1995). Precision farming data management using geographic information systems. *Photogrammetric Engineering and Remote Sensing*, *61*(11), 1383–1392.

Uyanık, G. K., & Güler, N. (2013). A study on multiple linear regression analysis. *Procedia: Social and Behavioral Sciences*, *106*, 234–240. doi:10.1016/j.sbspro.2013.12.027

Valdés-Vela, M., Abrisqueta, I., Conejero, W., Vera, J., & Ruiz-Sánchez, M. C. (2015). Soft computing applied to stem water potential estimation: A fuzzy rule based approach. *Computers and Electronics in Agriculture*, *115*, 150–160. https://doi.org/10.1016/J.COMPAG.2015.05.019

Van der Vorst, J., Beulens, A., & van Beek, P. (2005). 10 Innovations in logistics and ICT in food supply chain networks. *Innov. agri-food Syst,* 245.

van Engelen, J. E., & Hoos, H. H. (2020). A survey on semi-supervised learning. *Machine Learning*, *109*(2), 373–440. doi:10.100710994-019-05855-6

van Klompenburg, T., Kassahun, A., & Catal, C. (2020). Crop yield prediction using machine learning: A systematic literature review. *Computers and Electronics in Agriculture*, *177*, 105709. https://doi.org/10.1016/J.COMPAG.2020.105709

Vandana, C. P., Prakash, A., Bhowal, A., & Taj, D. M. (2020). Drone Assisted Effective Pesticide Sprayer. *International Journal of Scientific Research in Computer Science. Engineering and Information Technology*, *3*(6), 363–370.

Veena, S., Mahesh, K., Rajesh, M., & Salmon, S. (2018). The survey on smart agriculture using IOT. [IJRIREM]. *Int J Innov Res Eng Manag*, *5*(2), 63–66.

Vellidis, G., Tucker, M., Perry, C., Kvien, C., & Bednarz, C. (2008). A real-time wireless smart sensor array for scheduling irrigation. *Computers and Electronics in Agriculture*, *61*(1), 44–50. https://doi.org/10.1016/J.COMPAG.2007.05.009

Verdouw, C., Tekinerdogan, B., Beulens, A., & Wolfert, S. (2021). Digital twins in smart farming. *Agricultural Systems*, *189*, 103046. doi:10.1016/j.agsy.2020.103046

Veroustraete, F. (2015). The rise of drones in agriculture. *EC agriculture, 2*(2), 325-327.

Vibhute, A., & Bodhe, S. K. (2012). Applications of image processing in agriculture: A survey. *International Journal of Computers and Applications*, *52*(2).

Victer Paul, P., Ganeshkumar, C., Dhavachelvan, P., & Baskaran, R. (2020). A novel ODV crossover operator-based genetic algorithms for traveling salesman problem. *Soft Computing*, *24*(17), 12855–12885. doi:10.100700500-020-04712-2

Victor, O., Nic, J. L., & Xiaomeng, L. (2021). Factors affecting the adoption of mobile applications by farmers: An empirical investigation. *African Journal of Agricultural Research*, *17*(1), 19–29. doi:10.5897/AJAR2020.14909

Vikranth, K. (2021). An Implementation of IoT and Data Analytics in Smart Agricultural System–A Systematic Literature Review. *International Journal of Management* [IJMTS]. *Technology and Social Sciences*, *6*(1), 41–70.

Waheed, R., Chang, D., Sarwar, S., & Chen, W. (2018). Forest, agriculture, renewable energy, and CO2 emission. [doi:https://doi.org/10.1016/j.jclepro.2017.10.287]. *Journal of Cleaner Production*, *172*, 4231–4238.

Wallace, J. S. (2000). Increasing agricultural water use efficiency to meet future food production. *Agric. Ecosyst. Environ. 82*(1-3), 105–119. S0167-8809(00)00220-6. doi:10.1016/

Wang, A. X., Tran, C., Desai, N., Lobell, D., & Ermon, S. (2018). Deep transfer learning for crop yield prediction with remote sensing data. *Proceedings of the 1st ACM SIGCAS Conference on Computing and Sustainable Societies,* COMPASS. doi:10.1145/3209811.3212707

Wang, H., Ma, C., & Zhou, L. (2009). A brief review of machine learning and its application. *International Conference on Information Engineering and Computer Science.*

Wang, Q., Nuske, S., Bergerman, M., & Singh, S. (2013). Automated Crop Yield Estimation for Apple Orchards. 745–758. doi:10.1007/978-3-319-00065-7_50

Wang, X. (2011). Mixed reality and human-robot interaction (Vol. 47). Springer Science & Business Media.

Wang, F., Jiang, M., Qian, C., Yang, S., Li, C., Zhang, H., & Tang, X. (2017). Residual attention network for image classification. In *Proceedings of the IEEE conference on computer vision and pattern recognition* (pp. 3156-3164).

Wang, S.-C. (2003). Artificial neural network. In *Interdisciplinary computing in java programming* (pp. 81–100). Springer. doi:10.1007/978-1-4615-0377-4_5

Wang, X. P., Wang, M., Ruan, J. H., & Li, Y. (2018). Multi-objective optimization for delivering perishable products with mixed time windows. *Advances in Production Engineering & Management, 13*(3), 321–332. https://doi.org/10.14743/APEM2018.3.293

Wang, X., & Liu, N. (2014). The application of internet of things in agricultural means of production supply chain management. *Journal of Chemical and Pharmaceutical Research, 6*(7), 2304–2310.

Weisberg, S. (2005). *Applied linear regression* (Vol. 528). John Wiley & Sons. doi:10.1002/0471704091

Whitmire, C. D., Vance, J. M., Rasheed, H. K., Missaoui, A., Rasheed, K. M., & Maier, F. W. (2021). Using Machine Learning and Feature Selection for Alfalfa Yield Prediction. *AI, 2*(1), 71–88. doi:10.3390/AI2010006

Wójtowicz, M., Wójtowicz, A., & Piekarczyk, J. (2016). Application of remote sensing methods in agriculture. *Communications in Biometry and Crop Science, 11*(1), 31–50.

Wolfert, S., Ge, L., Verdouw, C., & Bogaardt, M. J. (2017). Big data in smart farming–a review. *Agricultural Systems, 153*, 69–80. doi:10.1016/j.agsy.2017.01.023

World Bank. (2011). ICT in agriculture: Connecting amallholders to knowledge, networks, and institutions. *WORLD (Oakland, Calif.), 64605*, 424. 64605

World Population Clock. (2020) 7.8 billion People. *Worldometer.* https://www. worldometers.info/world-population/#table-forecast .

Wynn, C. R. (2019). *Drone Technology: Is It Worth the Investment in Agriculture* [Doctoral dissertation, The University of Nebraska-Lincoln, USA].

XAG. (2019). Huawei cooperate on 5G-powered smart agriculture. *Geospatial News Portal.* https://www.thegeospatial.in/xag-huawei-cooperate-on-5g-powered-smart-agriculture .

Xu, D., Ouyang, W., Ricci, E., Wang, X., & Sebe, N. (2017). Learning crossmodal deep representations for robust pedestrian detection. In *Conference on Computer Vision and Pattern Recognition*, 4236–4244. IEEE.

Xu, D., Ouyang, W., Wang, X., & Sebe, N. (2018). PADNet: Multi-tasks guided prediction-and-distillation network for simultaneous depth estimation and scene parsing. In */CVF Conference on Computer Vision and Pattern Recognition*, 675–684. IEEE.

Yadav, S., Garg, D., & Luthra, S. (2020). Development of IoT based data-driven agriculture supply chain performance measurement framework. *Journal of Enterprise Information Management*.

Yadav, S., Sengar, N., Singh, A., Singh, A., & Dutta, M. K. (2021). Identification of disease using deep learning and evaluation of bacteriosis in peach leaf. [doi:https://doi.org/10.1016/j.ecoinf.2021.101247]. *Ecological Informatics*, *61*, 101247.

Yahya, A. (2020). *Emerging Technologies in Agriculture, Livestock, and Climate*. Springer. doi:10.1007/978-3-030-33487-1

Yang, M.-S., Lai, C.-Y., & Lin, C.-Y. (2012). A robust EM clustering algorithm for Gaussian mixture models. *Pattern Recognition*, *45*(11), 3950–3961. doi:10.1016/j.patcog.2012.04.031

Yazdani, M., Zarate, P., Coulibaly, A., & Zavadskas, E. K. (2017). A group decision making support system in logistics and supply chain management. *Expert Systems with Applications*, *88*, 376–392. doi:10.1016/j.eswa.2017.07.014

Yeasin, M., Dhandapani, A., & Ravichandran, S. (2021). Artificial Intelligence in Agriculture. In Ch. Srinivasarao (Ed.), *Agricultural Research, Technology and Policy: Innovations and Advances, ICAR-National Academy of Agricultural Research Management (NAARM), Hyderabad* (pp. 291–306).

Yost, M. A., Kitchen, N. R., Sudduth, K. A., Sadler, E. J., Drummond, S. T., Volkmann, M. R., & Yost MattYost, M. A. (2017). Long-term impact of a precision agriculture system on grain crop production. *Springer, 18*(5), 823–842. doi:10.1007/s11119-016-9490-5

You, J., Li, X., Low, M., Lobell, D., & Ermon, S. (2017). Deep Gaussian Process for Crop Yield Prediction Based on Remote Sensing Data. *Proceedings of the AAAI Conference on Artificial Intelligence*, *31*(1). https://ojs.aaai.org/index.php/AAAI/article/view/11172

Yousefi, M. R., & Razdari, A. M. (2015). Application of GIS and GPS in precision agriculture (a review). *International Journal of Advanced Biological and Biomedical Research*, *3*(1), 7–9.

Yu, H., Lee, H., & Jeon, H. Oct. 2017. What is 5G? Emerging 5G Mobile Services and Network Requirements. *Sustainability* *9*(10), 1848. su9101848. doi:10.3390/

Yu, H., & Son, C. (2020). Leaf spot attention network for apple leaf disease identification In *Conference on Computer Vision and Pattern Recognition Workshops (CVPRW)*, 229–237. IEEE .

Yu, X., & Wang, H. (2014). Improvement of Eclat Algorithm Based on Support in Frequent Itemset Mining. *Journal of Computing*, *9*(9), 2116–2123. doi:10.4304/jcp.9.9.2116-2123

Zaremba, W., Sutskever, I., & Vinyals, O. (2014). Recurrent neural network regularization.

Zhai, Z., Martínez, J. F., Beltran, V., & Martínez, N. L. (2020). Decision support systems for agriculture 4.0: Survey and challenges. *Computers and Electronics in Agriculture*, *170*, 105256. doi:10.1016/j.compag.2020.105256

Zhang, X., Wang, T., Qi, J., Lu, H., & Wang, G. (2018a). Progressive attention guided recurrent network for salient object detection. In *CVF Conference on Computer Vision and Pattern Recognition*, 714–722. IEEE.

Zhang, B., Xie, Y., Zhou, J., Wang, K., & Zhang, Z. (2020). State-of-the-art robotic grippers, grasping and control strategies, as well as their applications in agricultural robots: A review. *Computers and Electronics in Agriculture*, *177*, 105694. https://doi.org/10.1016/J.COMPAG.2020.105694

Zhang, X., Qiao, Y., Meng, F., Fan, C., & Zhang, M. (2018c). Identification of maize leaf diseases using improved deep convolutional neural networks. *IEEE Access: Practical Innovations, Open Solutions*, *6*, 30370–30377.

Zhang, X.-S. (2000). Introduction to artificial neural network. In *Neural Networks in Optimization* (pp. 83–93). Springer. doi:10.1007/978-1-4757-3167-5_5

Zhang, X., Yuan, Y., & Wang, Q. (2018b). Roi-wise reverse reweighting network for road marking detection in *British Machine Vision Conference*, United kingdom, 1–12.

Zhang, X., Zhang, J., Li, L., Zhang, Y., & Yang, G. (2017). Monitoring citrus soil moisture and nutrients using an IoT based system. *Sensors (Basel)*, *17*(3), 447. doi:10.339017030447 PMID:28241488

Zhang, X., Zhang, J., Li, L., Zhang, Y., & Yang, G. (2017, February). Monitoring Citrus Soil Moisture and Nutrients Using an IoT Based System. *Sensors (Basel)*, *17*(3), 447. https://doi. org/10.3390/s17030447

Zhao, G., Lianjun, Y., Yu, H., Wang, G., Sui, Y., & Zhang, L. (2016). Design and Implementation of Greenhouse Remote Monitoring System Based on 4G and. Virtual Network. 455–462.

Zhaogan, L., Taiyi, Z., Xiaodong, S., & Xiaohe, L. (2007). Limitations of Current 4G Systems and Its Substitute Schemes with TDD/TDMA. *WSEAS Transactions on Communications*, *6*(11), 850–860.

Zhao, H., Zhang, Y., Liu, S., Shi, J., Loy, C. C., Lin, D., & Jia, J. (2018). Psanet: Point-wise spatial attention network for scene parsing. In *Proceedings of the European conference on computer vision (ECCV)* (pp. 267-283).

Zhao, Z., Chow, T. L., Rees, H. W., Yang, Q., Xing, Z., & Meng, F. R. (2009). Predict soil texture distributions using an artificial neural network model. *Computers and Electronics in Agriculture*, *65*(1), 36–48.

Zhu, H., Chu, B., Zhang, C., Liu, F., Jiang, L., & He, Y. (2017). Hyperspectral Imaging for Presymptomatic Detection of Tobacco Disease with Successive Projections Algorithm and Machine-learning Classifiers. *Scientific Reports*, *7*(1), 1–12. doi:10.1038/s41598-017-04501-2

Zhuang, H., & Roth, Z. S. (2018). *Camera-aided robot calibration*. CRC press. doi:10.1201/9781315138725

Zou, J., Han, Y., & So, S. S. (2008). Overview of artificial neural networks. *Artificial Neural Networks*, 14-22.

About the Contributors

Pethuru Raj has been working in the Site Reliability Engineering (SRE) Center of Excellence, Reliance Jio Infocomm Ltd. (RJIL), Bangalore. My previous stints are in IBM Cloud center of Excellence (CoE), Wipro consulting services (WCS), and Robert Bosch Corporate Research (CR). In total, I have gained more than 17 years of IT industry experience and 8 years of research experience. Finished the CSIR-sponsored PhD degree at Anna University, Chennai and continued with the UGC-sponsored postdoctoral research in the Department of Computer Science and Automation, Indian Institute of Science, Bangalore. Thereafter, I was granted a couple of international research fellowships (JSPS and JST) to work as a research scientist for 3.5 years in two leading Japanese universities. Published more than 40 research papers in peer-reviewed journals such as IEEE, ACM, Springer-Verlag, Inderscience, etc. Have authored and edited 16 books thus far and focus on some of the emerging technologies such as IoT, Cognitive Analytics, Blockchain, Digital Twin, Docker Containerization, Data Science, Microservices Architecture, fog / edge computing, Artificial intelligence (AI), etc.

Kavita Saini (IEEE and ACM Member) is presently working as Professor, School of Computing Science and Engineering, Galgotias University, Delhi NCR, India. She received her PhD degree from Banasthali Vidyapeeth, Banasthali. She has 18 years of teaching and research experience supervising Masters and PhD scholars in emerging technologies. She has published more than 80 research papers in national and international journals and conferences. She has published 18 authored books for UG and PG courses for a number of universities including MD University, Rothak, and Punjab Technical University, Jallandhar with National Publishers. Kavita Saini has edited more than 12 books with International Publishers including Elsevier, IET Publisher, IGI Global and CRC Press, NOVA Science Publisher and published 25 book chapters with International publishers. Under her guidance many M Tech and Ph D scholars are carrying out research work. She has also published various patents. Kavita Saini has also delivered technical talks on Blockchain: An Emerging Technology, Web to Deep Web and other emerging Areas and Handled many Special Sessions in International Conferences and Special Issues in International Journals. Her research interests include Smart and Sustainable Agricultural, Blockchain Technology, Industry 5.O, and Cloud and Edge Computing, Digital Twin and Web-Based Instructional Systems (WBIS).

Vinicius Pacheco is an IT professional as well as an academic researcher. As a Brazilian federal public employee, he has over two decades of experience working at the IT department of several public offices, mainly in IT governance, cloud computing, unified communications, network management, and network security. He has also occupied the position of Chief Information Officer at two different federal organizations. As a researcher, he holds a Ph.D. in cloud security and has published several academic papers and book contributions, lately focusing on enabling privacy in the cloud computing paradigm.

* * *

Prerna Ajmani is a Senior Assistant Professor in department of Computer Science in Vivekananda Institute of Professional Studies. She has keen interest in computers and strongly believes in imparting quality education with emphasis on Computer Science. She has over 11 years of experience in research and academics. She is actively involved in research in the areas of IoT, Next Generation Technologies, cyber security, cloud computing, Ad-hoc networks, and blockchain.

Ganeshkumar C. is an Assistant Professor in management at Indian Institute of Plantation Management Bangalore (IIPMB), An Autonomous Organization of the Ministry of Commerce & Industry, Govt. of India. He has received his Ph.D. and MBA from Pondicherry Central University, Cleared UGC-NET in Management on 2010, He has 2 years of post-doctoral research experience in Indian Institute of Management Bangalore. Served as faculty at VIT University, Vellore. He has published 30 articles in National and International Journals, presented 15 research papers in various conference and organized different data analytics workshops. He has organized and acted as resources in various hands-on business analytics workshops using tableau, Python, R and Excel Solver software. His areas of research include Agri-food value chain, Big data analytics and Socially responsible supply chain. funded projects from Tea Board, IFPRI-USA and ADB. Apart from teaching and research, He has taken administrative roles like in VIT-Exam coordinator. In IIPMB, Boys hostel warden, Industrial visit coordinator and Analytics Research centre.

Anyway Chare is an Information Technology Lecturer in the Department of Marketing at Marondera University of Agricultural Sciences & Technology (MUAST), Zimbabwe. He is currently pursuing a Ph.D in Information Technology research degree at the Durban University of Technology (DUT) in South Africa. He holds a Master of Science in Information Systems degree from the National University of Science and Technology in Zimbabwe, a Bachelor of Science Information Systems Honours degree from the Midlands State University in Zimbabwe.

Chitrangada Chaubey is presently working as Research Scholar in department of Computer Science & Engineering. Earlier she has been associated with numerous academic institutions like GGSIP University-New Delhi, MDU - Rohtak and DCRUST - Murthal as Assistant Professor. She has got vast experience in Research and development with various research projects. Also, she has an excellent work experience in training and career counselling. She has an expertise in organizing and delivering guest lectures, workshops and conferences in both online/offline platforms. She is having great intrest in teaching subjects like Data Structures, Computer Organization & Architecture, Big Data, Blockchain, Cloud Computing and Enterprise Resource Planning. Apart from that she is currently working with various Government Undertaken projects for training and counselling students and academicians as well.

Garima Chauhan was born in Palampur, India in 1995. She obtained her Bachelor's and Master's degree in Agriculture and Agronomy in 2019 and 2021, respectively. Currently she is pursuing her doctorate degree in Agronomy at CSK Himachal Pradesh Agricultural University and has research work focused on the conjunctive action of nanofertlizers and various water regimes for higher crop productivity.

A. Chitty has completed her Post Graduation from Jawaharlal Nehru Technological University and she is currently working as Assistant Professor in CSIT Department in CVR College of Engineering, Hyderabad. She has more than 5 years of Teaching Experience. She has published several papers in National/International Conferences and Journals. Her research areas include Machine Learning, Cloud Computing, Wireless Sensor Networks, etc.

Arokiaraj David, PhD, is a Senior Assistant Professor at Jain University, Bangalore. He has received various awards - Young Scientist Award, Best Research Award, Young Academician Award and Best Teacher Award. He qualified for a Junior and Senior Research Fellowship (SRF/ JRF) and was awarded the National Eligibility Test for Lectureship (NET) conducted by UGC (Government of India). In the recent past, he has conducted various research programmes for the betterment of students as well as faculty. He has conducted his research in the area of Consumer Behaviour, HR Matrix, Sustainable Tourism, Environmental Management, Econometrics etc. He has published more than 40 research articles including 10 papers that were indexed in Scopus, 3 in ABDC, 2 WoS, 3 Patents, 2 Book chapters, one Copyright and one Book. He had a very good sound knowledge of various research software like SPSS, AMOS, STATA, and Smart PLS. Dr. Arokiaraj David has completed his doctorate in Management (International Business) and also master's degree in the area of Business Studies (Pondicherry Central University) and also Financial Management (Alagappa University) had more than 10 years of work experience including 5.3 years in research and 5.10 years in teaching. His future research was focused on green marketing, consumer behaviour, and policy analysis. He has supervised several Undergraduate and Postgraduate students' dissertations at the college levels.

Mohit Dayal is Technical committee Member of IEEE INDIaCom international conference, Delhi and Editorial member of International journal of Recent Advances in Science and Technology. He is currently working in Dr. Akhilesh Das Gupta Institute of Technology and Management as an assistant professor in computer science and engineering Department. He received his master's degree in Information Security from Ambedkar Institute of Advanced Communication Technologies & Research of Guru Gobind Singh Indraprastha University, Delhi. He holds a bachelor's degree in Computer Science & Engineering from Guru Gobind Singh Indraprastha University, Delhi. His research interests include machine learning, Internet of Things, Big Data, Web Application attacks and information security.

Pinalkumar Engineer is also working as Assistant Professor in Electronics engineering department at S.V. National Institute of Technology (SVNIT), Surat since 2007. He holds a PhD from Indian Institute of Technology, Bombay and a ME degree in electrical engineering with specialization in Microprocessor systems and applications (M. S. University of Baroda, 2002). His research interests include Edge Computing, Application Specific Processor Design, Energy-efficient Computing, VLSI architecture for real-time signal/image processing, High performance embedded computing.

Petros Venganayi Gavai is a lecturer at the Midlands State University, in the Department of Information and Marketing Sciences. He is a published author who has published research articles in Scopus indexed journals and non-Scopus indexed journals. He has also attended and presented at a number of international conferences. He holds a Post Graduate Diploma in Tertiary education from Midlands State University, a Bachelor of Science Honours Degree in Information Technology from Harare Institute of Technology, and a Master Science Degree in Information Systems Management from Midlands State University.

Jeganathan Gomathi Sankar is a member of SSM since 2015. Prior to this, he served as full time research scholar in Department of International Business at Pondicherry University. Dr. Sankar develops and teaches a course on Marketing Management, Services Marketing, Advertising and Sales Promotion, Sales Management, and Consumer Behaviour for post graduate students. A pure educationist by choice, he has been involved with teaching and research in marketing field. His wider areas of research include Internal Marketing, Services quality, Information diffusion, ICT, and MOOCs. His work has been published in Scopus indexed and UGC listed journals. Dr. Sankar holds a PhD from Pondicherry University, Masters in Business Administration from Anna University, Bachelors in Electronics and Communication Engineering from Anna University. He has secured UGC-NET.

Manvi Gupta is a final year student pursuing Bachelor's in Technology in Computer Science and Engineering.

Ritu Gupta is an experienced lecturer and researcher working at the school of computer science and engineering in Galgotias University, Greater Noida. She is having total 7 years of experience out which 4 years of industry and 3 years of academic experience. She has received the MTech in Computer Science from Uttar Pradesh Technical University. Her current research interests are IOT, Wireless sensor networks and Cloud computing.

Madhura N. Hegde is working as an Assistant Professor in Department of Information Science and Engineering in Sahyadri College of Engineering and Management, Mangalore. She has 8 years of Experience in Teaching and 1.5 year of Experience In Industry. She has research interest in the field of Bioinformatics, Machine Learning and Data analytics.

Deepa J. is currently working as Assistant Professor in the Department of Computer Science and Engineering in Easwari Engineering College, Chennai. She has completed her Under-Graduation (B.E) in Madurai Kamarajar University, Post-Graduation (M.Tech) in National Institute of Technology, Trichy and Ph.D in Anna University, Chennai. She has more than 13 years of experience in the teaching field and also published various papers in National/International Conferences and Journals.

Ganaraj K. is an accomplished Researcher Currently working as an Assistant Professor in Department of Information Science and Engineering, Sahyadri College of Engineering, Mangalore. Mr. Ganaraj K graduated from Vivekananda College of Engineering and Technology, Puttur and obtained his M.Tech from BIT Mangalore. He has 4 years of Teaching experience and 1 year experience in Software Industry. He published 5 conference papers and 4 book chapters and 3 journal papers. His area of interest includes Social Network Analysis, Machine Learning, Cloud Computing, Data Analytics and Big Data.

Jayapadmini Kanchan is an Assistant Professor in the Information Science & Engineering department. She has pursued B.E from HKBKCE, Bangalore, M.Tech from SCEM, Mangalore. Her area of interest includes Network Security, Data Analytics, Web Technology, etc. She has published various research articles in International journals and conferences.

Deepika Kataria is a lecturer working in the school of computer science and engineering in Galgotias University. She is having total 5 years of experience in teaching. Also she have registered a patent on drone based harvesting technology.

Sreeram Kattamuri is currently working as Associate Professor in IT Department in Sreenidhi Institute of Science and Technology, Hyderabad. He hold a PhD degree in Computer Science Engineering and has more than 15 years of experience in teaching field.

V. Gokula Krishnan is currently working as Professor in the Department of Computer Science and Engineering in Saveetha School of Engineering, Saveetha Institute of Medical and Technical Sciences (SIMATS), Thandalam, Chennai, Tamil Nadu, India. He has completed his Under-Graduation (BE) in Anna University, Post-Graduation (M.Tech) in Dr. MGR University and Ph.D in Sathyabama Institute of Science and Technology, Chennai. He has more than 16 years of teaching experience in various colleges in Chennai and Hyderabad. He has published several papers in SCI/Scopus/WoS indexed journals and also he has presented various papers in National/International Conferences. His area of interest includes Computer Networks, Computer Architecture, Data Structures, Software Engineering etc. He serves as the guest editor, editorial member and also as reviewer in many reputed international journals. He is a member of Professional Bodies like ISTE, IAENG, CSI, and IEEE.

Naveen Kumar is a Principal Scientist of Agronomy at CSK Himachal Pradesh Krishi Vishwavidyalaya, Palampur, Himachal Pradesh, India. He obtained his undergraduate degree (1986) in Agriculture, masters' degree (1988) in Agronomy from CSK Himachal Pradesh Krishi Vishwavidyalaya, Palampur and doctoral degree (1998) in Agronomy from Punjab Agricultural University, Ludhiana, Punjab. He holds the office of Head, Department of Agronomy, College of Agriculture, Palampur and Additional Director Research (Agriculture) at CSK Himachal Pradesh Krishi Vishwavidyalaya, Palampur. He is an expert on management and improvement of sub-tropical & temperate forages, pastures & grasslands. He has made 187 scholarly contributions which includes 65 peer-reviewed research papers, 12 book chapters, 42 popular articles and 65 symposium/conference papers. In addition to being co-investigator (CO-PI) in 12 research projects, he was the Principal Investigator (PI) of 6 research projects.

Pravin Kumar is the Lead Software Engineer at Cadence Design System, Bangalore. He is currently working in R&D Department as a Research and Product development Engineer. Prior to Cadence Design System, he worked as a senior Research Engineer for O-RAN and vRAN protocol stack development for 4G and 5G. He received his master's degree in Computer Science and Engineering from the Indian Institute of Technology (ISM) Dhanbad. His research interests include Emulation System on Chip (SoC), 4G/5G Protocol Stack Development, machine learning, Internet of Things(IoT), Algorithm Design And Analysis and Swarm intelligence.

Samuel M. is a lecturer at the Midlands State University, in the Department of Information and Marketing Sciences. He is a member of the ZIMCHE physical assessment for facilities and resources checklist for academic programs. The program included an assessment of the curriculum of academic programs for accreditation by ZIMCHE. This included the review of the degree programs at the University of Zimbabwe and the National University of Science and Technology in the domain of Computer Science, Information Systems, and Informatics. He has contributed to the 2019 Technology Transfer Workshop (Intellectual property (IP)) at Gweru Polytechnic. He is a published author who has published research articles in Scopus indexed journals and non-Scopus indexed journals. He has also attended and presented at a number of international conferences (ICT4D, Springer link organized and Elsevier organised) He holds a Post Graduate Diploma in Tertiary education from Midlands State University, a Bachelor of Science Honours Degree in Information Systems from Midlands State University, a Master Science Degree in Information Systems Management from Midlands State University. He is currently studying for PhD in Information Systems at the University of South Africa (UNISA).

Briget Munyoro is a lecturer at Midlands State University (MSU) in the department of Information and Marketing Sciences. She holds a Master of Science Degree in Information Systems with National University of Science and Technology (NUST) and a Bachelor of Technology Honours Degree in Computer Science with Harare Institute of Technology (HIT). She teaches Mobile computing, Operating systems, Software Project Management, Web technologies, Introduction to Informatics, Foundation of Information Systems and Information Informatics. Briget has been in the academic industry as an instructor for 8 years and has earned IBM badges in Data Science and Analytics, Artificial Intelligence, Cybersecurity, building a chatbot and Cloud computing to enhance her skills in the field of Computing. Her research interests are in the ICT application areas, Business Intelligence and Distributed Databases.

Ambika N. is a MCA, MPhil, Ph.D. in computer science. She completed her Ph.D. from Bharathiar university in the year 2015. She has 16 years of teaching experience and presently working for St.Francis College, Bangalore. She has guided BCA, MCA and M.Tech students in their projects. Her expertise includes wireless sensor network, Internet of things, cybersecurity. She gives guest lectures in her expertise. She is a reviewer of books, conferences (national/international), encyclopaedia and journals. She is advisory committee member of some conferences. She has many publications in National & international conferences, international books, national and international journals and encyclopaedias. She has some patent publications (National) in computer science division.

Bhumika Pal is Assistant Professor in the Department of Computer Science and Engineering at Dr. Akhilesh Das Gupta Institute of Technology and Management, New Delhi. She completed her M.Tech in Computer Science and Engineering from Guru Gobind Singh Indraprastha University, Delhi.

Priya Paliwal is Assistant Professor in the Department of Computer Science and Engineering at Dr. Akhilesh Das Gupta Institute of Technology and Management, New Delhi. She completed her M-Tech in Computer Science and Engineering from the Centre for Development of Advance Computing,(GGSIPU) Delhi.

Meghna B. Patel is associated with Ganpat University since 2007. She has Experienced with 15+ years and also completed PhD in Faculty of Computer Applications in 2018. She has a good reputation as a research dean, research coordinator, resolving problems, student-staff satisfaction, and driving overall operational improvements. During this tenure, she has experience in teaching various subjects in Bachelor and Master Courses of FCA. She also guiding the students of UG and PG for their industrial projects and guiding Research Scholars for their research work as a Research Supervisor and as a Research Coordinator also.

Bhupendra Prajapati works as a Professor in the Department of Pharmaceutics, Shree S.K.Patel College of Pharmaceutical Education and Research, Ganpat University, North Gujarat, India. He did his Ph.D. from Hemchandracharya North Gujarat University, Patan. He did his PG and UG from M.S.University, of Baroda. He has 19 years of experience in the academic/industry (17+2). He was awarded with the Carrier Award for Young Teacher by AICTE, New Delhi in 2013. He was also awarded Distinguished Associate Professor in TechNExt India 2017 by CSI, Mumbai, and President award of Staff Excellence in Research (2019) and Capacity Building (2020) by Ganpat University for the consecutive years 2021 and 2022 respectively. He claims in his name more than sixty national and international publications. He fetched grants for Research Projects, Staff Development Programs, Seminars, Conferences, and Travel Grants from National and State Government agencies. He is also given his guidance in industrial consultancy projects conducted at the institute. His two patents were published and Three applications were submitted to the Indian Patent office in the field of NDDS. He has delivered expert talks and invited scientific sessions at several national and international conferences and seminars. He is actively working in the field of lipid-based drug delivery and nanotech formulations. He guided 7 Ph.D. and 45 PG research scholars. 6 Ph.D. and 3 PG research scholars are currently working under his guidance in the field of Nanoparticulate Drug Delivery and Bioavailability Enhancement. In the last three-year 7 Prizes secured by his PG and Ph.D. research scholars work at national and international conferences. He is an active life member of IPA, APTI and CRSIC.

Jigna Prajapati has rich experience of 16 years in academia, research and IT industry, holding Doctorate in Computer Science. She claims on her name more than thirty international & national publication. She has secured many best paper awards in international & national Conferences in research category. Beside Assistant professor she is Motivator, Online and Offline Trainer & counsellor. Leading a team which involves work of Infrastructure/ Academic Planning & Development, Network Design & Management, Examination, Key Administration & Research. She is actively working in field of Artificial Intelligence & Machine Learning focusing on healthcare. She has delivered 11 invited session/hands-on in various conference, seminar and workshop. Area Of Interest: Software Fault Tolerance, Data mining, Artificial Intelligence, ML classification, Hands on for Weka 3.9.0.

Pinagadi. Venkateswara Rao obtained his Ph.D in Computer Science Engineering from Sathyabama University-Chennai, Tamilnadu, India in 2020.He had 14 years of experience in both industry and teaching profession. At present he is working as Associate Professor in ACE Engineering College (ACEEC) in the Department of Computer Science and Engineering. He has published various journals include Indian, International Journal and Conferences .His areas of interest are Machine Learning, Image Processing, Data Structures, and Compiler Design.

Hemamalini S. received the M.E., degree in Computer Science and Engineering from Sathyabama University, Chennai, Tamilnadu, in 2011 and pursuing Ph.D., degree in the domain of Image Processing at Anna University, Chennai, Tamilnadu. She is having 15 years of experience in teaching and presently working as Associate Professor in the Department of Computer Science and Engineering at Panimalar Institute of Technology, Chennai, Tamilnadu. Her area of interest includes Image Processing, Internet of Things (IoT), Cloud Computing, Artificial Intelligence, Machine Learning and Deep Learning. She has published many papers in International / National Journals and Conference Proceedings.

Manjunath Saginala is working as manager in Agrostar and completed his PGDM from Indian Institute of Plantation Management Bangalore (IIPMB) and B.Sc Agriculture from Acharya N.G. Ranga Agricultural University and his area of interest is drone application in the food and agriculture sector.

Pooja Saigal is presently working as Associate Professor, School of Information Technology at Vivekananda Institute of Professional Studies – Technical Campus (affiliated to GGSIP University), New Delhi, India. She received her PhD degree in Computer Science (Machine Learning) from South Asian University (established by SAARC), New Delhi, India. She holds a Master's degree and Bachelor's degree in Computer Applications. She is University Topper of PG program and was awarded Gold Medal by the erstwhile President of India, Dr. APJ Abdul Kalam, for her outstanding performance. She has total experience of over 19 years in teaching information technology and computer science subjects at post-graduate and undergraduate levels, with successful records of accomplishments. Her research interests include artificial intelligence, machine learning, optimization and image processing. She is UGC-NET qualified in computer science. She is reviewer of SCI-indexed journals: Neurocomputing (Elsevier), Neural Networks (Elsevier), IEEE Transactions on Cybernetics and many others. She has published over 25 research papers in international journals, books and conferences of high repute.

Esha Saxena is presently working as an Assistant Professor, in the Department of Computer Science & Engineering at Dr. Akhilesh Das Gupta Institute of Technology and Management, New Delhi. She is pursuing Ph.D from Jamia Hamdard University, New Delhi. She has received a Master of Technology's degree in Computer Science under the School of Engineering Sciences and Technology at Jamia Hamdard University, New Delhi and Bachelor's Degree from University of Rajasthan, Jaipur. Her area of interest includes Machine Learning and Deep learning. She has several publications in International / National Journals and Conferences. She also published a book on Database Management System.

Raghav Seth is pursuing his Bachelor of Technology degree in Computer Science Engineering at Dr. Akhilesh Das Gupta Institute of Technology and Management. He has completed several projects in the field of Web Development and has an interest in Natural Language Processing, Machine Learning, and Internet of Things.

Tarun Sharma was born in Hamirpur, India in 1996. He obtained his Bachelor's and Master's degrees in Agriculture and Agronomy in 2018 and 2020, respectively. His Master's thesis on the effect of different nutrient sources on cowpea was his first work as a researcher. He has published 11 research publications and is an ICAR-NET qualified scholar. Currently, he is pursuing his doctoral degree in Agronomy from CSK Himachal Pradesh Agricultural University, Palampur, India with the research on prospects of biofuel production from sweet sorghum-mustard cropping system.

Shwetha S. Shetty working as an Assistant Professor in Department of Information Science and Engineering in Sahyadri College of Engineering and Management, Mangaluru. She has 7 years of Experience in Teaching. She has research interest in the field of Networking, Cyber Security, Machine Learning and Data analytics.

Akashdeep Singh was born in Dharmashala, India in 1995. He obtained his Bachelor's and Master's degrees in Agriculture and Agronomy in 2018 and 2020, respectively. His Master's thesis on the crop diversification of the existing cropping system and its effect on soil fertility was his first work as a researcher. As of now, he has 11 publications to his credit which include 3 book chapters, 4 research papers, 2 review articles, and 2 popular articles. He is an ICAR-NET qualified Ph.D. Scholar in pursuing in the field of Agronomy from CSK Himachal Pradesh Agricultural University, Palampur, India with research on different crop establishment methods, conservation agriculture practices, and biofortification with zinc subsequently affecting soil health in the rice-wheat cropping system.

Deepti Singh is Assistant Professor in the Department of Information Technology at ABES Institute of Technology, Ghaziabad. She completed her M.Tech in Information Security from Guru Gobind Singh Indraprastha University, Delhi.

Bramha S. Tripathi is working as an Engineer-Firmware in the Depart. Of DBEB in Indian Institute of Technology Delhi. He received the Mtech in VLSI Design & B.E. in Electronics from RGPV University MP and has 9 years of experience in Prototype development, R&D, Product Designing, Project Lead, Teaching. His current research interests are in IIoT, Embedded System, Product Design Technology (3d Printing, Laser cutting, PCB Designing), wireless sensor network.

Kamal Upreti is currently working as an Associate Professor in the Department of Computer Science & Engineering, Dr. Akhilesh Das Gupta Institute of Technology & Management, Delhi. He completed is B. Tech (Hons), M. Tech (Gold Medalist), PGDM(Executive) and PhD in Computer Science & Engineering.He has published many patents, books, magazine issues and research papers in various reputed international conferences and journals. His areas of research interest include Machine Learning, Wireless Networking, Embedded System and Cloud Computing.

Index

S

T

U

V

W

Recommended Reference Books

IGI Global's reference books are available in three unique pricing formats: Print Only, E-Book Only, or Print + E-Book.

Shipping fees may apply.

www.igi-global.com

ISBN: 9781799822493
EISBN: 9781799822516
© 2021; 684 pp.
List Price: US$ **195**

ISBN: 9781799844082
EISBN: 9781799844099
© 2021; 279 pp.
List Price: US$ **195**

ISBN: 9781799884262
EISBN: 9781799884286
© 2021; 461 pp.
List Price: US$ **295**

ISBN: 9781799872504
EISBN: 9781799872528
© 2021; 297 pp.
List Price: US$ **195**

ISBN: 9781799875123
EISBN: 9781799875192
© 2021; 416 pp.
List Price: US$ **295**

ISBN: 9781799867098
EISBN: 9781799867111
© 2021; 405 pp.
List Price: US$ **195**

Do you want to stay current on the latest research trends, product announcements, news, and special offers?

Join IGI Global's mailing list to receive customized recommendations, exclusive discounts, and more.

Sign up at: **www.igi-global.com/newsletters.**

Publisher of Timely, Peer-Reviewed Inclusive Research Since 1988

IGI Global
PUBLISHER of TIMELY KNOWLEDGE

www.igi-global.com Sign up at www.igi-global.com/newsletters facebook.com/igiglobal twitter.com/igiglobal linkedin.com/igiglobal

Ensure Quality Research is Introduced to the Academic Community

Become an Evaluator for IGI Global Authored Book Projects

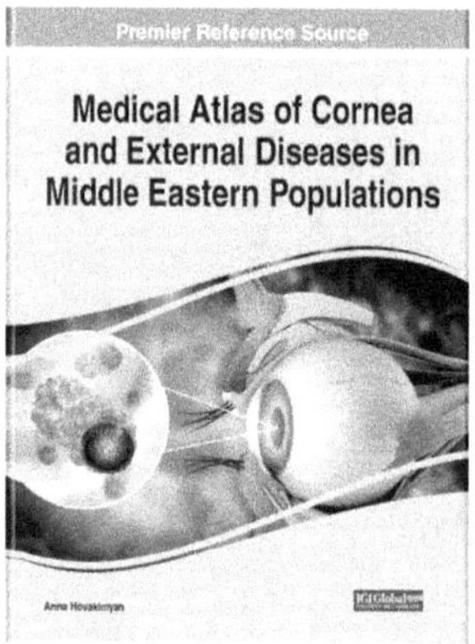

The overall success of an authored book project is dependent on quality and timely manuscript evaluations.

Applications and Inquiries may be sent to:
development@igi-global.com

Applicants must have a doctorate (or equivalent degree) as well as publishing, research, and reviewing experience. Authored Book Evaluators are appointed for one-year terms and are expected to complete at least three evaluations per term. Upon successful completion of this term, evaluators can be considered for an additional term.

If you have a colleague that may be interested in this opportunity,
we encourage you to share this information with them.

Easily Identify, Acquire, and Utilize Published
Peer-Reviewed Findings in Support of Your Current Research

IGI Global OnDemand

Purchase Individual IGI Global OnDemand Book Chapters and Journal Articles

For More Information:
www.igi-global.com/e-resources/ondemand/

Browse through 150,000+ Articles and Chapters!

Find specific research related to your current studies and projects that have been contributed by international researchers from prestigious institutions, including:

Massachusetts Institute of Technology

HARVARD UNIVERSITY

COLUMBIA UNIVERSITY IN THE CITY OF NEW YORK

Australian National University

- Accurate and Advanced Search
- Affordably Acquire Research
- Instantly Access Your Content
- Benefit from the InfoSci Platform Features

"It really provides an excellent entry into the research literature of the field. It presents a manageable number of highly relevant sources on topics of interest to a wide range of researchers. The sources are scholarly, but also accessible to 'practitioners'."

- Ms. Lisa Stimatz, MLS, University of North Carolina at Chapel Hill, USA

Interested in Additional Savings?

Subscribe to

IGI Global OnDemand *Plus*

Learn More

Acquire content from over 128,000+ research-focused book chapters and 33,000+ scholarly journal articles for as low as US$ 5 per article/chapter (original retail price for an article/chapter: US$ 37.50).

6,600+ E-BOOKS. ADVANCED RESEARCH. INCLUSIVE & ACCESSIBLE.

IGI Global
PUBLISHER of TIMELY KNOWLEDGE

IGI Global e-Book Collection

- **Flexible Purchasing Options** (Perpetual, Subscription, EBA, etc.)
- Multi-Year Agreements with **No Price Increases** Guaranteed
- **No Additional Charge** for Multi-User Licensing
- No Maintenance, Hosting, or Archiving Fees
- Transformative **Open Access Options** Available

Request More Information, or Recommend the IGI Global e-Book Collection to Your Institution's Librarian

Among Titles Included in the IGI Global e-Book Collection

Research Anthology on Racial Equity, Identity, and Privilege (3 Vols.)
EISBN: 9781668445082
Price: US$ 895

Handbook of Research on Remote Work and Worker Well-Being in the Post-COVID-19 Era
EISBN: 9781799867562
Price: US$ 265

Research Anthology on Big Data Analytics, Architectures, and Applications (4 Vols.)
EISBN: 9781668436639
Price: US$ 1,950

Handbook of Research on Challenging Deficit Thinking for Exceptional Education Improvement
EISBN: 9781799888628
Price: US$ 265

Acquire & Open

When your library acquires an IGI Global e-Book and/or e-Journal Collection, your faculty's published work will be considered for immediate conversion to Open Access *(CC BY License)*, at no additional cost to the library or its faculty *(cost only applies to the e-Collection content being acquired)*, through our popular **Transformative Open Access (Read & Publish) Initiative**.

For More Information or to Request a Free Trial, Contact IGI Global's e-Collections Team: eresources@igi-global.com | 1-866-342-6657 ext. 100 | 717-533-8845 ext. 100

Have Your Work Published and Freely Accessible

Open Access Publishing

With the industry shifting from the more traditional publication models to an open access (OA) publication model, publishers are finding that OA publishing has many benefits that are awarded to authors and editors of published work.

Freely Share Your Research

Higher Discoverability & Citation Impact

Rigorous & Expedited Publishing Process

Increased Advancement & Collaboration

Acquire & Open

When your library acquires an IGI Global e-Book and/or e-Journal Collection, your faculty's published work will be considered for immediate conversion to Open Access *(CC BY License)*, at no additional cost to the library or its faculty *(cost only applies to the e-Collection content being acquired)*, through our popular **Transformative Open Access (Read & Publish) Initiative**.

Provide Up To **100%** OA APC or CPC Funding

Funding to Convert or Start a Journal to **Platinum OA**

Support for Funding an **OA Reference Book**

IGI Global publications are found in a number of prestigious indices, including Web of Science™, Scopus®, Compendex, and PsycINFO®. The selection criteria is very strict and to ensure that journals and books are accepted into the major indexes, IGI Global closely monitors publications against the criteria that the indexes provide to publishers.

WEB OF SCIENCE™ Compendex Scopus PsycINFO® IET Inspec

Learn More Here:

For Questions, Contact IGI Global's Open Access Team at openaccessadmin@igi-global.com

IGI Global PUBLISHER of TIMELY KNOWLEDGE www.igi-global.com

www.ingramcontent.com/pod-product-compliance
Lightning Source LLC
LaVergne TN
LVHW062142300926
849161LV00010B/2605